Chemicals for Life and Living

Eiichiro Ochiai

Chemicals for Life and Living

 Springer

Eiichiro Ochiai
West 8th Avenue
V5Z 1E1 Vancouver, BC
Canada
eo1921@telus.net

ISBN 978-3-642-20272-8   e-ISBN 978-3-642-20273-5
DOI 10.1007/978-3-642-20273-5
Springer Heidelberg Dordrecht London New York

Library of Congress Control Number: 2011930819

Printed on acid-free paper

Springer is part of Springer Science+Business Media (www.springer.com)

This book is dedicated to Katsuko, Tomoyuki and Naoyuki

Preface

Congratulations! You seem to exist! What is more, you seem to be alive, for you are seeing and reading this sentence right now. You can read this page, because you have eyes to see and a brain to decipher, where a number of chemicals work together to change the image of letters to electric signals in your brain. These functions are attributes of being alive. But how come you exist and are alive? This is a loaded question, but consists mainly of two separate (though related) questions. One is where did you come from? From your parents, who came from their parents, and so on and on. Eventually, it leads to the question: how did life emerge and evolve on this planet? The other question is what is being alive and how (by what means) you are alive. You eat to maintain your activities or being alive. Your body converts what you eat into body parts and some forms of energy. *These Are All Carried Out By Chemicals*.

Well, where do you get food to eat? Bread from wheat, that is, a plant. Meat from, say, cow. Cow eats plants to produce its own body. How do plants produce our food (carbohydrates)? From water and carbon dioxide! *ALL THESE ARE CHEMICALS*. Water from soil and carbon dioxide from air. Air? Is it a chemical? In fact air consists mainly of two chemicals: nitrogen and oxygen; carbon dioxide is a rather minor component of the atmosphere.

You can eat because most of you earn money to buy food. You go to work to earn money. By what means? Maybe by a car. Car, that substantial thing, is made of hundreds of different chemicals: iron, plastics, etc. In the office, the first thing you do may be to sit down in front of a computer. You see some signs and messages on the monitor screen of the computer. How are these visible? Because the monitor emits light that hits your eyes. What is the monitor screen made of? Chemicals. Light itself is not a chemical, though.

Well, you see that anything that is "substantial" and "tangible" is made of chemicals and that *THERE IS NO MATERIAL THAT IS NOT CHEMICAL*. Of course, the messages themselves that are contained in diagrams and sentences on a computer screen or in this sentence are not chemicals, but they cannot be conveyed to the human eyes without intervention of chemicals. Chemicals are everywhere. *WITHOUT CHEMICALS THERE IS NO UNIVERSE, LET ALONE US, HUMAN BEING*.

So let us celebrate for the fact that we are here and now! *Let us enjoy the labor of chemicals, for we are chemicals. Indeed, this whole material world is a festival staged and played by chemicals.*

Chemicals, however, do not only good, but sometimes also bad things. There is no value judgment call in what chemicals do, though. Only, certain things that chemicals do are bad for us human beings or other organisms. As we are made of chemicals, and chemicals interact with each other, some chemicals work wonders for human health, but other chemicals sometimes do harm to our body's proper functions as well as the entire ecosystems on the Earth. Unfortunately, the word "chemicals," today, are often used to mean almost exclusively such chemicals that may harm human body and the environment. The word seems to be used to mean almost synonymous with "bad," and this is very unfortunate, because chemicals on the whole do a lot more wonderful things than bad things. After all, as said above, the universe would not exist, let alone human beings, without chemicals. So it is imperative that we understand the workings of chemicals and the proper handling of chemicals. And *this is what "chemistry" is all about.*

Chemistry is rather difficult a subject to be imparted to those who are not practicing chemistry. A typical conversation at a party goes like this: "What do you do for earning money? 'I teach chemistry at a college' 'Chemistry, eh?' I had to take it because it was required, but I hated it; I did not understand it." For one thing, most of the material chemistry deals with are mundane and not particularly exciting, but it also requires a completely new way of looking at things, which is not very obvious or intuitively understandable. In other words, we need to look at, say, sugar, in terms of concepts such as atom, molecule, and molecular structure and how atoms in a molecule might change (i.e., chemical reactions). You can see sugar, as powder, grains, or cubes, but you cannot see the atoms and molecules that make up the sugar grains. Atoms and molecules are so tiny and invisible that these may be regarded as just abstract concepts and hence foreign to human experience. And here lies the basic difficulty of chemistry. Today though, various techniques have been developed to render atoms and molecules visible and consequently have given credence to what chemistry has been preaching. Yet, to understand how our material world works requires much more than the visual images of atoms and molecules.

This book is not intended to be a systematic exploration of chemistry, but rather is a collection of stories about some of the interesting chemicals or everyday things. The emphasis is on "understanding of chemistry" of the workings of the chemicals and how chemists contribute to understanding them. It omits most of the material necessary for practicing chemists, though. Those materials require logical quantitative applications of the concepts and interpretations of data, thermodynamic, kinetic, spectral, or otherwise; the latter requires deeper understanding of chemical concepts and theories. In addition, chemistry is a practical science, i.e., dealing with chemicals, making them, decomposing them, allowing them to react with each others, and measuring their properties. These require practical lab techniques, which are also omitted in this book.

In this book, each chapter is intended to be and can be read independently if you have some knowledge of chemistry. If not, reading first the appendix Chaps. 19, 20, and 21, particularly Chap. 19, is suggested. Chapter 19 is not an easy reading. I would not expect everybody to understand it fully on first reading. Read once, twice, or more, and use the chapter as a reference as you read the other chapters. This chapter functions as a sort of glossary as well, but a glossary cannot give an overall picture of a discipline. Hence, I tried to present an overall picture of *chemistry (=how to understand the material world)* as concisely as possible in this chapter.

Vancouver, BC, Canada   Eiichiro Ochiai

About the Author

Eiichiro Ochiai was born in Tokyo, Japan and earned his PhD from the University of Tokyo. He has taught and carried out research at universities and colleges in Japan (University of Tokyo), Canada (University of British Columbia, University of Toronto), Sweden (University of Umeå) and the USA (Ohio State University, University of Maryland, Juniata College) and became Professor emeritus of Juniata College, PA, USA. So far, Prof. Ochiai has published about 110 research papers and articles on chemistry. Books published include *"Bioinorganic Chemistry, an Introduction"* (Allyn and Bacon, 1977), *"General Principles of Biochemistry of the Elements"* (Plenum Press, 1987), and *"Bioinorganic Chemistry, a Survey"* (Elsevier/ Academic Press, 2008). Other publications include about 100 short articles on education and 180 short articles on issues of economics, politics, peace, radiation problems, and civilization.

Contents

Part I

Essentials of Life

Essential chemicals for life include air (rather oxygen), water, and organic biocompounds such as carbohydrate, protein, and DNA. The bases of their workings are discussed here, particularly, with regard to the way the bioenergy is created in the life system and how the information in DNA sequence is translated into protein. Other materials such as clothes are required for our comfort; chemicals for those materials are also briefly talked about in this part.

Water

1

Life-sustaining Water! Water and Blue Planet-Earth

Water is everywhere: rivers, lakes, underground, and oceans, and rain. There is about 1.5×10^{18} t (1.5×10^{21} kg $= 3.3 \times 10^{21}$ lb) of water on this Earth. This is about 0.02% of the total weight of the Earth, and water covers 70% of the surface of the Earth. Indeed, the Earth is a planet of liquid water, which is largely responsible for the "blueness" of the planet and also for the presence of living organisms.

What properties of water are important? Important for what? Let us say "important for our lives, us ourselves and our living conditions." First of all, it is made of two most abundant elements in the universe and on the Earth, and hence it is abundantly available. Next, it is liquid at ambient temperatures and the temperature range for liquid is relatively large. It happens to be liquid at the pre-vailing temperatures on the Earth, and hence, it could have been used for the basis of living organisms, and also because it is a very good solvent for a large number of compounds. And the other important property is its high heat capacity. This has a temperature moderating effect, contributing to reducing temperature fluctuation of the Earth.

There is plenty of water alright, but the bulk of the water is salty and is not fit for the consumption by many organisms including human beings. The avail-ability of fresh water is limited, and the demand for it is ever increasing. Perhaps, one of the greatest challenges for chemists in the twenty-first century is to develop methods to produce usable fresh water cheaply (including cleaning of the polluted water).

As much as 70% of our body weight is water. The water content in many other organisms ranges from 70 to 95%. We are literally living with water. Water is one of the simplest compounds, but is quite unusual in many ways, as expounded above. Let us try to see why water is so unusual.

E. Ochiai, *Chemicals for Life and Living*,
DOI 10.1007/978-3-642-20273-5_1, © Springer-Verlag Berlin Heidelberg 2011

1.1   How Does Water Behave?

Let us first explore the basic nature of water. Take a look at Fig. 1.1 below. This represents the chemistry of water in a nutshell (multilayered, though).

Eighteen grams of water is a unit of the quantity of this substance. This amount corresponds to one "mole," which is the units of quantity used in chemistry (see Chap. 19). It is called the molar mass of water, 18 g/mol. Let us suppose that the container in the figure contains 18 g of water. It consists of 6×10^{23} particles of water molecule, which is shown in the figure as one red ball with two smaller blue balls attached. Here the red ball represents an oxygen atom and the blue a hydrogen atom. This is the smallest unit (molecule) of water. This is a very, very tiny entity. Since the very large number, 6×10^{23}, of it weighs 18 g, each molecule of water weighs only 3×10^{-23} g $(=18 \text{ g}/(6 \times 10^{23}))$. It is possible to pull the oxygen and the hydrogen atoms apart by some means. The simplest way of doing so is to put an electric current through water; this is called "electrolysis." Many of you may have seen it demonstrated in high school chemistry. Water is then decomposed into two volumes of hydrogen gas (H_2) and one volume of oxygen gas (O_2). This says that a water molecule consists of hydrogen and oxygen atoms in the ratio of 2:1. It cannot say, though, whether water is made of two hydrogen atoms and an oxygen atom or six hydrogen atoms and three oxygen atoms or other such combinations. You need other means to determine which is actually the case. For example, you can measure the weight of water gas at a certain high temperature and simultaneously its volume and pressure. From this kind of measurements, you can determine that the molar mass of water is 18 g. It had long been known that the molar mass of hydrogen is 1 g and that of oxygen is 16 g. Therefore, the molecule of water must consist of two hydrogen atoms and one oxygen atoms $(2 \times 1 \text{ g} + 1 \times 16 \text{ g} = 18 \text{ g})$. It has been determined from many other measurements as well that indeed a water molecule is made of two hydrogen atoms and an oxygen atom; we write this as "H_2O." A hostess at one of my favorite restaurants impressed me, when she wrote down "H_2O" when I ordered a glass of water.

How are these three atoms combined together? This is one of the fundamental issues in chemistry: structure of molecule. We omit the issue of how we determine the structures of molecules. The result of many, many experimental works is summarized in the figure. That is, two separate small hydrogen atoms attach to a

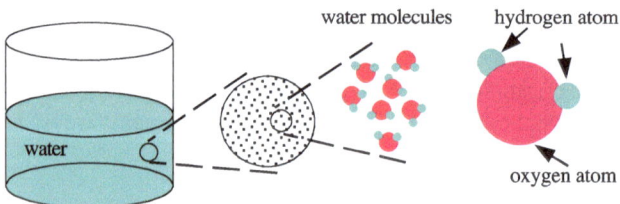

Fig. 1.1   Chemistry of water

larger oxygen atom in the manner of H–O–H, but the angle <H–O–H is not 180° (if it is 180°, it is straight), but about 106° (i.e., it is bent). The connection (bond in technical terms) between H and O is held by two electrons between the atoms (see Chap. 19 for a discussion of chemical bonding). These electrons are not equally distanced from the H and O atoms. Because oxygen is stronger in attracting electron(s) (more electronegative; see Sect. 19.6) than hydrogen, electrons are more distributed toward the oxygen atom than the hydrogen atom. As a result, the oxygen atom is slightly negatively charged and the hydrogen atoms are slightly positive. The technical expression for this situation is that the bond O–H is polar, and the water molecule as a whole is also polar (and is said to have a dipole moment). [Question: Would water molecule be polar if the structure is linear, i.e., H–O–H is straight?] This fundamental character is responsible for the many interesting and important chemical and physical properties of water.

Let us look at the water molecule a little more closely. The central atom oxygen has six valence electrons. An oxygen atom has eight electrons, but two of them are the so-called core electrons and usually are not involved in bonding and reactions, because they are very tightly held by the positively charged nucleus. The remaining six electrons are involved in bonding and chemical reactions, and these are called "valence" electrons. See Chap. 19 for a further discussion of this issue. Two of them are used to bind two hydrogen atoms. There are, therefore, still four more valence electrons on oxygen atom unused. They form two sets of paired electrons. Since they are not used for binding other atoms in water, each of these pairs can still be utilized to bind an atom that lacks electrons; in other words, oxygen and this other atom can form a bond by sharing one of these pairs. The oxygen in this situation donates a pair of electrons toward another atom that lacks electrons but is capable of accommodating a pair of electrons. The electron pair donor in this sense is called "BASE" (as defined by Jack Lewis and John E. Walke), and the one to accept the electron pair is termed as "ACID" (Lewis definition). Water in this sense is a "base." Water, or hydrogen in water to be more exact, can also act as acid, as it carries a partial positive charge and can separate out ("dissociate" is the technical word for this) as "H⁺" which can bind with a base. So you can see that a compound that can give off H^+ is also an "acid." Actually, "acid" is more commonly used in this sense; examples include hydrochloric acid HCl, and sulfuric acid H_2SO_4. Water becomes OH^- when it loses H^+. OH^- is called "hydroxide" ion. It is a much stronger base than water itself, because it has an extra pair of electrons (three pairs altogether now). Hence, its electrons can be more readily donated to an acid due to the fact that there are too many negatively charged electrons for the same positive charge on the oxygen nucleus. This whole situation can be described by the following chemical reaction equation:

$$H_2O + H_2O \rightleftharpoons H_3O^+ + OH^-$$

As you see, the proton H^+ that splits from a water molecule is actually bound with another water molecule, its oxygen atom to be exact, and exists as a hydronium ion (H_3O^+) in water. This (chemical) reaction is called "autodissociation" equilibrium.

As indicated above, OH⁻ is much stronger base than water itself, and hence, there are a lot more H_2O than H_3O^+ and OH^- in pure water, as these two, i.e., H^+ (of H_3O^+) and OH^-, tend to bind to form H_2O back. The extent of autodissociation is governed by the so-called mass action law: $[H_3O^+][OH^-] = K_w$, and K_w is called an equilibrium constant and known to be a very small number, 10^{-14} at room temperature. As you recall, $[H_3O^+]$ represents the molar concentration (mol/L) of the chemical species H_3O^+. More rigorously speaking, $[A]$ in an equilibrium constant expression represents "activity" which is a number and its magnitude related to the molar concentration. This relationship says that the product of the concentrations of hydronium ion and hydroxide ion in water is constant. So, if you add an acid (which gives off H^+) to water, you have increased the hydronium ion concentration, and accordingly the hydroxide ion in the solution will be reduced.

The oxygen atom in a water molecule, through its lone pair, approaches and can bind in a way to a hydrogen atom on another water molecule, without stripping the hydrogen (ion) away from the water molecule. This is very much like the acid–base reaction between two water molecules mentioned in the previous paragraph. The important difference is that no hydrogen (H^+) splitting occurs. This kind of bonding is called "hydrogen bonding" and occurs effectively between oxygen, nitrogen, or fluorine atom (in a compound) and hydrogen in another compound that is bound with nitrogen, oxygen, or fluorine (Fig. 1.2). The hydrogen bonds are moderately strong, somewhere around 18 kJ/mol. Compare this value with the regular chemical bonds whose bond energies range from 100 to 1,000 kJ/mol (see Chap. 19). Hydrogen bonding plays important roles in the chemistry of living, but this topic will be postponed to a later chapter.

Now we have listed all the important basic chemical properties of water molecules. Water as we know is a substance and is liquid at ambient temperatures. It turns into solid (ice) at lower temperatures below 0°C, or it boils at 100°C and changes into gas (vapor, steam in this case) under the standard condition. The standard condition here means that the atmospheric pressure is 1.0 atm (or 1.013 hPa or bar). Actually, the freezing point (temperature) and the boiling point of pure water under this condition are defined to be 0°C and 100°C, respectively. [Today, scientists try to redefine the standard condition to be 1.000 hPa or 1.000 bar (1,000 mbar) instead of 1 atm. For nonrigorous treatments, the old standard condition is still valid and is what is employed in this book].

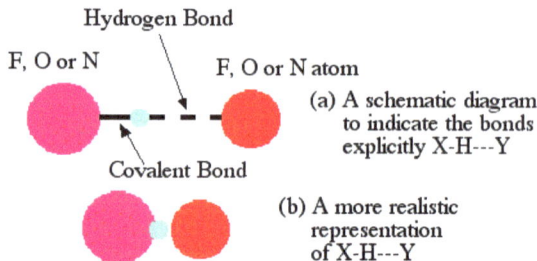

Fig. 1.2 Hydrogen bonding

At sufficiently low temperature any compound, i.e., aggregate of molecules, becomes "solid." The molecules line up more or less regularly and would not move from where they are. That is why they take a definite shape, solid. In reality, however, atoms in molecules are vibrating (over a short distance) around their respective positions even at these low temperatures. As you heat up a solid, the movements of molecules themselves in the solid become more and more vigorous, and beyond a certain temperature (melting point) molecules can move around almost freely. As there is still sufficiently strong force acting on each molecule from all other molecules there, the molecules remain being stuck together, but they would not form a definite shape. This is "liquid" state. As you heat it up further, the molecules become more and more agitated and gain enough energy to get out of the confines of liquid, and go out into the space. Molecules in the space move around almost independently; they hardly interact with each other. This is "gas" state.

Now you might have guessed that there is a good correlation between the strength of force that binds molecules together and how high temperature you have to go to melt or vaporize a compound. And you are quite right. In this sense water is quite unusual. Water is a simple and small molecule. Similar compounds of relatively simple and small molecules, such as ammonia (NH_3), hydrogen sulfide (H_2S), and methane (CH_4), are all gas at ambient temperatures; that is, they have much lower boiling temperatures (and melting points), compared to water. Water is exceptional. Why is water so unusual?

The forces between water molecules are relatively strong, because they are polar; i.e., they carry positive and negative charges on them, so that they attract each other strongly. They can also interact attractively through hydrogen bond (see a couple of paragraphs back). That is why! It can then be deduced that it would be harder to pull the water molecules apart from each other, as compared with the other ordinary compounds as suggested in the previous paragraph for comparison. Therefore, it would require a larger energy and a higher temperature to do so. The energy that is required to turn liquid water to vapor (steam) is called "vaporization energy (enthalpy)"; the value for water is unusually large.

The same basic properties, the polar structure and acidity–basicity, account for its being a good solvent for so many different compounds. Perhaps half of all the compounds that exist on the Earth and anywhere else are of ionic type. Take "table salt" for example. Its chemical name is sodium chloride, and consists of two ions, sodium ion (Na^+) and chlorine negative ion, chloride (Cl^-). The two ions, because of the opposite electric charge, attract strongly each other, and arrange themselves in an orderly fashion (crystal structure) in solid form. The ions interact so strongly that they are hard to be moved around and a high temperature is necessary to melt it. Compounds that are made of positive ions (the technical word for this is "cation") and negative ions ("anion") are called "ionic compounds" and are typically crystalline solids at ambient temperatures and have high melting points, as illustrated by table salt. However, when an ionic compound is brought into contact with water, ions come apart even at room temperature and the compound dissolves in water. Why? Where does the energy that is necessary to pull all the ions apart come from? That comes from the interaction between ions and water molecules. As a water

Fig. 1.3 Interactions
of a cation or an anion between
water molecules

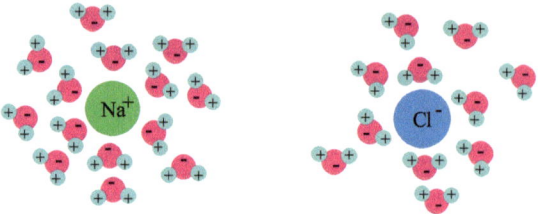

molecule has a negative charge centered on the oxygen atom and the positive charge on the two hydrogen atoms, it can attractively interact with ionic species. A cation interacts with the negative oxygen atom of water, and an anion with the positive hydrogen atoms (see Fig. 1.3). A cation is usually a Lewis acid, and hence the oxygen (a base in acid–base sense) on water would bind to the cation through acid–base interaction as well. There is no clear-cut distinction, though, between the ionic interaction (purely electrostatic interaction) and the acid–base interaction. The energy gained (through these attractive interactions) is called "hydration energy" and is often sufficiently large to overcome the large energy needed to separate cations from anions in a crystal. This is why many ionic compounds are soluble in water. This explanation is OK for ordinary ionic compounds, but unfortunately not quite sufficient for sodium chloride (among others). That is, the energy gained from hydration is not sufficient to pull apart cations from anions in the case of sodium chloride. It requires a little bit more detailed discussion of the so-called second law of thermodynamics and can be found in Sect. 19.7. This law says that a process, if accompanied by an increase in randomness, can occur more easily than otherwise. When the crystalline sodium chloride (or any other ionic compounds) dissolves and moves randomly in water, the randomness of the salt/water system will increase. This increase in randomness (entropy in technical terms) favors the dissolved state. By the way, the reason why a like substance dissolves in a like substance (for example, toluene in benzene) is this; that is, an increase in randomness when dissolved.

How about sugar, glucose, and alcohol? Your favorite beverage, beer, is an alcohol (ethanol, C_2H_5OH) dissolved in water. Sugar cube would easily dissolve in water when you stir. These are not ionic compounds like table salt. What you need to compare is the energy that is involved in the interaction between ethanol and ethanol molecules as against the interaction energy between water molecule and ethanol molecule. If the latter is not sufficiently large as compared to the former, ethanol would prefer to remain to be stuck together; that is, ethanol would not dissolve in water. Obviously, that is not the case here. There is strong enough interaction between water and ethanol. I am intentionally vague about the actual magnitude of the interaction energies here, because the interaction energy is only a part of the story, as we discussed in the previous paragraph. What kind of interaction? Hydrogen bonding between water molecule and the OH group portion of ethanol. Thus, water is a good solvent; that is, it can dissolve a large number of compounds that are either ionic or polar with groups that can form hydrogen bond. Ammonia, NH_3, for example, is slightly polar and can hydrogen-bond to water and hence is soluble in water, though it is not an ionic compound. Sugar has many OH groups that can hydrogen-bond to water and, hence, is very soluble in water.

Oil, in general, is nonpolar and has little ability to form hydrogen bond and hence is not soluble in water. When you try to mix oil with water, what you get is a suspension of small droplets of oil in water, which then separate out eventually as a separate layer. "Oil and water do not mix."

1.2   Acid–Base and pH

The idea of acid and base was mentioned above, but it needs to be explored further, because this is one of the most basic chemical properties and can be very useful in understanding some types of chemical reactions.

Let us review: water H_2O can release H^+ off and also accept H^+. When H^+ is released from water, it becomes OH^-. That is, $2H_2O \rightleftharpoons H_3O^+ + OH^-$; i.e., the H^+ given off is captured by another water molecule. Often, however, this is abbreviated as $H_2O \rightleftharpoons H^+ + OH^-$. A chemical species that gives off H^+ in water is defined as an "acid," whereas a chemical that accepts H^+ is a base. OH^- is called "hydroxide" (ion), and binds H^+ to form H_2O, and hence should be a base. As a matter of fact, hydroxide is the strongest base present in water solution. Sodium hydroxide, also called caustic soda NaOH, will dissociate completely into Na^+ and OH^- in water, and hence a strong base (sometimes called "alkali"). KOH (potassium hydroxide) does the same thing.

There are chemicals that release H^+ when they are dissolved in water. An example is hydrogen chloride, HCl, which will separate (dissociate) completely into H^+ and Cl^- in water; $HCl \rightleftharpoons H^+ + Cl^-$ (more rigorously, $HCl + H_2O \rightleftharpoons H_3O^+ + Cl^-$). So an aqueous (i.e., water) solution of HCl is called "hydrochloric acid," and is one of the strong acids, because HCl dissociates completely and hence provides a lot of H^+. Other common strong acids include sulfuric acid H_2SO_4 (turns into H^+ and HSO_4^- in water) and nitric acid HNO_3. What would happen if one mixes an acid with a base? A strong chemical reaction (between H^+ and OH^- to form water) will take place. For example, when hydrochloric acid is mixed with a sodium hydroxide solution, $HCl + NaOH \rightarrow H_2O + Na^+ + Cl^-$. This type of reaction is sometimes called "neutralization," because acid (H^+) and base (OH^-) react resulting in a neutral water molecule, if you mix them in a proper proportion.

To make "neutrality" more quantitative, it is useful to devise an expression of the strength of acid or base. This is "pH." You might have heard an expression such as "pH-balanced shampoo." Technically pH is equal to the negative of the logarithm of $[H^+]$, the molar concentration of H^+. [Suppose that you dissolve a chemical species A in water. You would then obtain a solution of A in water (aqueous solution of A). Notation [A] is called the molar concentration of A and indicates how many moles of chemical species A in a solution of 1 Liter (L) (technically in units of mol/L)]. So pH is related to the concentration of H^+ species (more precisely, H_3O^+) in a solution, but not in a straight way. As you saw before, a number can be expressed as something like $a \times 10^n$. A small number or a large number is often expressed in this manner. The concentration of H^+ in water is not very large, and often very small. So it is convenient to express it in the form of "$a \times 10^{-n}$." If the concentration is 0.5 mol/L, then it will be 5×10^{-1}. Or if it is 0.0000000013 mol/L, it is 1.3×10^{-9} mol/L. The pH of a solution whose $[H^+]$ concentration is 1×10^{-2} mol/L is 2 ($= -\log(1 \times 10^{-2})$). If it is 1×10^{-10} mol/L, the pH is 10. Since 1×10^{-2} mol/L is much greater than 1×10^{-10}, pH 2 is much stronger in acidity than pH 10. It turned out that pure water (with no added acid or base) dissociates (as mentioned above) into equal moles of H^+ and OH^-, and this concentration of H^+ (and OH^-) is 1×10^{-7} mol/L at an ambient temperature (25°C). Then what is the pH? Yes, it is pH 7. That is, pH of the pure water is 7. If pH of a solution is below 7, it contains more H^+ than that of pure water. So we can say that such a solution is acidic. When $[H^+]$ becomes lower than 10^{-7} mol/L or pH higher than 7, you have more $[OH^-]$ than $[H^+]$ and also more than that of pure water. So you can say that such a solution is basic (or alkaline). You may have noted in this description that when $[H^+]$ becomes smaller, $[OH^-]$ become larger and vice versa. As a matter of fact, the product of their concentrations $[H^+][OH^-]$ is always equal to 1×10^{-14}, as has been discussed earlier. As you also note, $[H^+]$ and $[OH^-]$ would be the same in pure water, as both come from the same H_2O molecule. Hence, $[H^+] = [OH^-] = 1 \times 10^{-7}$ in pure water. By the way, the degree of dissociation of pure water is extremely small as indicated the concentration of $[H^+]$ (and $[OH^-]$) being small as 10^{-7}; only about 0.0000002% of water molecules is dissociated into $[H^+]$ and $[OH^-]$. Or you can say that only two out of ten millions of water molecules are in the form of H^+ (H_3O^+) and OH^- in pure water.

A compound CH_3COOH, acetic acid, is the main ingredient of "vinegar" whose smell and taste are largely due to this ingredient. It would give off a little bit of H^+ when it is dissolved in water. The chemical reaction is $CH_3COOH + H_2O \rightleftharpoons CH_3COO^- + H_3O^+$. Unlike HCl, though, the extent of dissociation is much less than 100% (only several%). For this reason, this compound, acetic acid, is called a "weak" acid.

Another example of weak acid is carbonic acid (H_2CO_3). When you open a soda can, you get a lot of bubbles. That bubble is carbon dioxide, CO_2. Carbon dioxide gas dissolves a little in water, but dissolves more under high pressure of the gas. The can has been closed off under a high pressure of carbon dioxide. When you open the can, the gas can escape into air, and hence the pressure becomes less and the dissolved carbon dioxide comes out as bubbles. Still a significant amount of carbon dioxide remains dissolved in water and that reacts with water in this manner: $CO_2 + H_2O \rightleftharpoons H^+ + HCO_3^-$. This acid ($H^+$) gives that tongue-tingling sensation. But this sensation will be lost, as you leave a can open long enough. Why?

Soap is typically made of sodium salt of fatty acid(s). A main ingredient of animal fat is glycerol esters of fatty acids. A fatty acid is made of a long carbon chain plus an acid unit (COOH, as you saw in the case of acetic acid above) at the end; for example,

$$CH_3CH_2CH_2CH_2CH_2CH_2CH_2CH_2CH_2CH_2CH_2CH_2CH_2CH_2CH_2COOH$$
(palmitic acid)

Animal fat is made of a mixture of several different fatty acids (of different chain lengths). Soap can be obtained essentially by cooking the animal fat with sodium hydroxide (caustic soda, NaOH). It turns into:

$$CH_3CH_2CH_2CH_2CH_2CH_2CH_2CH_2CH_2CH_2CH_2CH_2CH_2CH_2CH_2COONa.$$

When this is dissolved in water as you do when you wash your hand, it dissociates to

$$CH_3CH_2CH_2CH_2CH_2CH_2CH_2CH_2CH_2CH_2CH_2CH_2CH_2CH_2CH_2COO^- + Na^+$$

Since this long chain with COOH at the end is a very weak acid and hence $RCOO^-$ is rather strong base, so it would pick up H^+ from water, leaving OH^- behind; i.e., $RCOO^- + H_2O \rightleftharpoons RCOOH + OH^-$, where R represents the long carbon chain. This means that a soap solution is basic (alkaline). A base can react with the esters of fatty acids and proteins. When you wash your hand, you feel your hand becomes a little slimy, or your hairs tend to become sticky when you wash hair with soap or a detergent. That is all due to this chemical action of the basicity of soap. "Hair conditioner" then would contain a counteracting chemical (what kind?) to neutralize the effect.

Why does the soap do what it does, wash a cloth, for example? That has little to do with acid–base issue. We figured above that water and oil would not mix. Hence, it is difficult to wash away oily stains with just water. A soap molecule (sodium palmitate for example), as you see in the chemical formula above, has two different characters. The long carbon chain is essentially an oil (the same kind as gasoline or wax), whereas the portion COO^-, being a negative ion, interacts with water and thus tends to dissolve in water. As a result, soap molecules form a sort of round shape (called "micelle") in water, with the oily portion (blue) stuck together inside and the COO^- portions (red) exposed to water (see Fig. 1.4). The oil stains will be embedded inside of micelles, because they interact well with the long chain of soap molecules (oil interacting with oil). Now the oil stains will be washed away as micelles.

Cell membranes of all the organisms are essentially made of double layer of long chain of hydrocarbons as seen in Fig. 1.5. The major ingredient of cell membrane is a compound called glycerol ester of fatty acids, consisting of two long hydrocarbon

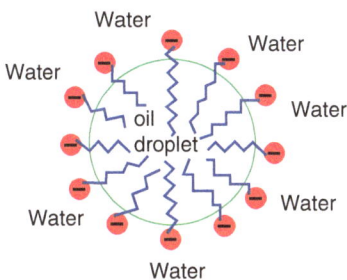

Fig. 1.4 Soap molecules wrap an oil droplet (formation of micelle)

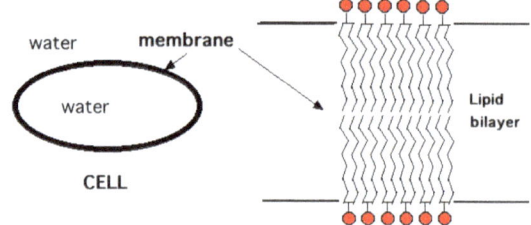

Fig. 1.5 Cell membrane is made of double layer of phospholipid

chains and a head portion which is made of phosphate group, negatively charged, and hence interacts with water. The double layer (lipid bilayer) is formed with the oily portion stuck back to back and the both surfaces are lined with the hydrophilic (water-liking) head portion (Fig. 1.5). This double layer forms the boundary of a cell, and the outer surface and the inner surface are exposed to water. The principle involved in cell membrane, thus, is the same as that involved in oil cleaning by soap.

1.3   Natural Water

Natural water that is found in lake, river, underground, and ocean is not pure water. Even the so-called pure water, rather potable water is not necessarily chemically pure water. Water also exists as solid ice in the Polar regions and on high mountains. Natural water contains a lot of different chemicals. Many of them are dissolved, but some others are simply floating (suspended). Many microorganisms inhabit natural waters, too. Larger organisms, algae and fish included, are regarded as separate from natural water body.

Let us first look at the terrestrial waters, lakes, and rivers. The total amount of the terrestrial waters on the Earth is about 5×10^{17} kg. With a few exceptions like Great Salt Lake and Dead Sea (in Jordan/Israel), the terrestrial waters contain relatively low levels of dissolved ionic species (low salinity). Though it varies widely, the salinity of river waters has been estimated on average as about 100 ppm (that is, 100 g of ionic compounds of various kinds per 1 million grams (1 m^3 volume) of water). On average, the contents of various elements (compounds) in rivers are HCO_3^- (bicarbonate anion) $>Si(OH)_4$ (silicate) $>Ca^{2+}>SO_4^{2-}>Cl^->Na^+$ $>Mg^{2+}$. These ions come dissolved from the rocks and soils through which the river water runs.

A high content of calcium ion (and other cations such as iron) in water causes some problems in its use. Such water is called "hard." For one thing, calcium carbonate scale forms and clogs a pipe, for example, when heated. You may have noticed white scales to form in your kettle when you use spring water. (Why does this happen, while the water is transparent before heating?) Calcium (Ca^{2+}) also binds with soap (sodium palmitate, for example) forming scam of calcium palmitate. Hence, it reduces the effectiveness of soap's cleaning power. How can you soften hard water (i.e., remove calcium)?

The largest natural water body is oceans. The total mass of the Earth's oceans has been estimated to be about 1.4×10^{21} kg (or 1.4×10^{18} t). The sea water is salty; it contains a lot of ionic compounds, the highest among which are sodium (Na^+), potassium (K^+), magnesium (Mg^{2+}), and calcium (Ca^{2+}) cations, and chloride (Cl^-) and sulfate (SO_4^{2-}) anions. Their concentrations are in the order of Na^+ (10 g (=0.4 3 mol)/L) $> Mg^{2+} > Ca^{2+} \sim K^+$, and Cl^- (19 g (=0.54 mol)/L) $> SO_4^{2-}$. In addition, seawater contains a large variety of elements (their cations and anions). About 70 elements have been identified in seawater, and certainly others are also present but may not be detected by currently available techniques.

Why is the seawater salty? The materials in seawater are supplied by the rivers that are not very salty. Well, the substances that have been put into ocean by rivers are removed (from the seawater) by a number of ways. As mentioned above, the predominant metallic ion in rivers is calcium (Ca^{2+}) but that in the seawater is sodium (Na^+). Why is this so? Calcium is removed (along with carbonate CO_3^{2-}) from the seawater by living organisms and some physical chemical effects. For example, seashells are all calcium carbonate ($CaCO_3$). In addition, millions of millions of minute algae use calcium carbonate as their shells. The chalk bed of the cliff along the Dover strait is made of the remains of such minute organisms. Silicate is removed likewise by microorganisms such as diatoms and others. Silicates, along with aluminate, also sediment to form clay minerals. On the other hand, sodium cation does not form insoluble compounds and hence remains dissolved and cannot be removed from the seawater easily. Therefore, sodium is concentrated in it. As a result, sodium (in the form of sodium chloride) has become the predominant species in the seawater.

1.4   Water Pollution

Water pollution is multifaceted. As there is no (chemically) pure water in nature, what constitutes "pollution" is a difficult issue. Shall we define "nonpolluted" water to be that fit for human consumption, i.e., "potable or drinkable"? Well, this is too narrow a definition. Seawater, polluted or not, is undrinkable. How about "nonpristine"? But can we really define the "pristine condition"? No, we cannot.

Today, we may define "water pollution" technically in terms of levels of chemicals and microorganisms contained in the water. Hazardous materials (chemicals and biological material) are listed and their tolerable levels are determined by the regulating agencies, such as EPA (Environmental Protection Agency), and any water that contains such a substance at a higher than the set level is defined "polluted." However, the tolerance levels are often determined politically, and their scientific rationale is often obscure. Besides, more importantly we, mankind collectively, do not yet know all the health effects of all chemicals and biological materials and hence we often do not have rational (scientific) bases to define the tolerance levels in many cases. As a contamination of natural water system is almost inevitable, consideration also needs to be given to the benefit of such a practice that has led to the contamination, as against its risk. That is, a "cost/benefit" analysis is necessary, and this is where political consideration would get in.

The causes of water contamination could be either natural or anthropogenic. Some natural processes may produce hazardous substances that can give adverse effects on the organisms in the water body or its surroundings. Toxic fumes containing toxic hydrogen sulfide are spewed out from hot spring, for example. However, this kind of situation is often considered not to be pollution. In other words, pollution seems to be defined as that caused by human activities (i.e., anthropogenic).

Pollution occurs most often in the wastewater released from domestic and industrial sources. Treatment of such a wastewater or how polluting substances may be prevented from escaping into the wastewater is an important critical issue. Chemistry and microbiology play important roles in these treatments for reducing water pollution. Analytical chemistry identifies and quantifies the polluting chemicals. Some chemical pollutants are talked about later in the book (Part V).

1.5   Why Is Seawater Blue?

Water in a glass looks colorless and transparent, does not it? However, seawater looks blue. The fact that a substance is colorless suggests that it does not absorb nor emit light in the visible range (320–800 nm in wavelength). Water molecules do absorb light in the visible range (red region), though very weakly. This is believed to be due to the overtone of vibrational frequency of water molecule. However, it absorbs so weakly that shallow or small quantity of water does look like not absorbing light and thus appears colorless. When light goes through a long path of water (such as deep water), a substantial amount of light (in the red region) is absorbed and thus such a water body looks blue. You may have seen even water in a swimming pool being blue.

In real situations, the apparent color of water body is determined by many other factors, besides the inherent light absorption by water. Reflection of the sky and presence of microorganisms that contain pigments of different colors are the major factors.

Air

2

Breath of Life! and Kiss of Death!

Air is taken for granted. We can neither see it nor smell it, and tend to ignore it. But without it we would not be able to live. Not only we, humans, but also the majority of living organisms on this Earth depend on it. Only a small number of organisms can live without it, or rather they will be killed by air. These organisms are called "anaerobic," whereas we are "aerobic." When we say "air" in these sentences, we mean "oxygen." Air is a mixture of nitrogen (N_2, about 78%) and oxygen (O_2, about 21%) plus minor components such as argon (Ar), carbon dioxide (CO_2), and water vapor (H_2O).

Of the two main components, nitrogen is relatively inert, but oxygen is reactive. Oxygen burns many substances; combustion is a chemical reaction in which oxygen binds with other chemicals. This reaction releases a lot of heat (energy). You will see and feel that in a fire. Oxygen can burn our body, but it requires certain conditions for oxygen to do so, and those conditions do not prevail ordinarily. That is why we are OK (not burned) in the presence of a lot of oxygen. However, oxygen does a lot of other kinds of damage to our body, the chemical basis of which is the same as combustion. As a matter of fact, oxygen can be regarded as toxic to organisms. Probably, this comes as a surprise to you. We talk about this issue later.

2.1 Where Did "Air" Come From?

Where have they (nitrogen and oxygen) come from? Nitrogen, being nonreactive, has remained more or less unchanged in the atmosphere throughout the history of the Earth. It does not mean that no change has occurred to nitrogen. Some reactions have taken place and do take place with nitrogen. Nitrogen, for example, can react with oxygen in the atmosphere under, say, lightning conditions. As a result, some nitrogen oxides (NO, NO_2, etc.) form. More importantly, nitrogen is fixed as

E. Ochiai, *Chemicals for Life and Living*,
DOI 10.1007/978-3-642-20273-5_2, © Springer-Verlag Berlin Heidelberg 2011

ammonia by a number of microorganisms; the reaction is: $N_2 + 6H^+ + 6e \rightarrow 2NH_3$. Ammonia ($NH_3$) is utilized by organisms, and it eventually turns back to nitrogen (N_2) or nitrate (NO_3^-). Nitrate is also used by organisms. However, the quantity that undergoes these reactions is miniscule compared with the entire nitrogen (N_2) present in the atmosphere, which is 4×10^{18} kg. The amount of nitrogen biologically fixed is estimated to be 2×10^{11} kg/year.

Oxygen, all of it, on the other hand, has come from an entirely different route. By the way, the oxygen in the air exists as dioxygen (O_2) molecules, but we often use simply oxygen to mean dioxygen. The current prevailing idea about the ancient Earth asserts that no significant free oxygen (O_2) was present in the earlier atmosphere. A main reason for this thinking is that oxygen, being reactive, would have reacted with a number of compounds available at the time of formation of the Earth, and thus, would not have remained as free molecule long in the atmosphere, even if there were free oxygen at the beginning. If so, then where has the free "oxygen" come from and why would not "oxygen" be consumed and hence disappear? On the present Earth, we, animals and others, are consumers of oxygen, and who are the producers? Yes, green plants and a lot other minute organisms, phytoplankton (algae), are the producers. They produce carbohydrates using carbon dioxide (CO_2) and water (H_2O), with the aid of sunlight; that is, they carry out photosynthesis. The reaction can be written essentially as $6CO_2 + 6H_2O \rightarrow C_6H_{12}O_6$ (carbohydrate) $+ 6O_2$. Reactions involved in photosynthesis are very complicated, and of multiple steps, but oxygen comes from the decomposition of water. So plants produce oxygen, which animals consume, and a steady state has been established. And so, the current 21% figure (oxygen in the air) seems to be more or less constant (see also Chap. 3).

There is no other good way of making free oxygen in nature. An alternative is a direct decomposition of water by sunlight. It does occur, but it is not very significant. Photosynthesis then must have been the only significant means to create the free oxygen in the atmosphere throughout the history of the Earth. So when did photosynthesis start or rather when did the first photosynthetic organisms emerge? By the way, there are other types of photosynthesis where water is not decomposed; for example, some photosynthetic bacteria use hydrogen sulfide (H_2S) instead of water. Hence, what we are interested in is "water-decomposing" photosynthetic organisms. The earliest such organisms are believed to be similar to the contemporary cyanobacteria (often called bluegreen algae). When they emerged has not been determined unequivocally, but many people believe that it was sometime around 2.7–3 billion years ago. By the way, the Earth is 4.6 billion years old, and the earliest organisms are believed to have emerged around 3.5–3.8 billion years ago.

Cyanobacteria kept proliferating and releasing an increasing amount of oxygen ever since about 3 billion or so years ago. However, the oxygen content in the atmosphere stayed quite low for a long time, up until about 2.2 billion years ago (a current hypothesis says so). There were a vast amount of oxygen-consuming materials (oxygen sink) in the ocean; the most important was iron. The result was the formation of a vast amount of iron ores, known as "banded iron formation." The main iron ore in Minnesota, for example, is of this type. (This issue is discussed further in Chap. 14).

The oxygen content of the atmosphere started to go up about 2.2 billion years ago, perhaps, because the consumable iron in the ocean had been exhausted about that time. Up until then, the majority of organisms were living without oxygen, that is, "anaerobic," including most of cyanobacteria. They could not live in the presence of oxygen, because they were made of compounds that readily react with oxygen and thus would be destroyed when exposed to oxygen. So the increase of oxygen in the air was a grave pollution threat for the majority of the organisms then living, and most of them did actually perish. Some organisms found ways to defend themselves against this terror, and they and their descendants including us have survived to this day.

2.2   Biological Functions of Oxygen

An oxygen atom binds fairly strongly with other atoms, and as a result, an oxidation of a compound by oxygen tends to produce stable end products. Hence, such an oxidation reaction will release a lot of energy (heat) and is a preferred reaction, that is, thermodynamically preferred. For example, the oxidation of methane (CH_4) will produce very stable compounds, carbon dioxide (CO_2) and water (H_2O), and will release energy of 890 kJ/mol (or 13,300 kcal/1 kg). Methane is the major component of the natural gas; hence, this heat is what we use for cooking and house heating. Likewise, we burn our food with oxygen in our body to extract energy for our lives (see Chap. 3). This is the positive side of oxygen.

2.3   Toxicity of Oxygen

Oxygen (O_2), more precisely dioxygen, is a rather abnormal molecule, as mentioned in Chap. 19. The majority of stable molecules including H_2 (hydrogen molecule), proteins, carbohydrates, and DNAs have even numbers of electrons, and all of the electrons in these molecules are paired up. [Just a brief review: An electron is like a tiny magnet, which can be positioned in two ways. In one position the north (of the magnet) is up, and in the other the north is down. In ordinary molecules, electrons combine in such a way that an electron of the north up pairs up with another of the north down. As a result, the magnetic effects of electrons are canceled in such a molecule. This situation is called "diamagnetic" (not magnetic)]. On the other hand, O_2 has two unpaired electrons in its most stable form (the ground state). It can be expressed as ˙O=O˙, where the dot represents a single electron. A molecular entity with unpaired electron(s) behaves "paramagnetically" (like a magnet), and is also called a "free radical." In general, a free radical is very reactive and tends to acquire an electron to pair up. However, dioxygen has two unpaired electrons (and hence called "biradical"), and it is relatively nonreactive toward ordinary compounds with no unpaired electrons. For example, when hydrogen (H_2) is mixed with O_2, the reaction to form water ($2H_2 + O_2 \rightarrow 2H_2O$) should occur potentially, but it doesn't, at least not at an appreciable rate (see Chap. 19). That is why dioxygen would not readily react with us, animals and plants, though oxygen potentially can burn us.

But if you produce a free radical (with a single unpaired electron), oxygen immediately reacts with it. When you strike a match, for example, you are producing a small amount of such free radicals. Once free radicals form, they react with oxygen, and this reaction results in the formation of free radicals. Let us see that in the form of chemical equations:

$$F^{\cdot} \text{(a free radical formed)} + O=O \rightarrow F-O\,O^{\cdot}$$

$$F-OO^{\cdot} + H-R \text{ (gasoline or wood)} \rightarrow F-OO-H+R^{\cdot}$$

then $R^{\cdot}$ will carry on similar reactions.

In the first step, the odd electron on $F^{\cdot}$ pairs up with one of the unpaired electrons on the oxygen. In the second step, the free radical $FOO^{\cdot}$ abstracts a hydrogen atom from another molecule. This kind of reaction can automatically produce reactive entities (free radical in this case), and hence it will continue, once started. Besides, the reaction with oxygen produces a lot of heat, and the temperature will rise, accelerating the reaction; thus combustion takes place. The chemical equations given above represent only a few important reactions involved in combustion. The whole process of combustion is very complicated. Because the cell membrane is made of essentially the same type of compounds as gasoline (called hydrocarbons), as we see in Fig. 1.5 (Sect. 1.2), that is RH in the above equations, the cell membranes will be attacked by oxygen once a free radical is somehow produced. A result of such a reaction is the formation of ROOH entities (they are called hydroperoxides). Hydroperoxides further decompose resulting in damages in the structural integrity of cell membranes. This is considered to contribute to the aging processes of cells.

Oxygen, that is, dioxygen can obtain one more electron: $O_2 + e$ (electron) $\rightarrow OO^{\cdot-}$ ($O_2^{\cdot-}$). The resulting entity has now a single unpaired electron, and it is called "superoxide" (free radical). This is much more reactive than oxygen (dioxygen) itself. For example, it can abstract a hydrogen atom as in the second step of the equation shown above. Therefore, superoxide is much more damaging than the ordinary oxygen. Superoxide forms as a byproduct in several enzymatic reactions involving oxygen and is intentionally produced in some immunological cells to attack and kill the invading bacteria.

A hydroperoxide ROOH (e.g., CH_3OOH) will be one of the first products of free radical chain reaction of O_2 or O_2^- with hydrocarbons. A hydroperoxide is reactive and will react in the following manner with iron (Fe(II)), for example: $ROOH + Fe\,(II) \rightarrow RO^{\cdot} + OH^- + Fe(III)$. A similar reaction can happen with hydrogen peroxide $HOOH + Fe\,(II) \rightarrow HO^{\cdot} + OH^- + Fe\,(III)$. The free radicals formed in these reactions, $RO^{\cdot}$ and $HO^{\cdot}$ are extremely reactive and damaging to cells and tissues. These reactions of oxygen, superoxide, hydroperoxide, hydroxy radical ($HO^{\cdot}$), or alkoxy radical ($RO^{\cdot}$) are the reasons for the toxicity of oxygen. The reactive oxygen-free radicals (such as $HO^{\cdot}$) are considered to be involved in damaging DNAs as well, and the damage may lead to mutation and hence, a cancerous state.

A special kind of iron compound is contained in a number of proteins and enzymes involved in respiration. The iron compound takes the form of either Fe_2S_2

or Fe_4S_4 (called "iron–sulfur cluster" – see Chap. 6). The iron in some of these clusters can readily be oxidized by oxygen, and the protein or the enzyme that contains it then loses its biological activity. This is another source of oxygen toxicity. Some organisms take advantage of this fact, and use such an iron–sulfur cluster as a monitoring device for the presence of oxygen.

2.4   Biological Strategies Against Oxygen Toxicity: Antioxidants, Etc.

As expounded above, the rise of the oxygen content of the atmosphere starting around 2.2 billion years ago created a panic among the then-living organisms. Many of them did not survive. The rise was not abrupt, and hence the organisms had chances to evolve to develop the defense mechanisms against this toxicity.

One of the most basic ways to do this is to consume oxygen in the cell all the time so that the free oxygen level in the cell may be maintained at a very low level. Some organisms invented the "aerobic respiration." The energy that is potentially available in foodstuff was not fully utilized in the "anaerobic" organisms. In other words, carbohydrates are only partially oxidized in the anaerobic metabolism (it is called "fermentation" or glycolysis). Most of the energy inherent in carbohydrates is unused or wasted. You can further oxidize the products of fermentation, but that requires oxygen. This is what some organisms devised. They succeeded in developing ways and means to oxidize them further using oxygen and extracting much more energy from the same amount of food, and simultaneously keeping the intracellular oxygen level very low. This is the aerobic respiration (see Chap. 3).

Of course, this may not be sufficient to counter the toxic effects of various types of oxygen-derived entities. Therefore, organisms have developed various means to destroy some very toxic oxygen entities, superoxide and hydroperoxide (and hydrogen peroxide). They have not developed very effective specific means to combat very reactive $HO^{\cdot}$ or $RO^{\cdot}$ radicals.

Hydrogen peroxide (HOOH) forms as a product of certain enzymatic reactions and is a stronger oxidizing agent than oxygen (O_2) itself. It can be decomposed by an enzyme called catalase: $2HOOH \rightarrow O_2 + 2H_2O$. You might have seen the action of catalase yourself. When you cut your hand, sometimes your mother applied a solution that contained hydrogen peroxide to the cut. It hurts but is supposed to disinfect. You might have observed that bubbles formed on the applied spot. The blood on the cut contains catalase, and that decomposes the hydrogen peroxide and forms oxygen gas that forms bubbles. This oxygen is slightly different from the oxygen in the air and has a stronger reactivity (thus, called "active oxygen") and hence is supposed to kill bacteria. Anyway, catalase is widely distributed in your body, as well as among different organisms. Catalase has iron in it, which acts as the catalyst for decomposing hydrogen peroxide.

In recent years, another group of enzymes has been discovered that quickly decomposes hydrogen peroxide. It is called peroxiredoxin and is found both in humans and bacteria. In this enzyme, the catalytic element that decomposes hydrogen peroxide is sulfur, the sulfhydryl group of cysteine amino acid.

Hydroperoxides (ROOH) can be decomposed by enzymes called peroxidases. Most of peroxidases are dependent on iron as catalyst, but a few of them use manganese. A very interesting peroxidase, called glutathione peroxidase, uses selenium as its catalytic element. Selenium is one of the most toxic elements, but a very small quantity of it is required by almost all organisms. It is used for glutathione peroxide as well as a few other enzymes. This enzyme is much more efficient in destroying certain types of hydroperoxides than the other more common iron-dependent peroxidases.

A more toxic superoxide free radical O_2^- is taken care of by enzymes called superoxide dismutases. They catalyze this reaction: $2O_2^- + 2H^+ \rightarrow O_2 + H_2O_2$. That is, it converts superoxide into less toxic hydrogen peroxide and oxygen.

The derivatives of oxygen such as superoxide and hydroxyl radical are damaging to cells, and hence they are often called "active oxygen." Another type of active oxygen is the so-called singlet oxygen. At the beginning, we talked about the fact that O_2 (dioxygen) has two unpaired electrons. That is, the most stable form of dioxygen. However, these two electrons can pair up, and the O_2 molecule in such a situation is called "singlet oxygen." The energy of singlet oxygen is higher than the regular dioxygen. This means that the singlet (di)oxygen is less stable than the regular dioxygen. Such an oxygen molecule can form, for example, as a result of radiation of ultraviolet ray on the regular oxygen. The singlet oxygen has different reactivities from that of regular oxygen and sometimes leads to damage of DNA and other cell components. Hence, the singlet oxygen is often considered to be one of the active oxygen (species).

Some naturally occurring compounds act as scavenger or quencher for these (bad) active oxygen species. These compounds are often called "antioxidants." Vitamin C, E, and K react readily with the free radical entities, superoxide and hydroxyl radical, and make them nonradical, nonreactive compounds. These are called scavengers. Red wine has been recognized to be good for health. Some of the chemicals (particularly, polyphenols) contained in red wine are believed to act as free radical scavenger. Vitamin A and its relative carrotene (the orange pigment of carrot) are known to convert the singlet oxygen to the regular form. This process is a kind of quenching.

2.5   Air Pollution

The oxygen in the air is in fact quite toxic as mentioned in the previous sections. But it is not considered to be a pollutant, as it is essential as well to the living organisms. Any extraneous compound that is put into the atmosphere can be potentially a pollutant. Three different kinds of pollutant can be recognized. One kind affects the physiology of plants and animals and causes ill effects on their health. Acid rain-causing and smog-causing compounds are such examples. The second kind is to increase the greenhouse effect of the atmosphere. Carbon dioxide is one of the important greenhouse effect enhancing gases, but it does not cause any significant physiological harm to living things at the current ambient level. It must be noted, though, that at a high level carbon dioxide can suffocate people, as some people in Cameroon were killed by an explosive release of a large amount of carbon dioxide from a lake. Carbon dioxide is not the only greenhouse enhancing gas; almost any organic compounds as well as water vapor can act as a greenhouse gas. The third kind causes depletion of ozone in the stratosphere.

2.5.1 Acid Rain and Smog

Modern human living produces a lot of different chemicals. For example, we need electricity for our living. How do we produce electricity? Most of electricity comes from power generation stations, though a significant amount is obtained from hydroelectric generators in some countries. In the power generation station, coal or petroleum is burned to produce heat that produces steam that drives the electric generator. Coal is chemically mostly carbon, and petroleum is made of mainly carbon and hydrogen. Therefore, the burning of coal or petroleum should produce only carbon dioxide and water. Carbon dioxide may contribute to the greenhouse effect but is not toxic to organisms.

The problem is that the fossil fuels contain small quantities of sulfur and nitrogen compounds (and others). After all, they have arisen from living organisms that were made of carbon, hydrogen, nitrogen, oxygen, and smaller quantities of phosphorus and sulfur (plus many microelements – see Chap. 6). Therefore, all fossil fuels contain small quantities of sulfur and nitrogen, though the actual content varies widely depending on the location of the source.

When coal or petroleum is burned, sulfur dioxide and some nitrogen oxide will be produced. Sulfur dioxide (SO_2) can react with oxygen in the air under a certain condition and forms sulfur trioxide (SO_3). Sulfur trioxide then reacts with water in the air or rain and turns into sulfuric acid (H_2SO_4) and the reaction is: $SO_3 + H_2O \rightarrow H_2SO_4$. This is a simplified picture of how acid rain occurs. Sulfuric acid is one of the most corrosive acids and affects the physiology of plants and animals, if inhaled, and corrodes buildings and statues made of marble, concrete, or metal. Sulfur dioxide itself forms sulfurous acid ($SO_2 + H_2O \rightarrow H_2SO_3$). Sulfurous acid is also corrosive but less so than sulfuric acid. Nitrogen oxide can also form acid when it reacts with water. And, this acid also contributes to making acid rain. Why a strong acid like sulfuric acid is corrosive is illustrated by its effect on a marble statue. Chemically, it is due to reactions such as:

$$CaCO_3 \,(\text{marble}) + 2H_2SO_4 \rightarrow Ca(HSO_4)_2 \,(\text{water soluble}) + H_2O + CO_2$$

$$2CaCO_3 + 2H^+ \,(\text{from the acid}) \rightarrow Ca(HCO_3)_2 \,(\text{water soluble}) + Ca^{2+}$$

That is, the marble will be washed away by acidic rain.

The internal combustion engine in our car burns gasoline fuel. Gasoline is a mixture of hydrocarbons that are made of elements carbon and hydrogen. So when gasoline is burned, carbon dioxide and water will form as the end products, neither of which is toxic. A little of gasoline, unburned or partially burned, may be spewed out from the exhaust pipe. The car uses air for the combustion. Air consists of nitrogen and oxygen. Oxygen is used to burn gasoline. No problem! Nitrogen, though quite unreactive as mentioned earlier, can react with oxygen at high temperatures in the engine. This reaction will produce nitrogen oxide(s). Nitrogen oxide may consist of several different chemical species: N_2O, NO, NO_2, but mostly "NO." These compounds will form acids, when bound with water.

Hydrocarbons or their partially burned compounds will rise to the higher level in the atmosphere and so will nitrogen oxides. At higher level under the strong effect of sunlight (particularly, ultraviolet portion of sunlight), hydrocarbons (or partially burned compounds) react with a reactive chemical species OH (you recall this is called "hydroxyl radical") in the atmosphere. This reactive species forms under the effect of sunlight. Some of the compounds that form in these reactions are aldehydes. (Aldehyde has a chemical formula RCHO in general; R can be H, CH_3, or others.) Some of them are toxic themselves. "NO" reacts with OH radical (or O radical) and turns into NO_2 (nitrogen dioxide). NO can also turn into NO_2 by just reacting with O_2. NO_2 is colored yellow to brown. This is the cause of the brown haze in the smog. Smog is usually caused by the effect of sunlight, and so it is more properly called "photochemical smog." NO_2 and an intermediate in the hydrocarbon's reaction then react and form peroxyacetyl nitrate (PAN) (chemical formula is CH_3COONO_2). PAN and aldehydes are the eye- and throat-irritating components of the smog. In the lower level of atmosphere, the reactive species OH reacts with oxygen (dioxygen) and forms ozone (O_3). Ozone is also toxic and has a pungent smell. All these chemical species that form under the strong sunlight contribute to the smog.

2.5.2 Greenhouse Effect

Global warming is a topic talked about every day and everywhere these days. It may be contributing to the increase in the formation of tornadoes and other severe weather phenomena. Carbon dioxide (CO_2) is blamed mainly for its cause, though there are other causes natural and anthropological. Carbon dioxide comes from a number of sources. We, living organisms, burn our foodstuff (in physiological sense) and turn the organic compounds in the food into water and carbon dioxide, which we exhale. We cannot help but to release carbon dioxide. The quantity of carbon dioxide released by all living organisms is significant, but carbon dioxide is also consumed by another type of organisms, that is, plants. Plants use carbon dioxide and convert it to carbohydrate (food) through photosynthesis (see Chap. 3). Release and consumption of carbon dioxide seems to have been well balanced before the civilization of mankind started to release a lot of carbon dioxide into the atmosphere. We burn fossil fuel to energize most of things: automobile, production of electricity, and so forth. This will produce carbon dioxide (and water). As a result, the content of carbon dioxide in the atmosphere is rising. Why does this cause the global warming? It is because carbon dioxide is a greenhouse gas.

What is the greenhouse effect? This is more of a physics question, but relevant here. So we will talk about it a little. A greenhouse lets sunlight in. Sunlight is mostly light of wavelength of visible range and just outside of visible range. The light shone on any material excites the motions of atoms and molecules in the material; this will show up as temperature rises. That material reradiates energy as heat or rather light in the infrared range. The night vision camera or eyeglasses detect infrared radiation coming from an object. The infrared radiation is dependent on the temperature of the object. So a living human body will show up clearly in the night vision camera, though it is dark. [Let us review the wavelength of different types of light (electromagnetic wave). A light of wavelength range of 200–350 nm (nm = nanometer = 10^{-9} m) is called ultraviolet; the visible range (i.e., light the human eyes can detect) is 350–800 nm; the range from about 1,000 nm (1 mm = micrometer) to 50 mm or so is called "infrared."]

Scientists have derived an equation that relates the wavelength of radiation to the temperature of the radiating body. This is known as the blackbody radiation equation. Using this equation, we can calculate how radiation from a body will distribute in terms of wavelength. For example, the surface temperature of our sun is about 6,000 K (K = Kelvin is degrees in Celsius plus 273). The radiation from this body (Sun) can be estimated to be centered around 480 nm or so (this is in the visible range). This agrees with the observation of sunlight. The radiation from a body whose surface temperature is 15–25°C (60–77°F) centers around 10 μm, which is in the middle of infrared range.

Now let's get back to the greenhouse effect. The material inside a greenhouse becomes heated by sunlight that gets in. Suppose that the temperature becomes 25°C; then the inside surface of the greenhouse radiates infrared light centering around 10 μm. Now the crux of the matter is that the glass (or plastic) window of the greenhouse can let visible sunlight through, but that the glass absorbs the infrared light, that is, would not let it through. The glass then radiates infrared light, most likely about the half of it to the outside and the rest to the inside. So, only about half of the heat generated inside can escape to the outside. Hence, there is a further heat up inside the greenhouse. Heat can also escape through the glass by heat conduction, which is dependent on the temperature difference between inside and outside.

On the Earth, the atmosphere functions as the glass of greenhouse. Sunlight comes through the atmosphere. A number of compounds in the atmosphere absorb some portions (particularly, in the ultraviolet region) of sunlight, but a substantial portion of it reaches the surface of the Earth. We live on that light. It also heats the Earth. Now the Earth's surface radiates back infrared light. And because the surface temperature is somewhere around 15°C on the average, the infrared radiation centers around 10 μm. A portion of it then escapes into the space, and a steady state has been established between the incoming energy flux and the outgoing energy flux. Hence the surface temperature has been more or less constant overall.

Atmosphere contains carbon dioxide and other compounds as minor components. Carbon dioxide absorbs infrared light. The major components of the air, nitrogen and oxygen, do not absorb infrared light. So, the infrared light radiated from the Earth's surface is absorbed by carbon dioxide. Carbon dioxide will reradiate the heat about half into the outer space and about half is radiated back onto the Earth's surface. As the concentration of carbon dioxide in the atmosphere increases, the heat trapped inside the atmosphere will then increase and heat up the Earth's surface. This is the mechanism of greenhouse effect of carbon dioxide, though quite simplified, and is considered by many to be the major cause of the current global warming.

Carbon dioxide is not the only greenhouse gas. As can be inferred from the description above, any compound that absorbs the appropriate portion of infrared range will act as greenhouse gas. Methane, the major component of natural gas, is one such example. Methane comes from microorganisms (methane bacteria), and also cows and other animals produce methane in their digestive systems. Some people even think that methane may be the major cause of greenhouse effect. Some chlorofluoro carbons that are discussed below in connection with the ozone depletion are also effective greenhouse gases. No matter what, carbon dioxide is believed to be the most important greenhouse gas, as we human beings burn fossil fuels as well as any material containing carbon (wood, for example) as an energy source, and that inevitably produces carbon dioxide. It might be pointed out that the temperature of the surface of Venus is about 430°C, partially due to the predominant presence (about 90%) of carbon dioxide in its atmosphere.

2.5.3   Ozone Depletion

In the previous section we talked about how sunlight is partially blocked by the Earth's atmosphere. What blocks the sunlight? A number of minor components in the atmosphere absorb light in the ultraviolet region. Examples include some of the organic compounds such as aldehyde(s) and benzene, and inorganic compounds such as N_2O, NO, and NO_2. The most important of all such compounds is ozone (O_3); that is, it is made of three oxygen atoms. The composition tells you that it will form when O_2 (dixoygen, the form found in the air) reacts with O (oxygen atom). How ozone is produced and decomposed is a crucial issue.

In order to understand this matter, we need to introduce the concept of the energy of light. Light is a wave and hence is characterized by its wavelength λ or frequency v. The energy of a light with frequency v is given by hv, where h is Planck's constant (a universal physical constant). As wavelength l and frequency v are related by an equation $c = \lambda v$, the energy of a light is related to its wavelength in the form of hc/λ, where c is the speed of light. This says that a light of a shorter wavelength has a larger energy than that of a longer wavelength. Hence, ultraviolet light (shorter than 320 nm; nm = nanometer = one billionth of meter = 10^{-9} m) has larger energy and is more damaging than visible light (320–800 nm).

Now, the sunlight consists of lights of varying wavelength centered about 480 nm as mentioned earlier. It contains light with shorter wavelength than that as well as light with longer wavelength. When it hits the upper layer of the Earth's atmosphere, it cleaves the bond of O_2 (dioxygen, i.e., the regular oxygen molecule in the air). This bond is difficult to split, as it requires a lot of energy. So it occurs only when dioxygen absorbs light with wavelength shorter than 240 nm (way into the ultraviolet range). As a result, dioxygen splits into two oxygen atoms, and simultaneously the sunlight shorter than 240 nm will be consumed or stopped there and would not reach the lower part of the atmosphere. The oxygen atom produced then reacts with dioxygen and forms ozone (O_3). The O–O bond in an ozone molecule is more easily split than that in a dioxygen molecule. So the decomposition of ozone occurs with light of longer wavelength and is known to occur with light of 320 nm or shorter. This means that light of wavelength shorter than 320 nm (i.e., ultraviolet light) will be consumed to decompose ozone, and not significant amount of ultraviolet light reaches the Earth's surface. So ozone layer will act as a filter of ultraviolet light.

The reactions involved in these processes are:

$$O_2 \rightarrow 2O \ \text{(by light with wavelength shorter than 240nm)}$$

$$O_2 + O \rightarrow O_3$$

$$O_3 \rightarrow O_2 + O \ \text{(by light with wavelength shorter than 320nm)}$$

$$O_3 + O \rightarrow O_2 + O_2$$

These reactions constantly generate and decompose ozone, and a steady-state ozone layer has been established in the upper level of atmosphere. The maximum level of ozone occurs at about 18–20 km above the Earth's surface. This layer protects all living organisms from being exposed to the dangerous ultraviolet light. As a matter of fact, the creation of the ozone layer is believed to have made possible the emergence of terrestrial (land) organisms.

A satellite Nimbus has on board a spectrometer to measure ozone level in the atmosphere. Data from the satellite showed that the ozone level above the Antarctic continent during the winter season started to decrease noticeably from about 1980. It developed further to form almost a hole (no significant ozone there); hence, this phenomenon is known as "ozone hole." There is a sign that the range of ozone hole is further spreading. Even in the Northern Hemisphere, a region of low ozone level has developed over the Arctic ocean and the Northern Siberia. Is the decrease of ozone level occurring only above the Polar regions? No, it is spreading to the southern mid-latitude. A decrease of ozone in the upper atmosphere means that an increased amount of ultraviolet will reach the Earth's surface.

The critical question is what is causing the decrease of ozone level in the atmosphere. Scientists, particularly atmospheric chemists, tried to figure out this problem for the last 2 decades, by lab works and computer modeling (i.e., simulating what might be happening in the atmosphere), and measuring the levels of various chemical species in the atmosphere. The consensus of these investigations is that ozone is being decomposed chemically by free radical species. As mentioned somewhere else, a free radical is a chemical species with an unpaired electron. A free radical X reacts with ozone (O_3) in the following way:

$$O_3 + X \rightarrow O_2 + XO$$

$$XO + O \rightarrow O_2 + X \text{ (then X will repeat the same cycle of reactions)}$$

X can be almost any free radical species including NO and OH, but chlorine (and bromine) has been recognized to be the major culprit.

Where does "chlorine" come from? It is believed to come from chlorofluoro-carbons (CFCs; a collective name for chloro/fluoro derivatives of hydrocarbons). CFCs have been manufactured since the 1930s. CFC-12 (chemically CF_2Cl_2, dichloro difluoro methane) was introduced originally as refrigerant. CFC-12 is non-toxic and nonflammable, making the refrigerator safe enough. Other CFCs have also been used for refrigerators, as an expanding agent for foams, propellants for aerosol sprays (CFC-11, $CFCl_3$), and cleaning agents for microelectronic compo-nents (chiefly CFC-113, CF_2ClCF_2Cl). CFCs in the troposphere were about 50 parts per trillion (ppt) by volume in 1971 and had risen to about 270 ppt by 1993.

CFC-11 and -12 are not reactive in the troposphere; they go up to the stratosphere where they are susceptible to a slow decomposition by ultraviolet light (shorter than 250 nm):

$$CF_2Cl_2 \rightarrow CF_2Cl + Cl \text{ (by light shorter than 250nm)}$$

The chlorine atom (Cl) thus produced then reacts with ozone, destroying it. (C–F bonds would not be split easily by the ultraviolet light.) Other CFCs behave simi-larly. The problem with CFCs is that they are quite inert compounds and are slow to decompose to form Cl and hence they persist in the stratosphere. Their lifetimes have been estimated to range up to 100 years.

Life Itself (A): How Do We Get Energy to Live by?

3

(The Chemical Logic of Life on Earth)

Life! This wonderful thing! What is it to be alive? Where did life come from? How do living things operate themselves? There are a number of different kinds of living organisms on this planet. How are they related and different from each other? How have they evolved? These are "biology" questions. But they require some answers from chemistry as well. We do not have all the answers yet, though. The chemical bases of the organisms currently living on the Earth are fairly well understood, thanks to the scientists' efforts in the last hundred years or so, particularly in the last half a century. We have now a discipline called "molecular biology" or "chemical biology," which strives to understand the biological phenomena, particularly those associated with genes (DNA) in terms of chemistry. Yet we have not come to a full understanding of even the lowly bacterium, say, an *E. coli*, in so much as we may recreate it. We have no more than the vaguest ideas about how life might have begun on the Earth.

Let us now look at ourselves. What do you see in a mirror? Head, hair, face, hands, and maybe teeth, etc. They are all made of proteins, except the teeth. The bulk of our body is in fact made of proteins. Our body is, anatomically speaking, made of tissues and organs that are made of cells. The cells are wrapped by membranes whose major component is called "phospholipids." So the skin you see is made of phospholipids and proteins. The other two important types of chemicals for organisms are not directly obvious. When you observe a cell of your body under a microscope, you will observe two main types of small bodies in a cell. One is "(cell) nucleus" and the other "mitochondrion"; there is only one nucleus but several mitochondria in a cell. The latter is where cells produce energy by burning our food, carbohydrates. The nucleus is where all the information cells need is stored, that is, where DNAs are located. Protein(s), carbohydrate(s), DNA (and RNA), and lipid(s) (fat) are considered to be the four major types of biocompounds. In addition, vitamins and hormones are essential but required in very small quantities. Each of these is a collection of various types of compounds. These are treated in the discipline of biochemistry.

To talk about these compounds and their behaviors fully requires a large amount of time and space. As a matter of fact, a typical textbook of biochemistry is huge, and often more than a thousand pages long. A proper learning of biochemistry

E. Ochiai, *Chemicals for Life and Living*,
DOI 10.1007/978-3-642-20273-5_3, © Springer-Verlag Berlin Heidelberg 2011

requires at least two semesters in college. I would not attempt to tell you all the details of these interesting chemicals and their behaviors. Instead, I will present a fairly simplified picture of the chemistry of life on Earth, not so much as the material of life as how life processes are carried out in the sense of chemical reactions.

We are here talking about mostly organic compounds. Chemical reactions they undergo are mostly of two different types. One can be called "acid–base" type and the other "oxidation–reduction" type. In fact, these are the two basic types of chemical reactions, whether organic or inorganic (see Chap. 19). Some reactions, however, may not easily be classified into either of these. One exception is reactions of free radical type. One such example is discussed in Sect. 5.3.3.

3.1   Reactions of "Acid–Base" Type

"Acid and base" was talked about in other chapters (for example, Chaps. 1 and 19). The sourness of vinegar is due to an acid, acetic acid CH_3COOH, which gives rise to H^+ (proton) that is responsible for the acidity. Caustic soda $NaOH$ is a typical base, which gives rise to HO^- (or OH^-). [OH^- is called "hydroxide ion."] According to Jack Lewis (a British chemist), a base is defined to be a donor of a pair of electrons and an acid is an acceptor of such a pair (from a base). Acid/base thus defined is called "Lewis acid/base." This pair of electrons on a base is then used to form a bond to an acid. Recall that a bond is formed typically by a pair of electrons occupying the space between two atoms. The simplest example is $HO^- (\text{base}) + H^+ (\text{acid}) \rightarrow HO - H$. An essential feature here is a movement of a pair of electrons. In this case, a pair of electrons on oxygen of OH^- moiety is moved toward H^+. In a reverse reaction, a pair of electrons occupying the space between O and H will move toward the oxygen atom, and as a result: $H - OH \rightarrow H^+ + OH^-$. A base can be any entity which can provide easily a pair of electrons. A good example is ammonia, NH_3; its nitrogen atom has a pair of electrons (called "lone pair"). Hence it can, for example, do the following: $H^+ + NH_3 \rightarrow {}^+NH_4 \, (\text{ammonium ion})$.

We will extend this idea to other more common types of reaction. For example,

$$H_3C - Br + {}^-OH \rightarrow H_3C - OH + Br^-. \tag{3.1}$$

The reaction of this type is called "substitution reaction," as the OH^- entity substitutes for Br^-. In this reaction, the pair of electrons on OH^- entity approaches the carbon atom and displaces Br^-. In this process, the two electrons between C and Br end up with Br. That is:

As in the above simpler acid–base reaction, a pair of electron comes from only one of the atoms involved in a bond formation or a pair of electron ends up on only one of the atoms in bond breaking. This type of bond cleavage or formation is called "heterolytic." In other words, reactions that involve "heterolytic" bond cleavage or bond formation are defined to be of "acid–base" type. A heterolytic bond cleavage and formation are simultaneously taking place in substitution reaction. The formation of water from an acid (H^+) and a base (OH^-) is a very strong reaction in the sense that the reaction gives off a lot of heat (energy) or more properly that the (Gibbs) free energy change is a large negative value. Formation of a bond always releases energy. Substitution reactions of "acid–base" type such as reaction (3.1) are accompanied by a relatively small change in the (Gibbs) free energy. The reason is that such a reaction involves substitution of an entity by another similar to it and ends up in compounds that are similar to the starting compounds, as exemplified above (3.1). Or we may say that the bond formation is energy releasing, but the bond splitting is energy absorbing, and so that the energy change in a substitution reaction is small, as it involves both bond formation and bond cleavage.

The majority of biochemical reactions that are involved in the degradation (digestion) of foodstuffs (polysaccharides (such as starch), proteins, etc.) and the building of biologically important compounds such as DNAs, RNAs, and proteins are of "acid–base" type.

The chemical reactions involved in the digestion of foodstuff are essentially "hydrolysis." Let us take an example of proteins (i.e., meat). A protein is made of a number of small molecules called amino acids. There are about 20 amino acids in nature. You might have heard "glutamic acid," "histidine," "tryptophan," etc.; these are examples of amino acids. Their general chemical formula is R-$CH(NH_2)$-$COOH$; different amino acids have different R groups. Two amino acids, say, $R^1CH(NH_2)COOH$ and $R^2CH(NH_2)COOH$ combine in the following manner:

$$R^1CH\left(NH_2\right)COOH + \left(NH_2\right)CH\left(R^2\right)COOH$$

$$\rightarrow H_2O + R^1CH\left(NH_2\right) - CONH - CH\left(R^2\right)COOH$$

In the process, one molecule of water is removed, and a new connection through a "–CO(NH)–" bridge (called "peptide bond") is formed between the two amino acids. This kind of reaction is called "condensation." Condensation is a reaction in which two small molecules combine to form a larger molecule, usually shedding off a small molecule such as water as in the case above. If one continues to add on and on amino acids in this manner, one gets a protein. I hasten to add, though, that this is not the way proteins are actually produced in cells.

On the contrary, you can chop the bridge bond by adding water to a protein: $-CO(NH) - + H_2O \rightarrow -COOH + NH_2 -$. This process is the opposite of condensation and is called "hydrolysis" (decomposition by water), and this is what happens when meats are digested in the stomach and other parts of your body. Starch and other carbohydrates are also hydrolyzed into smaller units (monosaccharide), and DNAs and RNAs, too, are degraded by hydrolysis. These reactions are accompanied by a relatively small change (usually negative) of (Gibbs) free energy. [This is because a hydrolysis typically will form more fragments and hence the randomness (entropy) increases; and that helps make the free energy more negative (see Chap. 19)]. That means that a hydrolysis will proceed without added energy. All these reactions, however, need to be catalyzed by enzymes to proceed at appropriate speeds, as are all other biological chemical reactions.

To build the necessary compounds such as proteins and DNAs, you have to combine smaller units by condensation reactions. Condensation reactions are typically accompanied by positive free energy changes, though of relatively small magnitudes. Condensation is opposite of the hydrolysis above and is accompanied by the formation of a more rigid entity and hence by the decrease in randomness (entropy). That means that such a reaction would not proceed without an added (negative) energy.

All organisms on the Earth use a single chemical compound as the energy source for most of such biochemical reactions. That is a molecule called "ATP" (*Adenosine Tri Phosphate*). It is a sort of energy carrier. ATP has a condensed phosphate group (see Fig. 3.1), which will yield a substantial amount of free energy upon hydrolysis (-50 kJ/mol at pH 7 and at 5 mM concentration; the so-called standard value is -35.7 kJ/mol). The organisms use the trick to combine the hydrolysis of ATP with a difficult condensation reaction (such as combining amino acids), so that the overall free energy change of the combined reactions becomes negative. All these reactions are of "acid–base" type. Even mechanical

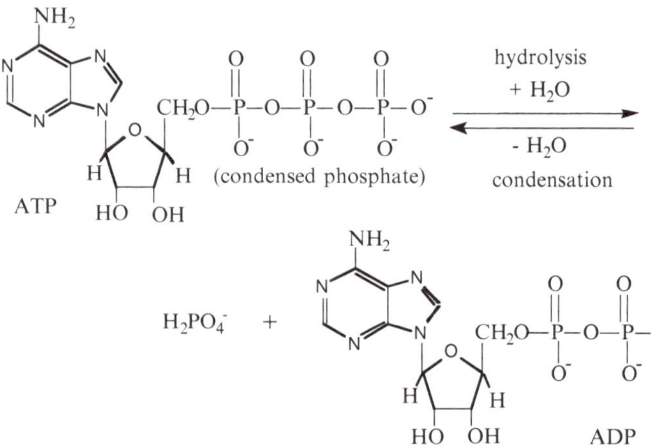

Fig. 3.1 ATP (adenosine triphosphate) – its hydrolysis and formation

works, such as muscle contraction, are carried out by use of the energy of hydrolysis of ATP.

How much of ATP do we need for a day? An interesting estimate of ATP requirement was made by the textbook authors, Garrett and Grisham ("Biochemistry," Saunders, 1995). Suppose that an average adult of 70 kg (150 lb) weight takes in 2,800 kcal/day, which is about 12,000 kJ/day. Assume that about 50% of that energy is converted into the energy of ATP. 6,000 kJ (half of 12,000) is stored in about 120 mol of ATP (=6,000 kJ/50 kJ/mol). One mole of ATP weighs 551 g; this is its molar mass. Therefore, 120 mol × 551 g/mol = 66 kg. That is, as much as one's body weight of ATP will be necessary. But, of course, not 66 kg of ATP is actually produced. As ATP is hydrolyzed, ADP is produced, and ADP is swiftly converted back to ATP, and thus recycled. In fact, only about 50 g of ATP is present in our body at any moment. If this were not the case and, instead, 66 kg of ATP was purchased, the cost of energy per person per day will be 66 kg × $10/g (current rate) = $660,000!.

Now, organisms need to produce ATP, a lot of it, as shown above. Not really a large amount at one time, but organisms have to keep producing it over and over again. In the adult humans illustrated above, they have to produce 120 mol of ATP over 24 h; this means 5 mol of ATP every hour. Since they are doing it with a supply of only about 0.1 mol of ATP material (50 g/551 g/mol), they have to reproduce in every minute all ATP they have. How? It involves oxidation–reduction reactions; i.e., the other type of chemical reactions. These reactions produce energy, and the energy is used to drive the reverse of reaction shown in Fig. 3.1 utilizing enzymes collectively known as ATPases.

3.2   Reactions of Oxidation–Reduction Type

3.2.1   Oxidative Metabolism (Catabolism)

We can remove electrons from an atom or molecule. Removal of electron in the chemical reaction is defined as "oxidation." For an oxidation to occur, there must be an acceptor of the electron(s). Accepting electron(s) is "reduction." Oxidation and reduction thus happen simultaneously. In order to see how and where electrons move in chemical reactions, it is convenient to define "oxidation state" of an atom in a molecule or compound.

For example, sodium in sodium chloride NaCl (Na^+ and Cl^-) (table salt) is in +I (+1) oxidation state and chlorine in −I (−1) oxidation state. Magnesium bromide $MgBr_2$ (Mg^{II} or Mg(II) and $2Br^-$) has Mg in +II oxidation state and Br in −I oxidation states. These are obvious cases. How about magnesium carbonate $MgCO_3$? (Mg^{II} and CO_3^{2-}) What are the oxidation states of C and O in carbonate CO_3^{2-}? There is a convention (rule 1) in which oxygen atom in all compounds except a very few is assigned −II oxidation state. Let us apply this rule. Since each O atoms carries −II and the overall electric charge of carbonate is −2, then the carbon atom in carbonate must be assigned "oxidation +IV." [Throughout this book, we would use Roman numerals to indicate the oxidation state of an atom in a chemical compound, and Arabic numerals for the electric charge of a chemical entity. A convention used in many textbooks is to use Arabic numerals to express both the oxidation state of an atom and the electric charge of a chemical species; and this is confusing].

Another convention (rule 2) is that hydrogen atom in all compounds (except when it binds to a metallic element) has +I oxidation state. So the oxidation state of carbon in methane, CH_4, must be $-IV$. How about the oxidation state of the carbon in methanol, CH_3OH? There are four hydrogen atoms, which contribute $+4$, and the oxygen carries $-II$; therefore, the carbon must carry "$-II$." How about the oxidation state of carbons in ethanol C_2H_5OH? [the answer is "$-II$"]. The carbon atoms in acetic acid, CH_3COOH, the ingredient of vinegar, should be "0, zero" in oxidation state [try it]. Carbon dioxide CO_2 has its carbon in $+IV$ oxidation state. As you see here, carbon atoms in compounds can take different oxidation states, from $-IV$ to $+IV$. "$-IV$" is the most reduced state and "$+IV$" is the most oxidized state of carbon. So, for example, in the process of reaction: $CH_4 + (1/2)O_2 \rightarrow CH_3OH$, the carbon atom can be regarded to have lost two electrons (try to see it) and hence to have been oxidized. Addition of oxygen to a compound is usually regarded as "oxidation" reaction. The chemical agent that oxidizes others is defined as "oxidizing agent." Of course, oxygen (O_2), the oxidizing agent itself, is reduced, starting from "zero" and becoming "$-II$"; so the other chemical that reacts with oxygen can be said to be a "reducing agent."

[Note: The oxidation state does not necessarily represent the actual electric charge, though it does so in certain cases including that in ionic compounds. It is a convenient way to keep track where and how electrons move in oxidation–reduction reactions].

Let us look at nitrogen cases, using rules 1 and 2. N in ammonia NH_3 is $-III$, the most reduced. N in N_2 is obviously zero. N in nitrate, NO_3^-? Try it. ["$+V$" is the right answer]. So the reaction to form ammonia: $N_2 + 3H_2 \rightarrow 2NH_3$ is a reduction as far as nitrogen is concerned, and the reducing agent here is hydrogen which is formally oxidized. To produce nitrate from ammonia you have to oxidize NH_3; that is, $NH_3 + 2O_2 \rightarrow HNO_3 + H_2O$. The oxidation state of N in NH_3 is $-III$ and that in HNO_3 is $+V$. Therefore, eight electrons have moved from the nitrogen atom to oxygen atoms in this process.

Your car can rust; it is the oxidation of the iron of your car by oxygen in the air to form iron oxide. It can be expressed formally as $4Fe + 3O_2 \rightarrow 2Fe_2O_3$. In this process, iron changes its oxidation state from "zero" to $+III$. Thus, iron loses electrons (be oxidized) to oxygen. Iron can be also in $+II$ oxidation state; that is, $2Fe + O_2 \rightarrow 2FeO$ is also possible. As a matter of fact, iron can change very readily its oxidation state between $+II$ and $+III$. And this change is very widely used in biological systems (see Chap. 6).

[Note: Rule 3 (of determining oxidation states) Halogen, F, Cl, Br, or I takes $-I$ oxidation states in compounds bound with other elements; for example, C is C^{-II}, and H^{+I} and Cl^{-I} in CH_3Cl. Exception is when they bound with each others; in this case, the lighter element carries $-I$ oxidation state. For example, in FCl F is F^{-I} and Cl^{+I}, whereas Cl^{-I} and I^{+III} in ICl_3].

Oxidation–reduction reactions tend to be accompanied by larger free energy changes than substitution reactions of acid–base type mentioned earlier. Oxidation reactions, particularly reactions with oxygen, give off a lot of energy. Biological systems convert this energy to that stored mainly in ATP, but also in carbohydrates in certain organisms. The major energy-producing foodstuff is carbohydrates, which is exemplified by glucose $C_6H_{12}O_6$. What is the oxidation state of carbons in this compound? "Zero." It (carbon in this molecule) can be oxidized eventually to carbon dioxide CO_2 (+IV oxidation state). This reaction is not a simple single step reaction. It can be defined as "energy-yielding metabolism" (of carbohydrates) and consists of many steps.

The earlier part of this entire process does not actually require oxygen; this part is called "glycolysis" (a fermentation). Oxidation in this part is done by removing hydrogen atoms. That is, removal of hydrogen can also be regarded as "oxidation." Fermentation of glucose produces pyruvic acid $CH_3(CO)COOH$. The reaction can be represented formally as: $C_6H_{12}O_6 \rightarrow 2CH_3(CO)COOH + 2H_2$. Let us use the rules outlined earlier and figure out the oxidation state of carbons in this product. It is +2/3 on each of carbon. Do not worry about the fractional value; the oxidation state of an individual atom should be an integer, and different carbon atoms in this molecule should be assigned different oxidation states. But all we need now is an average value, which in this case is a fractional value. OK, you started with $C_6H_{12}O_6$ where the oxidation state of carbons is "zero," and now ended up with $CH_3(CO)$ COOH in which the oxidation state of carbons is +2/3. So the carbon atoms must have lost electron(s); i.e., this is an oxidation. This reaction would give off energy of about −76 kJ (per mole of glucose). The net yield of ATP in glycolysis is 2 mol of ATP from 1 mol of glucose.

As the oxidation state of carbon in pyruvic acid is +2/3, it can be oxidized further (up to +IV). This is done in the subsequent respiratory process using oxygen as the ultimate oxidizing agent. First, it goes through a process called "TCA" cycle (tricarboxylic acid cycle) or "citric acid cycle," which produces $FADH_2$ (through succinic acid) and NADH in addition to some ATPs. FAD is flavin adenine dinucleotide and similar to NAD; $FADH_2$ is the reduced form of FAD. The special hydrogen atoms in $FADH_2$ and NADH are further oxidized in the final process of respiration (called "electron transport/oxidative phosphorylation") to be turned completely into water ($FADH_2$ and NADH turns to FAD and NAD^+, respectively). A number of ATPs are produced in this last process. Overall, 38 mol of ATP are produced from 1 mol of glucose. However, only 36 mol of ATP will be produced from 1 mol of glucose in brain and muscle cells, because of a slightly different type of metabolism predominates in these cells.

Let us examine the relationship between energies gained in the respiratory process and the simple combustion. Suppose that you conduct the following combustion reaction in water solution: $C_6H_{12}O_6(glucose) + 6O_2 \rightarrow 6CO_2 + 6H_2O$, the associated free energy change is −2,872 kJ per mole of glucose (180 g). As argued before, 1 mol of ATP will produce about −50 kJ of free energy; therefore, the total free energy converted in the form of ATP amounts to $40 \times (-50) = -2,000$ kJ (2ATPs produced in glycolysis is included). This is 69.6% of the simple combustion energy.

In other words, the efficiency of energy conversion in the mitochondria is 70%, more than two thirds, which is very good.

In another process similar to citric acid cycle, NADPH is produced instead of NADH. NADPH is used not to produce ATPs but to reduce some important biocompounds, such as sulfate and nitrate, and ribonucleotides (to produce the raw material of DNA) and produce some important biological compounds such as lipids.

Proteins and lipids are also degraded and partially used to make ATP. A lipid, for example, fatty acid $CH_3CH_2CH_2CH_2CH_2CH_2CH_2CH_2CH_2COOH$, has the carbons in low oxidation state [try to calculate it; $-16/10$]. So when it is oxidized, it will give off more energy (per carbon atom) than carbohydrates. That is why fat gives more calories.

A small number of organisms use compounds other than oxygen, as the oxidizing agent. Nitrate (NO_3^-) is a good oxidizing agent (see Chap. 8 on the fireworks) and is used for example by *E. coli* when not enough oxygen is available.

The things are getting messier. So let us summarize here what we talked about so far. The free energy in oxidizing glucose (net reaction: $C_6H_{12}O_6 + 6O_2 \rightarrow 6CO_2 + 6H_2O$) is utilized to convert ADP to ATP, the biological energy carrier. ATPase, one of the crucial enzymes, that is involved in the process of converting ADP to ATP (the reverse reaction in Fig. 3.1) was studied among others by Paul D. Boyer of UCLA, and John E. Walker of Research Council of Molecular Biology in Cambridge, UK. These two scientists shared a Nobel Prize in 1997, with Jens C. Skou who studied a related enzyme, Na(I),K(I)-ATPase. The details of how ATPases work are beyond this discourse.

3.2.2 Reduction

We obtain carbohydrates, our energy source, from plants. Plants produce them from carbon dioxide and water. The overall reaction can be expressed as $6CO_2 + 6H_2O \rightarrow C_6H_{12}O_6 + 6O_2$. This is opposite of the oxidative metabolism of glucose mentioned above. The reaction can be regarded to be a reduction of carbon dioxide by hydrogen that comes from water. Since oxidation of glucose is energy releasing, the formation of glucose from CO_2 and H_2O must require a lot of energy (energy absorbing). Plants use "energy from sunlight" to accomplish this feat, synthesis of carbohydrates. Hence, the process is called "photosynthesis." The details of photosynthesis are too much to talk about here. But essentially, sunlight forces electrons out of the green pigment, chlorophylls in the first part (called "photosystem I") of the photosynthetic machinery in a green leaf of plants, and the electrons are to be added to carbon dioxide (remember that an addition of electrons to a compound is "reduction"). Another part (photosystem II) of the machinery contains a mechanism to decompose water to produce oxygen O_2 and electrons. This process also uses the energy from sunlight. And the electrons produced in the photosystem II are then transferred to photosystem I to replenish the consumed electrons. The photosynthetic apparatus is contained in a vesicle called "chloroplast" in cells of green leaves and produces some ATP molecules in addition to carbohydrates.

A small number of bacteria use hydrogen sulfide (H_2S) instead of water (H_2O) as the electron source. These are called "sulfur photosynthetic bacteria" and produce carbohydrates, as well as ATP, and deposit sulfur (S) in the surroundings. Organisms that use sunlight to produce their food and energy, i.e., both water-decomposing ones (cyanobacteria and plants) and sulfur photosynthetic bacteria, are called "photoautotrophs." Some special microorganisms conduct photosynthesis using Fe(II) as the source of electrons.

Yet other organisms use chemical energy (instead of energy from sunlight) to produce carbohydrates and ATP. For example, some microorganisms oxidize iron, Fe(II) to Fe(III) by oxygen. This oxidation produces some energy, and the organisms use this energy to synthesize carbohydrates from carbon dioxide and water. Others oxidize ammonia (NH_3) (to N_2 and NO_2^-, NO_3^-) or some organic compounds using oxygen as the oxidizing agent. These organisms are called "chemoautotrophs," as they use the energy of chemical reactions to produce their own food and ATP.

There are several other important reduction processes. One is the reduction of sulfate (SO_4^{2-}) to produce hydrogen sulfide (H_2S). Hydrogen sulfide is then incorporated into organic compounds such as cysteine (one of the amino acids) and glutathione. Nitrate (NO_3^-) likewise is reduced to ammonia in most organisms. Ammonia is then incorporated into organic compounds such as proteins and DNAs. Ammonia can also be produced by reducing the nitrogen molecule in the atmosphere. The overall reaction $N_2 + 3H_2 \rightarrow 2NH_3$ does not require energy input; i.e., the free energy change is slightly negative. However, both the industrial and biological processes require a large amount of energy input to effect this reaction. The industrial process operates at high temperature and high pressure, and it requires energy. The biological process is catalyzed by an enzyme called nitrogenase, which operates under an ordinary condition, ambient temperature and pressure, but requires a lot of energy in the form of ATP to carry out the reduction. Nitrogenase (see Fig. 21.12 for the molecular structure) is found in a number of microorganims such as *Azotobacter*, *Chromatium* (photosynthetic), anaerobic *Clostridium*, and bacteria called *Rhizobium* found in the nodules on the roots of plants such as alfalfa, clover, peas, and beans, as well as some fungi. These processes contribute to the global cycling of nitrogen.

3.3 Chemical Logic of Life on Earth

Let us now see the overall picture of the chemistry involved in the life process discussed in the last two sections. Figure 3.2 provides such a picture. The ultimate source of energy of our lives and all our activities is Sun. It drives the

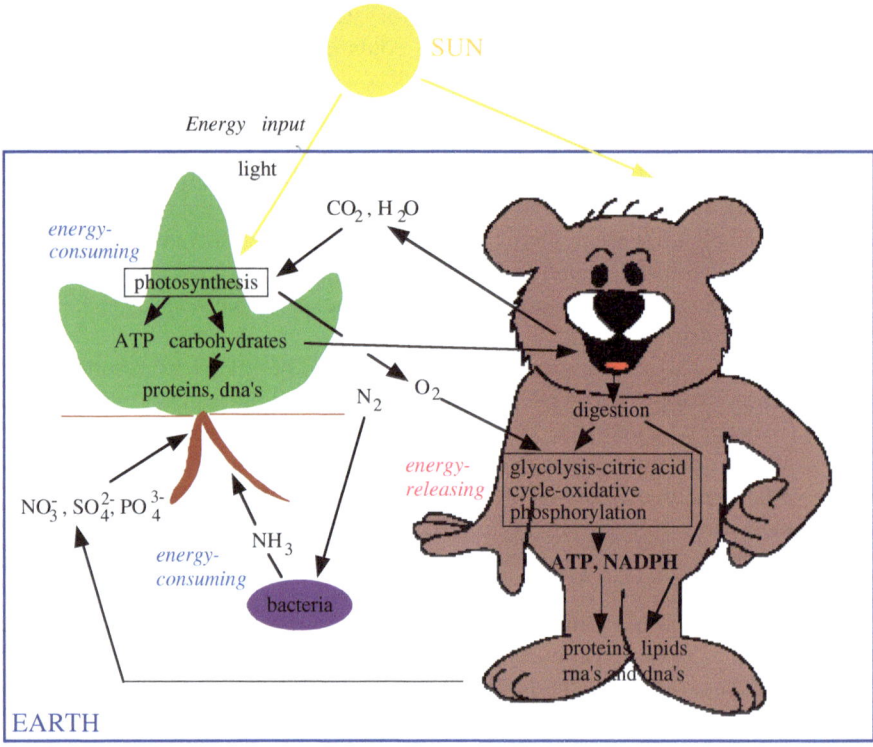

Fig. 3.2 Chemical logic of life on earth

energy-consuming photosynthesis, which produces ATP, the chemical energy carrier, and carbohydrates. Carbohydrates are the energy source for animals. Carbohydrates are degraded with use of oxygen (oxidation), and the energy released in the process is converted into the chemical energy in the form of ATP. Other accessory reactions are also indicated.

Figure 3.3 is a diagram with essentially the same information, but gives a little more details and is intended for audience more chemically inclined. It shows the overall life process on the earth. Three types of organisms are indicated: autotrophs, heterotrophs, and some special types of bacteria and fungi as shown in different colors. Major inorganic compounds are shown as cycled throughout this system. Biological functions of some inorganic compounds are talked about in Chapter 6.

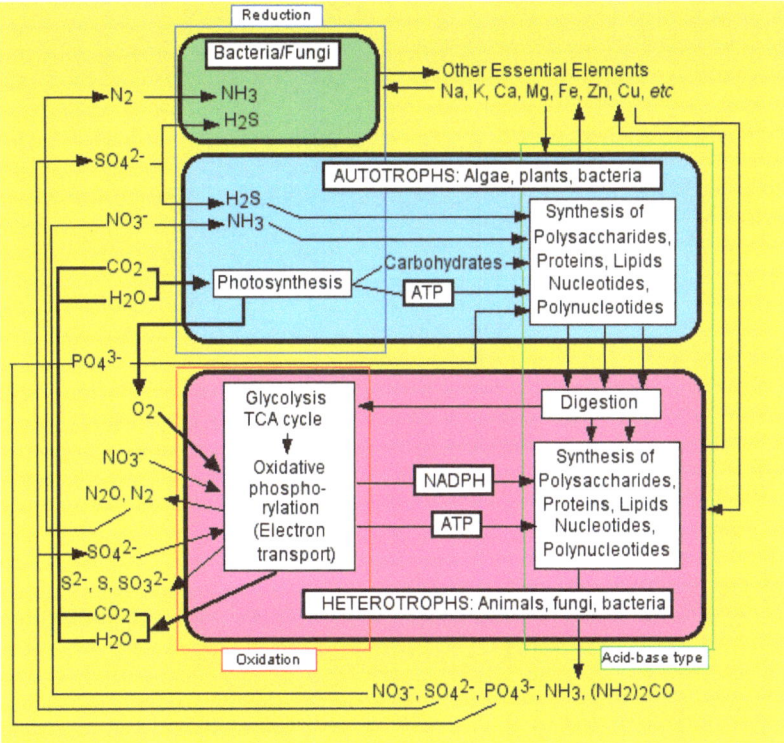

Fig. 3.3 Types of chemical reactions involved in life process on Earth (chemical logic of life on Earth)

The main biological processes are shown in boxes of thin line. The three categories of biochemical reactions are indicated as boxed in different colors. The yellow colored background represents solar energy.

Life Itself (B): Why Are We Like Our Parents?

4

Besides "being alive," the characteristic of living organisms is to produce progenies. And the progenies are like their parents. A human being begets human beings; a frog begets tadpole(s), which eventually become frog(s). A single *E. coli* bacterium cell divides into two identical *E. coli* bacterial cells. Something must be transmitted from a parental organism(s) to its progeny to direct the progeny's cells to become similar to its parent(s). This something is a "gene," and the gene will govern the characters of the organisms. This is the central dogma of "Life on the Earth." There are two things that a gene must do. One is that it should be able to duplicate itself or rather should be able to dictate its own duplication (replication). The second is that it must contain enough information to replicate the whole organism of which the gene is a part. Let us see how this is done in terms of chemistry.

It turned out that the chemical principle behind these functions of a gene is very simple and of a very fundamental chemistry. The gene had been known to be a chemical molecule called DNA. DNA is an acronym for *Deoxyribo*N*ucleic* A*cid.* However, it took a long time and an enormous amount of human endeavors to unravel the secrets of how DNA works as a gene. The efforts by scientists in this respect had culminated in a discovery of the so-called double helix structure of DNA by two (then) young scientists, James Watson and Francis Crick in 1950s. They were awarded a Nobel Prize for the discovery in 1962 (along with M. H. F. Wilkins). The story of this discovery is very well known and has been told over and over again by many people including Watson himself. So we will not repeat it. We will look at the very fundamentals.

4.1 Structure of DNA

A DNA molecule is a long chain-like chemical consisting of four units called "nucleotides" labeled as A, C, G, and T. You may visualize it as, say, a necklace made of beads of four different colors, aquamarine (A), green (G), cobalt blue (C), and tan (T) (Fig. 4.1a). A necklace may consist of at most hundreds of beads, but a DNA molecule may be made of hundreds of thousands or even millions of these

E. Ochiai, *Chemicals for Life and Living*,
DOI 10.1007/978-3-642-20273-5_4, © Springer-Verlag Berlin Heidelberg 2011

Fig. 4.1 DNA model

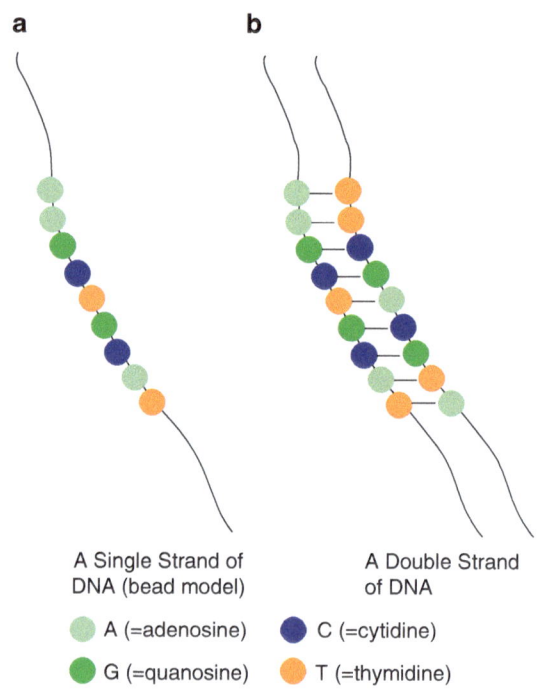

A Single Strand of   A Double Strand
DNA (bead model)   of DNA

○ A (=adenosine)   ● C (=cytidine)

● G (=quanosine)   ● T (=thymidine)

beads. It is a very, very long strand. Remember that the actual DNA molecule is made of atoms that are very, very, very small. For example, the gene (a chromosome) of a bacterium *E. coli* is made of about five million beads (nucleotides), which is about 1.6 mm (one sixteenth of an inch) long when stretched. This is visible to human eyes. You may consider this to be short in view of everyday things, but it is enormously long in the world of individual molecules. This strand of beads functions as a sort of recording tape, and the information on the tape is thus written in those four letters, A, C, G, and T. In contrast, the information on a digital recording tape, CD, or computer diskette is written by two letters, i.e., 0 and 1.

It turned out that the functioning DNA is a double strand. That is, it consists of two complementary strands, and this is the basis of the functions of DNA. Bead A specifically binds laterally with bead T, and G with C. Suppose that beads (a portion of a DNA) are arranged in the order of AAGCTGCAT on a strand, then on the other will be the complementary sequence TTCGACGTA, as seen in Fig. 4.1b.

The well-known structure of a DNA is double helix (there are other types of structures known.). The "double" portion means "double-stranded" as shown in Fig. 4.1b and is essential to the functioning of DNA. The "helix" portion is nonessential to the functioning of DNA, but is a result of chemical nature of the molecule and is also important in storing DNA. A DNA in the nucleus of eukaryotic cells coils up and then coils up further (supercoiled). As a result, its overall volume is reduced so that it can be confined in a small volume of nucleus. DNA cannot do this by itself. Some kinds of protein assist DNA in coiling up.

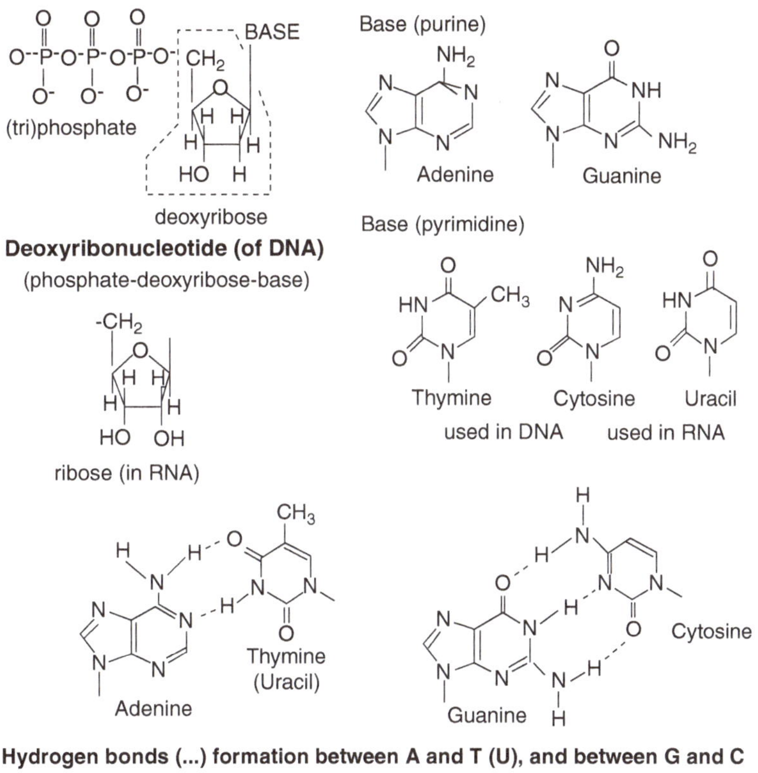

Deoxyribonucleotide (of DNA)
(phosphate-deoxyribose-base)

Hydrogen bonds (...) formation between A and T (U), and between G and C

Fig. 4.2 Structures of nucleotides

So the crucial question is why those beads (nucleotides) bind specifically. This is a straight chemistry. We have to see the chemical structures of those components (called deoxyribonucleotides; nucleotides for short): A, C, G, and T of DNA. Each of these consists of three parts: (deoxy) ribose, phosphate, and base (Fig. 4.2a). The first two are common to all nucleotides and make up the backbone of DNA. There are four different bases used in DNA. They are A=adenine, C=cytosine, G=guanine, and T=thymine. They have chemical structures as shown in Fig. 4.2b. Adenine and guanine belong to a class of compounds called purine base and have similar structures. Cytosine and thymine are pyrimidine bases. These compounds use nitrogen and oxygen, in addition to carbon and hydrogen. Oxygen and/or nitrogen atoms in a compound can bind relatively strongly to oxygen and/or nitrogen atoms in another compound through "hydrogen bond." "Hydrogen bond" is typically formed between water molecules (H_2O), as was talked about in Chaps. 1 and 19. It turned out that adenine binds with thymine most comfortably through hydrogen bonds and that guanine does so with cytosine, as shown in Fig. 4.2(b). The other combinations such A–C or G–T are possible, but these combinations are not as strong as the standard combinations A–T and G–C.

Nucleotides (represented by single alphabets A, C, G, ant T) bind through the phosphate as shown in Fig. 4.3. If you combine a large number of nucleotides by this means, what you obtain is a DNA molecule (i.e., a polymer of nucleotides). You may regard the chain of (deoxy) riboses bound through phosphate as the thread (in Fig. 4.1) and the four bases as the beads in Fig. 4.1.

The bases take planar shapes, and these planar molecules tend to stack parallel on top of each other. This tendency of ring aromatic compounds is seen, for example,

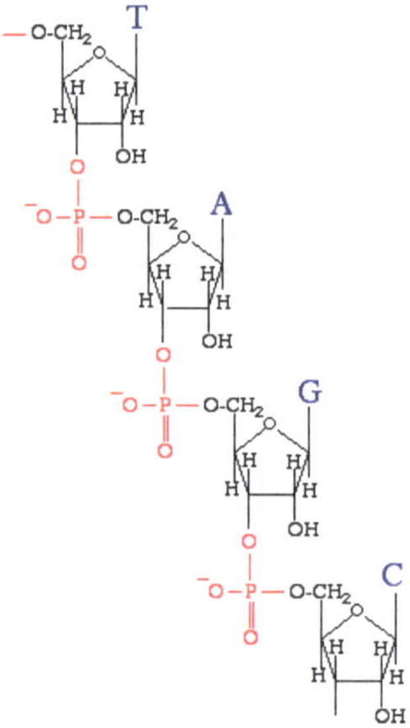

Fig. 4.3 Chain of nucleotides

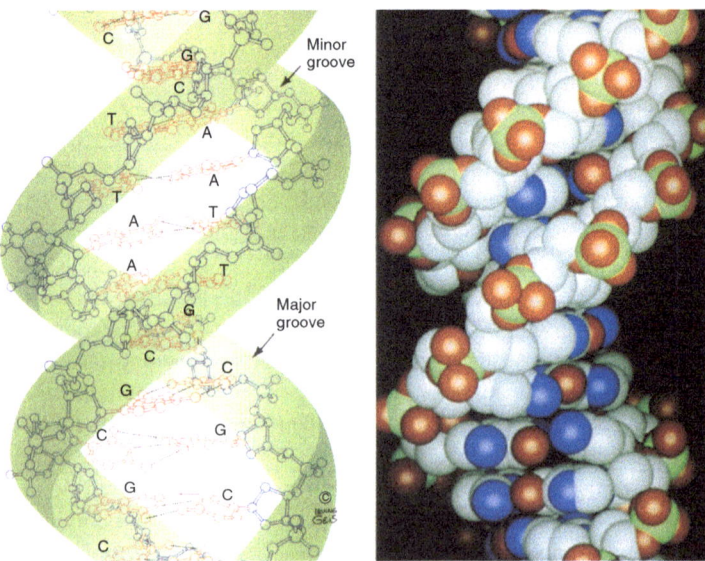

Fig. 4.4 Double helix [from D. Voet and J. G. Voet, "Biochemistry, 2nd ed" (J. Wiley and Sons, 1995)]

in the structure of graphite – see Chap. 11. As a result of these several properties of nucleotides, a DNA has strong tendencies to form a double strand and a helix structure, as seen in Fig. 4.4. The stacking tendency between base rings is partially responsible for the helix.

4.2   How Is a DNA Replicated?

This is quite clear at least in principle by now. It is based on the specific interaction between A and T, and between G and C. That is, take, for example, the double helix in Fig. 4.1. Let us label the left strand as "l" strand and the other "r" strand. (This is the complementary strand of "l"). Suppose that you separate the two strands and the "l" strand is isolated. Then you provide a pool of components A, C, G, and T and a means to bind nucleotides (enzyme called DNA polymerase) for the "l" strand. This enzyme binds nucleotides one by one sequentially. The top bead A on the "l" strand binds a bead T (laterally through hydrogen bond), and next another bead on "l" binds laterally a bead T. Beads T and T are then connected through the phosphate group by the enzyme. Next the bead G on "l" binds a bead C, and the bead C then is connected to the previous T on the right hand by the enzyme. This is repeated; then you see that an "l" strand will reproduce the complementary "r" strand. The reverse will also be true; i.e., an "r" strand will reproduce the corresponding "l" strand. Thus, a double strand will have been replicated (see Fig. 4.5).

How this is accomplished, i.e., mechanics of these chemical reactions are currently very intensely studied, is beyond the level of this book. Hence, this topic will not be pursued further here. But, the very basic reason why we are like our parents or in other words why a gene molecule (DNA) is (almost) faithfully replicated and transmitted to a progeny can be understood as in the previous paragraph. This replication mechanism of DNA, however, applies to only cell division.

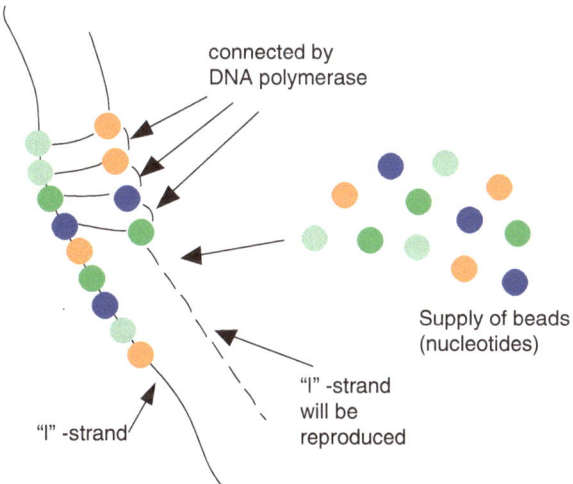

Fig. 4.5 Replication of DNA double helix

The issue of inheritance in sexual organisms like us is a little more complicated, because we get half of the gene from mother and the other half from father. But again we are not able to elaborate on this issue here. The issue is more of biology (so-called genetics) than chemistry. The chemical principles are about the same.

We said, "DNA is (almost) faithfully replicated" in the paragraph above. The qualification "almost" implies that replication may not always be exact. In other words, a cell may make mistakes in replicating a DNA. It happens not very often, but frequently enough. If this happens, a wrong DNA may form, which would give wrong information. Mistakes can be caused by some factors (some cancer causing factors, for example) or without any particularly cause. The distinction between the right combination A–T/G–C and wrong combinations such as A–C/G–T is not quite definite. Chemically speaking, the difference in interaction energy between the right and the wrong combination is not very great. Hence, there is some chance that the DNA-making mechanism may simply connect wrong nucleotides occasionally. This may be disastrous to the organism. Therefore, many DNA-making mechanisms (DNA polymerases) contain in it three functions. One is polymerizing nucleotides (making DNA chain), of course. The other two are monitoring and repairing mechanisms. It monitors what nucleotides are connected and can identify a wrong one. When it has recognized a wrong one, the repairing mechanism snips off the wrong one. And then the polymerase portion reconnects another; this time a right one, hopefully. There are many other mechanisms known in organisms that repair "damaged" DNAs. All these are chemical reactions, but too complex to be talked about here. It is also to be noted that these occasional changes in DNA are the ultimate cause of change of species, i.e., evolution.

4.3   What Do DNAs Do?: Protein Synthesis

When we say "you look like your mother," we are not talking about your DNA being similar to your mother's DNA (though this is true). We are talking about your body features, face complexion, hair color, the color of eyes, etc. What we get from our parents are their genes (DNAs). DNAs then create us, our body. DNAs dictate or provide information for creating all organs and tissues, i.e., body. How?

Or, what kind of information is stored on the DNA tape (strand)? A number of things are stored, but the most important kind of information is the instruction on how to construct proteins. Specifically, it dictates the sequence of amino acids that are to be bound in a protein. A protein is made of a series of amino acids and can function properly only when the amino acids are bound in a specific sequence. The proteins make up your body: muscle and others. Many proteins also function as enzymes that control all the chemical reactions necessary for life. Thus, proteins are manifestation of the genetic information.

There are about 20 different amino acids in nature, and hence the information written in the four letters A, C, G, and T should be able to specify each of these 20 or so amino acids. Obviously, you cannot specify all 20 or so names by just a single one of the four letters. How about using two letters for each amino acid?

Well it can distinguish $4 \times 4 = 16$ names. It is still not enough. It turned out that a set (called "code") of three nucleotides (of A, C, G, and T) specifies an amino acid. Code is like a three-letter word. Sixty-four different codes or words can be obtained by using three letters out of the four letters; this is more than enough. There is some redundancy; that is, several different sets specify a same amino acid. Questions such as "how has the genetic code developed" or "is there any chemical basis for the genetic code" are very interesting, but have not yet been answered.

To talk about the genetic code in detail, we need to delve into further details of protein formation process. DNA does not directly dictate the sequence of amino acids in a protein. The information (i.e., sequence) in DNA first needs to be transcribed into a nucleic acid of another type, that is, RNA (ribonucleic acid). The DNA is a permanent copy, and the corresponding RNA is a sort of temporary copy from which the cell makes a protein. Chemically, RNA is very similar to DNA, but it has more varied functions than DNA. The details are not important here. The differences between DNA and RNA are twofold: (1) RNAs use ribose instead of deoxyribose (see Fig. 4.2); and (2) the bases used are A, C, G, and U. That is, U, uracil, is used instead of thymine. Uracil has a very similar structure to thymine, and it combines with A, adenine, like in DNA.

A portion of DNA (usually only a very small portion of a very large (long) DNA) specifies a protein. When that protein is needed, the portion of DNA will be copied onto an RNA molecule (called messenger RNA, m-RNA). This portion of a DNA is called the gene for the protein. Therefore, a large DNA molecule has on it a number of genes for different proteins.

Like the replication process of DNA, the A, G, C, and T sequence of that portion of DNA is then duplicated using A, C, G, and U (of ribonucleotides) this time, and the connection (polymerization of A, C, G, and U) is made by an enzyme called RNA polymerase. This process is called "transcription." The next step in making a protein is to use this copy of an m-RNA and to "translate" the information on the m-RNA into the sequence of amino acids in a protein. This is done in the following manner.

A set of three ribonucleotides out of A, C, G, and U specifies an amino acid of a protein. Such a set is called a "codon." Therefore, the sequence of m-RNA can now specify the sequence of amino acids in a protein. Examples are as follows. UUU and UUC are codons for phenylalanine; CUU, CUA, CUG, and CUC (all CUX) for leucine; CAU and CAC for histidine. Glycine is specified by GGX (X = A, C, G, and U). GAU and GAC specify aspartic acid. Interestingly, some codes UAA, UAG, and UGA are used to indicate where to stop. The AUG code is used to specify where to start, but it specifies an amino acid methionine, if it occurs in the middle of a gene.

Now how are these codes translated into amino acids? An amino acid binds to a special kind of RNA called transfer-RNA (t-RNA for short); there is a specific t-RNA for each and every amino acid. A t-RNA for an amino acid has in it a set of three consecutive nucleotides called anticodon that binds specifically to the codon for that amino acid on m-RNA. The binding between a codon on m-RNA and the corresponding anticodon on t-RNA is again due to the specific hydrogen bonding similar to that in the formation of double helix of DNA.

One problem in this process is that more than two codons are used for an amino acid in m-RNA. A t-RNA has to recognize all different codons for a single amino acid as such. As you recall, codons for an amino acid are made of three nucleotides, the first two of which remain the same for that amino acid and yet the third nucleotide is different. So both UUU and UUC have to be recognized as the code for phenylalanine. The third nucleotide can be variable. A t-RNA uses regular nucleotides for the first two of the anticodon and a special kind of nucleotide (other than regular A, C, T, and U) for the last. For example, the anticodon for phenylalanine (codon is UUU and UUC) is AA(Gm) (see Fig. 4.6). Gm is a slightly modified guanine (G), in which 2'-position of ribose of guanosine is methylated. Gm then binds either U or C; the proper partners for U and C are A and G, respectively. Therefore, Gm plays the role of either A or G. But in other t-RNAs different nucleotides are used for the third nucleotide. Three of the codons (on m-RNA) for an amino acid alanine are GCU, GCC, and GCA. The first two of the anticodon for alanine are CG. The third component of the anticodon has to recognize the three different ones. In this case, a nucleotide called inosine (I) is used, which is a derivative of adenine (A). In this case, the third component is not critical in specifying an amino acid.

However, aspartic acid uses codons GAU and GAC, while glutamic acid's codons are GAA and GAG. The third component distinguishes the two amino acids. It turned out that G is used for either U or C (of codon) in the third position of the anticodon and U is used for A or G (of codon). In other words, the anticodon for

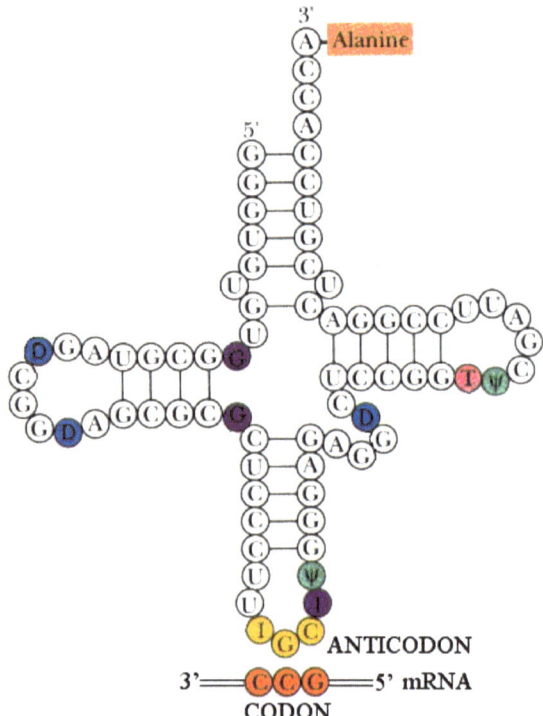

Fig. 4.6 A t-RNA has this shape (called "cloverleaf") and it has the anticodon in yellow at the bottom portion, which binds the corresponding codon on m-RNA. t-RNA has several special nucleotide, shown here as colored. D in *blue* is dihyrouridine, F in *green* = pseudouridine, G in *purple* = either methyl or dimethylguanosine, I in *yellow* = inosine, I in *purple* = methylinosine, and T in *pink* = robothymidine [modified from R. H. Garrett and C. M. Grisham, "Biochemistry" (Saunders 1995)]

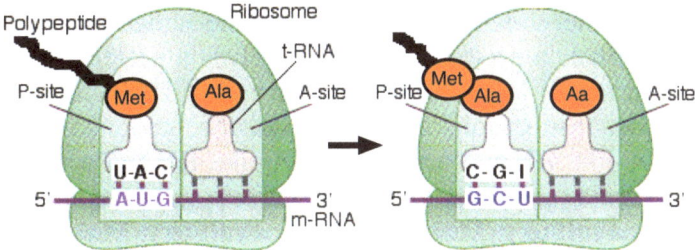

Fig. 4.7 Sequence of protein synthesis (*Met* methionine, *Ala* alanine, *Aa* amino acid)

aspartic acid is CUG and that for glutamic acid is CUU. These choices make a very good chemical sense, as A and G are chemically of the same type (purine base), while C and U are of pyrimidine base. That is, nature knows chemistry very well.

Now, the whole process of making a protein is shown in Fig. 4.7 and goes like this. A gene for a protein (a portion of a DNA) is transcribed into the corresponding m-RNA. Codons on m-RNA specify amino acids, and so the message on m-RNA can be translated into a sequence of amino acids that constitute the protein. An amino acid methionine, for example, will bind the end of its specific t-RNA (t-RNA^{met}). The t-RNA^{met} has an anticodon UAC that corresponds to a codon (AUG) for that amino acid. Hence, t-RNA^{met} that now has methionine at the end binds to the codon on m-RNA at site P on the ribosome. A t-RNA^{asp} (with anticodon AGG) with aspartic acid at the end now binds to the next triplet GAU at site A. Then methionine at the end of t-RNA^{phe} and the adjacent aspartic acid at the end of t-RNA^{asp} are then connected by an enzyme system on the ribosome. The aspartic acid residue is now at site P, and it is the growing end of the polypeptide. Another amino acid bound at the end of t-RNA now binds to site A, and the connection of amino acids thus continues.

The details of mechanisms of how these chemical reactions are effected may be difficult to understand. However, the crux of how the information of a gene on a DNA is used to make a protein, i.e., how the sequence of A, C, G, and T specifies and is translated into the sequence of amino acids is quite simple. It is based on specific interactions through hydrogen bonds between the bases (of nucleotides), as exemplified in Figs. 4.2 and 4.5.

4.4   Chicken–Egg Issue Regarding the Origin of Life

There are three different types of chemicals essential for life: carbohydrates, proteins, and nucleic acids (DNA/RNA). The duplication of DNA requires an enzyme DNA polymerase among others. DNA polymerase is a protein. However, as outlined above, a protein is produced according to the blueprint on a DNA. So, DNA needs protein(s) for its own duplication, and proteins need DNA as the source of information on their structures. So, in the beginning, which came first, DNA (gene) or protein? This is the ultimate chicken–egg issue.

Some people thought that proteins came first, perhaps without the instruction by a gene, and indeed some proteins have been shown to form spontaneously. One of the problems is this: does the protein have a specific sequence so that it functions in biologically meaningful manners? Is not a blueprint (DNA) necessary to make such a specific protein after all? These ideas were discussed and arguments were made when the enzymes that catalyze any bioreactions were considered to be "proteins"; i.e., only proteins can be enzymes. This was a kind of dogma hypothesized or even believed but not really proven.

Then in 1980s two scientists (among others), Thomas Czech of University of Colorado and Altman of Yale University, discovered that one of nucleic acids, RNA, does exhibit catalytic activities. These two scientists shared a Nobel Prize in chemistry in 1989 for this discovery. Though the gene is DNA in the majority of organisms, the gene in some organisms (including the AIDS virus) is RNA. Therefore, an idea was put forward that RNA acted in the beginning as both enzyme and gene. This idea called "RNA world," if proven, will make moot the chicken–egg issue.

In mid-2000, a report came out, which demonstrated in detail that the ribosomal RNA is indeed the site where proteins are formed. t-RNA with an amino acid bound

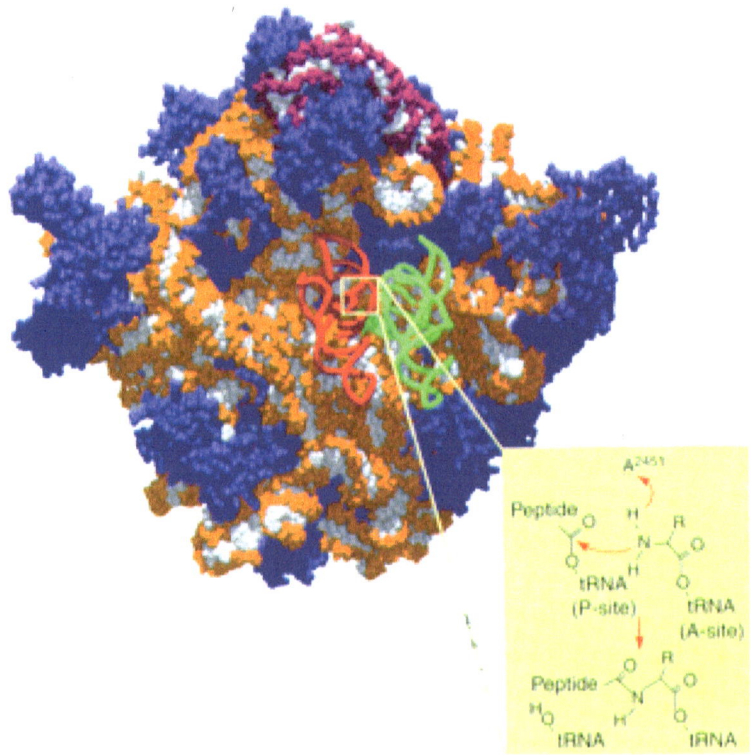

Fig. 4.8 Protein synthesis on ribosome. The chemical formulas indicate how the peptide bond is formed (from Cech T (2000) Science 289:878)

bind to the ribosomal RNA, which acts as catalyst to form peptide bonds, and hence a protein. The structure and the mechanism of peptide bond formation by the ribosomal RNA are shown in Fig. 4.8.

Would this study settle the issue of "chicken–egg issue" once and for all? Well, not quite yet. There is an issue about how RNA was created first without organisms. No plausible idea has been put forward about this issue, let alone being proven.

Clothing and Shelters: Polymeric Material

5

5.1 Necessity for High Molecular Weight: What Makes a Compound a Solid?

Many organisms need external shell or similar things for mechanical protection and strength. Examples include shells of clams and oysters, eggshells of chicken, corrals, extra skeleton of insects such as beetles, and bark of trees. Organisms also need scaffold to maintain their statures. This is "bone" and similar tissues. We will talk about "bone" in another chapter.

We, human beings, do have our skin, but require more than the skin for comfort and protection. That is, we need clothes and associated goods (shoes, etc.) and houses or some kind of building for our lives and social lives. Obviously, the materials for these purposes should take certain forms (solid) with enough mechanical strength and other suitable properties including resistance against elements. Examples of materials used for these purposes include brick, concrete, wood, cotton, silk, leather, plastics, and others. One common property of all these materials is that they are solid at ordinary temperatures; "solid" in technical sense, not in the ordinary sense, i.e., "hard/firm/rigid." What factors are necessary for a compound to be solid at room temperature? First let us look at this issue.

Well, from your experience, you can tell that the first two among the materials mentioned above, brick and concrete, are quite different from the rest. They are hard and brittle, whereas the others are rather soft and plastic, although some wood and plastics can be quite hard. Brick and concrete are typical inorganic material similar to rocks, whereas the others are made of organic compounds. We focus on the organic material in this chapter. Another chapter (Chap. 14) is devoted to the issue of rocks and the related material.

As we have seen throughout this book, chemical compounds can be gas, liquid, or solid at certain temperatures, and all the chemical compounds take all three forms

E. Ochiai, *Chemicals for Life and Living*,
DOI 10.1007/978-3-642-20273-5_5, © Springer-Verlag Berlin Heidelberg 2011

at different temperatures. You are familiar with the case of water. Water is solid (ice) below 0°C, and turns into liquid above it; it boils at 100°C turning into gas (steam). You have seen dry ice; it is the solid form of carbon dioxide (CO_2). Carbon dioxide is a gas under an ambient condition. If you warm dry ice under an open condition, it turns directly into gas. You do not see the liquid form like water. However, you can see it to turn into liquid, if you warm it carefully in a closed container. Iron is a solid at room temperature. It melts (turns into liquid) at 1,535°C. It comes out as liquid from the furnace when iron is produced from the iron ores. It can boil at 2,750°C. "Boiling" is turning into a gas whose pressure is one atmospheric pressure.

What factors determine whether a compound is a gas or liquid or solid at say 25°C (room temperature)? They are the cohesive forces that bind molecules together. If the force is weak enough (or the interaction energy is low enough), the molecules may be able to move around freely independent of each other; that is, such a substance is a gas. When the force is of an intermediate strength, the material behaves as a liquid, in which molecules are stuck together but loosely enough to move past each other. In the solid form, molecules are interacting so strongly (relative to the disturbing force) that they cannot move freely from each other and more or less sit tight in their positions.

There is always a force counteracting the cohesive force; that is, heat (temperature). Heat (higher temperature) encourages movements of molecules. If this disturbing effect (high enough temperature) overcomes the cohesive force between molecules, a compound can change (from the initial solid state) to liquid and then to gas.

The material that is solid around room temperature must have a strong cohesive force (high attractive interaction energy). The intermolecular force among molecules and ions are electrostatic in nature; that is, the attractive force between a positively charged entity and a negatively charged entity or the repulsive force between entities of like charges. So, when a molecule or entity has a localized electric charge and another has a similar electric charge but with the opposite sign, then the interaction between them is strong. Typical ionic inorganic compounds are held together by strong ion–ion interactions, and hence, they are solid at room temperature and require a lot of heat to melt it (therefore, high melting point). A familiar substance, table salt, is an example, in which ions Na^+ and Cl^- are arranged in an orderly fashion (crystal structure).

Typical organic compounds, on the other hand, have relatively weak intermolecular interaction, as they do not have large localized electric charges. As a matter of fact, the simplest organic compounds, hydrocarbons, have no localized electric charge; they are said to be "nonpolar." For example, the smallest hydrocarbon, methane (CH_4) is a gas at room temperature. Methane is the main component of natural gas.

However, another kind of attractive force does exist even between non-polar molecules. That is called "London (dispersion) force." This is often called by other names such as van der Waals force or hydrophobic interaction. The electrons in a molecule are not sitting still; they are always moving. The fact that a molecule is nonpolar can only be true when the distribution of all electrons is averaged over time.

At one instant, the electrons may not be evenly distributed in a molecule. If so, at that moment, the molecule has localized electric charges in it. This molecule affects then the electron distribution in another molecule nearby, creating (inducing) an instant localization of electrons in the latter molecule. These instantly created localized electric charges then contribute to attractive interaction between these molecules. This is what Dr. London proposed to explain relatively weak attractive force between nonpolar molecules.

Well, then you can see that the more electrons there are in a molecule and the more easily movable the electrons in it are, the stronger the London force would be. That is, the London force between the next simplest hydrocarbon ethane (C_2H_6) is stronger than that between methane. This is reflected in their boiling temperature; $-164°C$ with methane but much higher $-88.6°C$ with ethane. That is, it requires more heat (thus higher temperature) to break the interaction between ethane molecules to turn liquid to gas than to do the same with regard to methane. The hydrocarbon with five carbon atoms, pentane (C_5H_{12}), has now enough number of electrons that makes pentane a fairly volatile liquid at room temperature (boiling temperature is $36.1°C$). Still larger heptane (C_7H_{16}), octane (C_8H_{18}), and nonane (C_9H_{20}) are liquid at room temperature; they constitute gasoline. If you continue lengthening the hydrocarbons, the London force would become stronger. Indeed beyond about 15 carbons ($C_{15}H_{32}$=pentadecane), the (linear) hydrocarbons are solids, though melting at relatively low temperatures. These are wax. You can make very long chains of CH_2 units (tens of thousands of them) by artificial means. This is made from a large number of ethylene ($CH_2=CH_2$) molecules connected in a linear fashion. Such a large molecule is called a "polymer," and the unit material, ethylene in this case, is called "monomer." ("poly" means "many" and "mono" is "one"). The polymer made from ethylene is called polyethylene and is used for wrapping plastic film in everyday life.

Benzene C_6H_6 is another kind of hydrocarbons, called "aromatic compounds." It takes a hexagon shape, whose chemical structure resembles a unit of beehive. It is liquid at room temperature. That is, it has a sufficiently strong interaction among the molecules to be liquid, but not strong enough to be solid. When two benzene molecules are combined, a compound called "naphthalene" results. It is a solid with a special odor at room temperature and used as mothball. Why is "naphthalene" solid, whereas benzene is liquid at room temperature? Think about it in terms of London force. It must be noted that the interaction between benzene molecules or naphthalene molecules is not only due to London force but also due to another kind.

If we replace one or two (or more) hydrogen atoms in a hydrocarbon by chlorine atoms, the resulting molecule would have a much higher boiling temperature than the hydrocarbon. For example, let us compare methane (CH_4) and monochloro methane (CH_3Cl). The former has a boiling temperature of $-164°C$, while the latter's boiling temperature is $-98°C$. In Chap. 19, you learn that a chlorine atom attracts electrons to itself more strongly than a carbon or a hydrogen atom. As a result, the electrons that are binding carbon atom and chlorine atom are not equally distributed between the two atoms. They are more densely populated toward the chlorine atom. Hence, a (partial) negative electric charge appears on the chlorine atom and, accordingly, a positive charge on the carbon atom. The bond C–Cl is thus said to be polar, and the monochloro methane molecule as a whole is polar. These partial electric charges on this molecule make their interaction stronger.

Diamond is the hardest solid; diamond is talked about in Chap. 15. To summarize the reasons for it to be a solid, diamond is made of carbon atoms that are strongly bound (by covalent bonding) and make up a three-dimensional structure. In other words, in diamond the interparticle (atom) interaction is very strong. It has a very high melting temperature and boiling temperature; 3,550 or so and 4,827°C, respectively.

Let us talk about another type of compound. Glucose is made of 6 carbon atoms, 12 hydrogen atoms, and 6 oxygen atoms ($C_6H_{12}O_6$) and is solid at room temperature. This is a typical example of carbohydrate which means "hydrated carbon," as the chemical formula $C_6(H_2O)_6$ indicates. In terms of (molar) mass (which roughly represents the number of electrons), glucose is similar to a hydrocarbon of 12 carbon atoms (dodocane$=C_{12}H_{26}$), which is liquid at room temperature. Well, this means that the intermolecular interaction in glucose is much stronger than that in dodecane. It turns out that glucose has five OH groups in it. As we mentioned in the chapter on water (Chap. 1), the OH groups interact with each other relatively strongly through hydrogen bonds. Hence, glucose has a relatively high melting temperature of 146°C for the size. What would happen if we make a compound that is made by combining two glucose-like molecules? Such a compound is sugar (sucrose), though one of the two C_6 carbohydrate units is not glucose. Its melting temperature is yet higher, 186°C, understandably.

However, glucose and sugar are soluble in water unlike the hydrocarbons (oil) mentioned above. Well, if the material is soluble in water, it would not be suitable for clothes or shelter, would it? It turns out that if you combine a large number of glucose in a certain linear fashion, you obtain "cellulose." Cellulose can be regarded as a polymer of glucose. Such a polymer made of glucose or similar compounds is called in general "polysaccharide" (polymer of sugars). The plant cell walls are made of cellulose and lignin. A special plant, cotton plant, produces cotton, i.e., cellulose in chemical terms. (By the way, glucose and similar compounds are the initial products of "photosynthesis" conducted in green leaves. This is talked about in another chapter).

Well, if glucose and sugar are soluble in water, would not "cotton" (cellulose) be also soluble in water because its constituent, glucose, is soluble? It would be so, because it has a lot of "OH" groups that can interact with water. The fact of the matter is that, however, cellulose (cotton) is not soluble in water. Why? We come to that later. But the fact that cellulose would interact with water, i.e., it has an affinity toward water, is suggested by the observation that paper towel, which is made of cellulose, sucks up water.

Starch is another polymer of glucose, like cellulose. But you know that starch is quite different from cellulose. Plant starch typically consists of two components: water-soluble amylose and water-insoluble amylopectin. Amylose is a linear polymer of glucose, but its connection is different from that in cellulose. The structure of amylopectin is different from that of amylose. We discuss these issues later.

Our body is shaped mainly by proteins. Silk is a protein produced by silk worm, and our hair is made of some proteins, one of which is keratin. A protein is a special kind of polymer. It is a linear polymer of 20 or so monomers called amino acids.

We hope we have shown you that many organic building and clothing material are polymers. There are a number of organic compounds that are not polymers but are solids at ordinary temperature and insoluble in water. The compounds of this type, however, are not used for extra skeleton, shell, or clothing. Why would they not be used for such purposes? Polymeric material can easily be made into fibers or films because of their chemical structures, but nonpolymeric material cannot be made into these forms.

5.2   Natural Polymeric Material

Important natural polymers (biopolymers) include cellulose, chitin, proteins, and nucleic acids. Nucleic acids, DNA and RNA, are discussed in Chap. 4 and are not used as mechanical supports for organisms/cells/tissues. All of these materials can in principle be produced by connecting small molecules (the repeating unit called monomer) through condensation reactions. A reaction to connect a large number of monomers is called "polymerization." In the case of these biopolymers, the reaction should further be characterized as condensation polymerization to contrast with another type of polymerization described later. It can generally be written as follows:

$$\text{H-M-OH} + \text{H-M-OH} \rightarrow \text{H-M-M-OH} + \text{H}_2\text{O} \,(\text{condensation with removal of } \text{H}_2\text{O})$$

$$\text{H-M-M-OH} + \text{H-M-OH} \rightarrow \text{H-M-M-M-OH} + \text{H}_2\text{O}, \text{ and so on.}$$

It is not a simple matter to accomplish this kind of reaction in the biological system. All of these materials require elaborate reaction systems to make. We would not talk about the processes of making them in the biological systems, and we will only look at their structures and some properties.

5.2.1   Cellulose, Starch, and Chitin

These are representatives of polysaccharides and are polymers of glucose or a glucose derivative. However, they are very different, as you know. Why are they so different? Well, let us see how they are constructed. Figure 5.1 shows the structures of glucose (the monomer), cellulose, starch 1 (amylose), and starch 2 (amylopectin). You may not concern yourself with the details of the structures. You should, however, note the two different forms of (d)-glucose, α and β, depending on the way the OH group is attached to the carbon 1, and accordingly the different ways of connecting the glucose units in cellulose and amylose. Figure 5.2 shows in a space-filled model how four glucose groups bind in these two ways, through

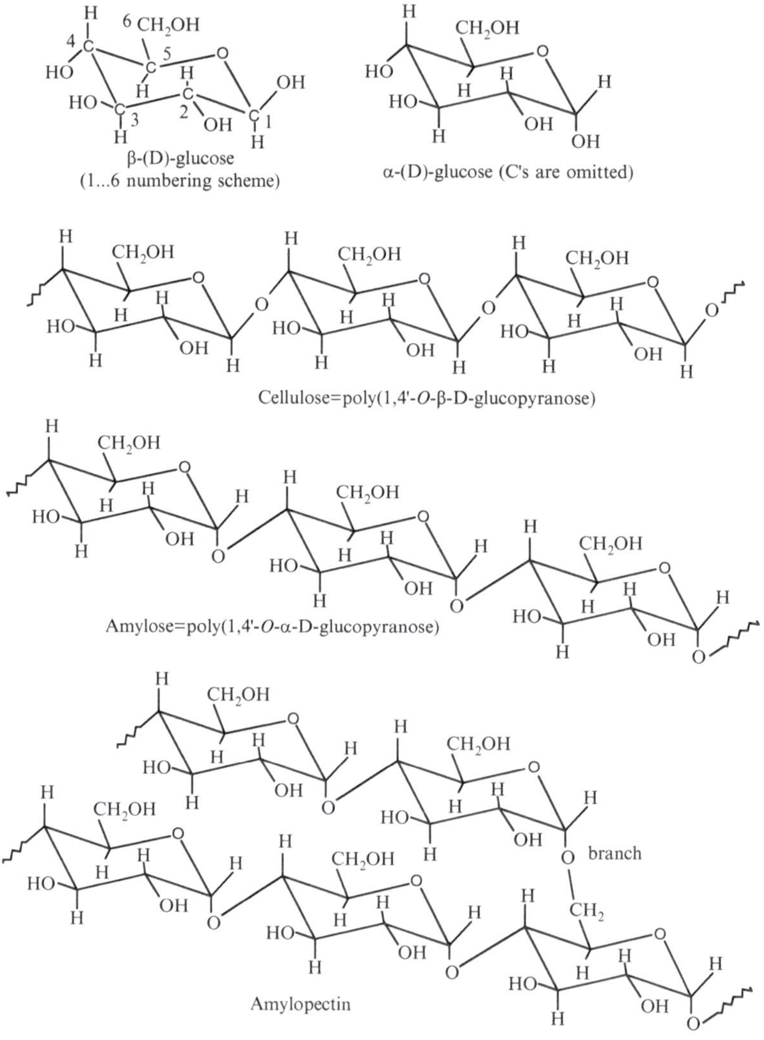

Fig. 5.1 Glucose, cellulose, amylose, and amylopectin

β-(cellulose) or α-linkage (amylose). Note that in the case of β (1→4)-linkage, the four glucose groups arrange themselves in a linear fashion, suggesting that cellulose molecule is linear with each glucose group oriented in the same way. In the case of α-linkage (1→4), on the contrary, the glucose groups orient themselves differently from each other, and the figure suggests that the long chain of α-linked glucose (amylose) may form a spiral (helix).

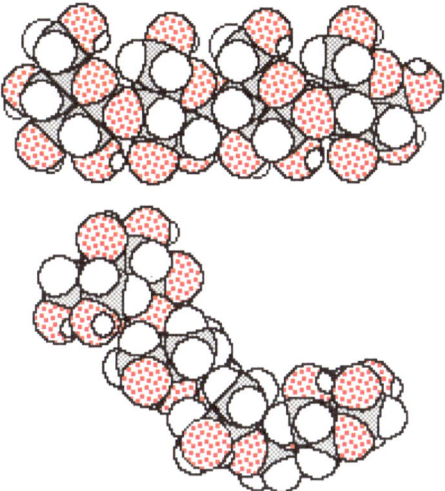

Fig. 5.2 Space-filling structures of tetramers of glucose; the upper structure is obtained when glucose groups are connected through b(1→4), i.e., a partial model of cellulose, and the lower is a partial model of amylose (starch 1)

The difference in structure is reflected in their properties. The polysaccharide chains in cellulose are bound strongly together through hydrogen bonding between OH groups of glucose units. As a result, they form linear, fibrous, sturdy polymers. Hence, it provides a good material for clothing and paper. The interactions between the OH groups among cellulose molecules leave little OH that interacts with water molecules. That is, it is insufficient to make cellulose soluble in water, but it makes cellulose (cotton and paper) a good absorbent of water as is used in paper towel.

The arrangement of glucose through α-linkage as shown in Fig. 5.2 makes the amylose polymer chain into a helical structure. Because of the helical structure, the OH groups in the glucose units can hydrogen-bond to water more effectively (than those in cellulose), and the chain–chain interaction is weaker here than that in the linear cellulose. Hence, amylose is soluble in water. Thus, the amylose portion of starch is called "soluble starch." Amylopectin is amylose chain intercrossed here and there (through 1→6 connection), and the polymer molecules entangle themselves extensively. Hence, amylopectin is not soluble in water.

You might have used iodine to detect the presence of starch. The iodine molecule (I_2 and other forms such as I_3^- and I_5^-) can fit in the helix of amylose. This complex (of iodine and starch) exhibits the characteristic deep purplish blue color. [The website of Dr. S. Immel of Darmstadt Technical University (http://caramel.oc.chemie.tu-darmstadt.de/~lemmi/graphics/polysaccharides.html) gives a variety of pictures of models of cellulose, amylose, and others as well as the interaction between them and water molecules and iodine].

We, animals, can digest starch. This is the basis of the food chain; plants are eaten by animals. Starch itself cannot be used by us. We have to digest it. The process of digestion is the reverse of polymerizing glucose. That is, you have to add a water molecule to split the glucose–glucose connection. This is called "hydrolysis." Animals accomplish this reaction, using enzymes called amylases, which attack the α-linkage in the glucose polymer. More specifically, there are two enzymes: exoglycosidase and endoglycosidase, and each splits different positions of the long chain of amylose. By the way, we store the extra glucose or starch in the form of glycogen, which is similar to amylopectin in structure. In other words, we polymerize glucose back into polymers in order to store.

Well, what about "cellulase" an enzyme to chop cellulose into glucose molecules? The enzyme attacks β-linkage. Most animals do not have cellulase; so we cannot digest cellulose. Some animals such as cow, sheep, and goat can eat grass and straw, which contain a lot of cellulose in addition to starch. They digest cellulose as well? Well, yes and no. No, they do not, to be exact. However, they harbor a lot of bacteria and other microorganisms in their rumen and they do have cellulase that can decompose cellulose into smaller units (down to glucose), and the animals use the digested products. This is an example of symbiosis. Why have most animals not developed cellulase so that they can utilize cellulose that is present on the Earth much more plentifully than starch? If we can develop cheap sources of cellulase, the food-shortage problem for mankind may be solved. On the other hand, that may completely change the ecosystem on the Earth.

Beetles have fairly hard armors, and crustaceans such as shrimp have also shells. The material constituting these shells is called chitin. Chitin is similar to cellulose. The only difference is that the OH group at carbon 1 on glucose unit is replaced by acetyl amine group ($-NH(CO)CH_3$). This unit is called N-acetyl-β-D-glucosamine. So chitin is a $\beta(1\rightarrow4)$-linked polymer of glucosamine. Chitin is also found in bacterial cell walls, fungi, and yeast. The hard shell of lobster or crab is made of chitin impregnated with calcium carbonate.

5.2.2   Lignin

Lignin is another important component of plant cell walls. It comprises 15–25% of the dry weight of wood. It occurs between the small fibers of cellulose. It even chemically binds to cellulose. Trees have to resist both stretching tension and compression. Cellulose is responsible for resistance to stretching, and lignin to compression. Lignin is a three-dimensional polymer of phenolic compound called coniferal alcohol and other similar compounds (see Fig. 5.3). [Phenol itself is a compound in which one of the hydrogen atoms of benzene (C_6H_6) is replaced by an OH group; i.e., the chemical formula is $C_6H_5(OH)$]. These phenolic compounds bind each other (i.e., polymerize) in many different ways and form a three-dimensional network as shown in Fig. 5.3. Because of this structure and the presence of a large number of the benzene rings, lignin is fairly rigid and very insoluble in water.

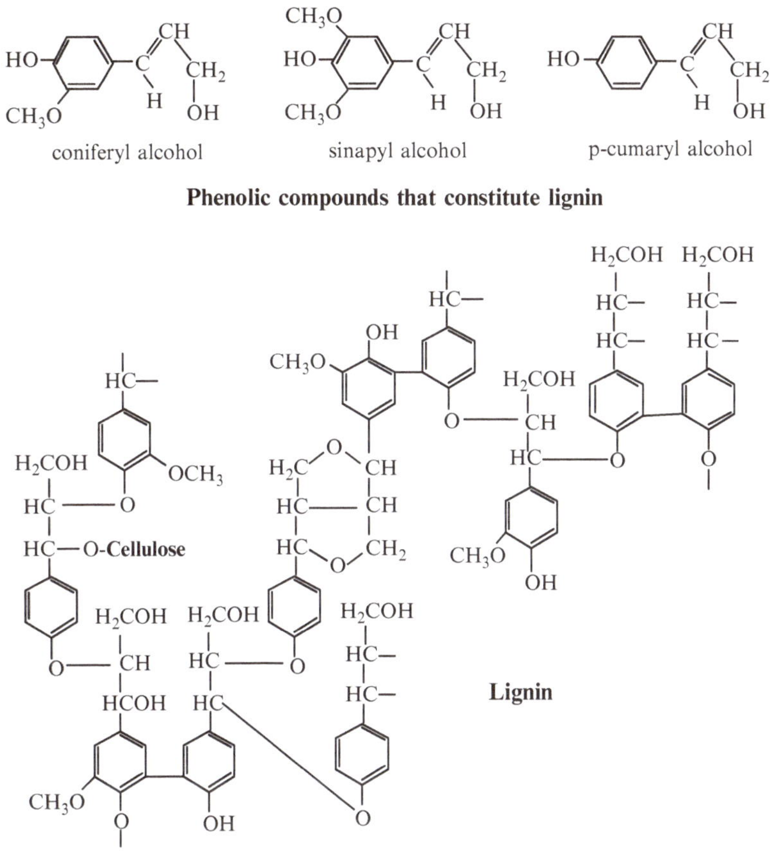

coniferyl alcohol   sinapyl alcohol   p-cumaryl alcohol

Phenolic compounds that constitute lignin

Fig. 5.3 Lignin

5.2.3   Proteins: Silk, Keratin, Etc.

Proteins are polymers, but are quite different from those discussed above. In cellulose and starch, the constituting unit (monomer) is a single chemical entity, glucose, and the polymer consists of a chain of a large number of this same unit connected. Not a single but about 20 different monomers called amino acids constitute a protein (polymer). The formation of proteins in our body was talked about in the previous chapter.

Amino acid has a general chemical formula $(NH_2)CH(R)COOH$, and its chemical identity is defined by the group R. Some of amino acids are glycine wit R=H, alanine with $R=CH_3$, serine with R=OH, cysteine with R=SH, histidine with R=imidazole $(C_3N_2H_3)$, glutamic acid with $R=CH_2CH_2COOH$, and tyrosine with

R=CH$_2$(C$_6$H$_4$)OH. Two amino acids, (NH$_2$)CH(R^1)COOH and (NH$_2$)CH(R^2)COOH, condense to form a dipeptide connected through a peptide bond (–CONH–):

$$\left(NH_2\right)CH\left(R^1\right)COOH + \left(NH_2\right)CH\left(R^2\right)COOH$$
$$\rightarrow \left(NH_2\right)CH\left(R^1\right) - CONH - CH\left(R^2\right)COOH$$

The third amino acid (NH$_2$)CH(R^3)COOH then attaches itself to the end, and so on. The resulting long chain of amino acids is a polypeptide, and a protein can be a polypeptide or an ensemble of polypeptides.

The functions of proteins are enormously varied, including enzymes, oxygen-carrier (like hemoglobin), muscle, and switch of genes. Proteins can also take a variety of shapes. Some examples of proteins are shown in Fig. 21.10–21.12. These are examples of the so-called globular proteins. Silk produced by silkworm is a typical example of fibrous proteins and functions as a cocoon-forming material. Chinese people devised a method to spin fibers out of this cocoon a long time ago. Spiders construct webs; the material for the web is also silk. Silk is not a single protein; it consists of two major proteins: fibroin and sericin.

The relative spatial arrangement of amino acids in a polypeptide can take, approximately speaking, three forms: α-helix, β-strand, and random coil. They are sketched in Fig. 5.4. By the way, these arrangements of amino acids are said to be of secondary structure of a polypeptide. The primary structure is the ordering of amino acids in the polypeptide chain. The tertiary structure is the overall three-dimensional structure of a polypeptide. Some proteins like hemoglobin consist of more than two polypeptides. This is the quaternary structure. A protein of a single polypeptide can take helix structure in some portions and β-strand structure or random coil in other portions, as illustrated in Fig. 5.4.

Fibroin is a very special protein, consisting of a sequence of repeated arrangements of GAGAGS (G = glycine, A = alanine, and S = serine). When fibroin is exuded from a nozzle of a silkworm or spider, a strong shearing force is applied to it and the polypeptide chain is stretched out and forms a β-strand. A number of such strands together form a structure called β-pleated sheet (Fig. 5.4), and this provides silk with its mechanical strength. Another protein sericin, a polypetide in α-helix fashion but randomly oriented, surrounds the crystalline fibroin. This combination gives silk its flexibility (through sericin) and strength (through β-pleated sheets of fibroin) (see Fig. 5.4).

Another important natural fibrous material is "wool." Wool is mainly made of proteins called "keratins." Not a single kind, but several different keratins are involved: acidic and basic keratins and keratin-associated proteins. Keratins are related to silk fibroin mentioned earlier. Both α-helix and β-pleated structures of keratin are involved. The keratin-associated proteins contain high level of the sulfur-containing amino acid, cysteine. The sulfhydryl (–SH) group of cysteine can readily be oxidized and combine with another sulfhydryl sulfur atom of another cysteine residue on another polypeptide. The result is the formation of sulfur–sulfur

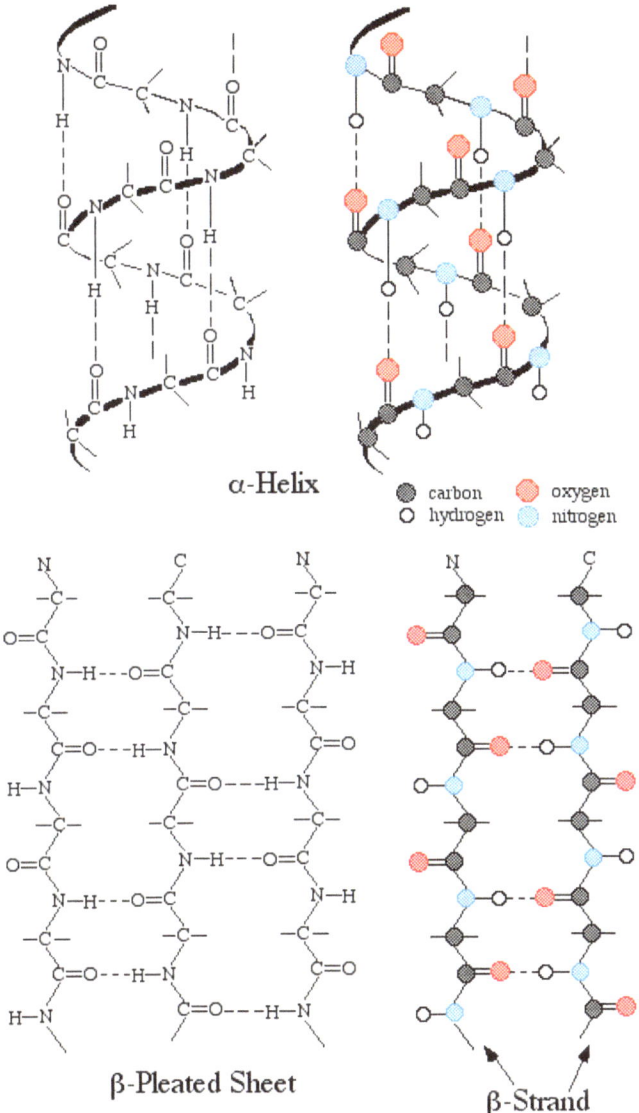

α-Helix

carbon   oxygen
hydrogen   nitrogen

β-Pleated Sheet   β-Strand

Fig. 5.4 Secondary structures of polypeptdides

(−S−S−) bridge between two polypeptides. Such cross-linking makes the protein stiffer, increasing mechanical strength of wool.

Human hair is essentially the same as wool. It is made of keratins and keratin-associated proteins. When you have your hair set, the chemistry mentioned in the last paragraph is made use of. First your hair will be treated with a smelly jerry kind of staff. That is a kind of compound containing SH group (ammonium thioglycolate is commonly used), which reduces the –S–S– bridges and breaks them down to –SHs. Now the cross-linkages have been removed, and the hair has become more flexible. You will have it shaped in any way you want. When this is done, the next step is to recreate the –S–S– bridges in new locations (new setting). Now the hair shape has been fixed. This last step is chemically speaking "oxidation" and is usually accomplished by oxygen coming from hydrogen peroxide; occasionally an oxidant such as sodium bromate might be used.

5.2.4 Natural Rubber

Rubber can be obtained from a number of trees, but commercial production is done with tropical plant *Hevea brasiliensis*. Rubber is a polymer of a single monomer isoprene (2-methyl butadiene). Isoprene itself can chemically (i.e., artificially) be polymerized by various means to produce polyisoprene, but this is not the way the natural rubber is produced in rubber trees. The biological process to produce rubber (i.e., biological polymerization) is shown in Fig. 5.5. This process is a kind of condensation with removal of a pyrophosphate group (P_2O_6).

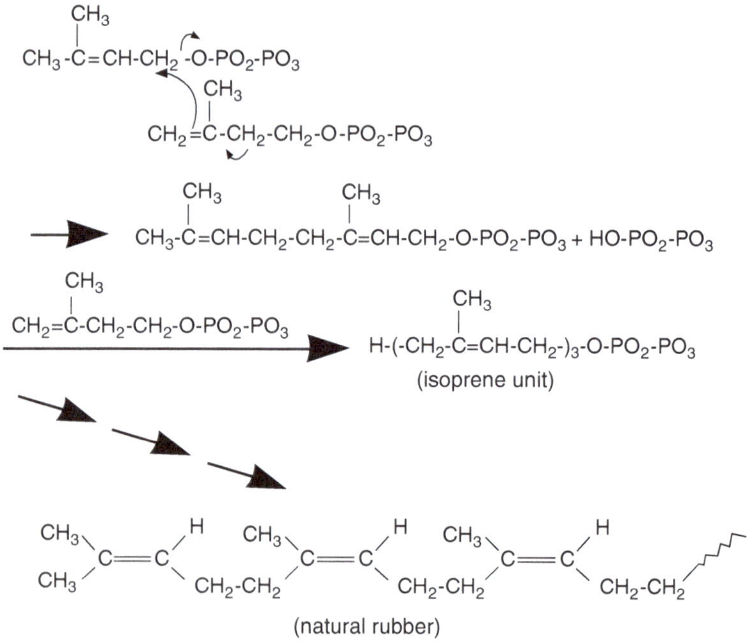

Fig. 5.5 The biological synthesis of polyisoprene

5.3   Synthetic Polymers

A large number and quantity of polymers are synthesized and used today. They are used for a variety of purposes: clothing, wrapping film, plastics, bottle for soda and others, water piping, all kinds of small parts for automobiles and other machinery, and CD/DVD.

An earlier motivation to produce polymers artificially was to produce natural fibers like silk and/or to modify the natural material to desirable fibers. Cotton is a very good fiber, but silk has some appealing characters such as luster and its soft feel upon touch. But silk was much more expensive. People tried to convert cellulose which was much more abundantly available to silk-like fiber. From this effort came the first synthetic fiber, "rayon."

In 1845, a Swiss chemist tried to treat cotton with nitric acid (and sulfuric acid). The resulting material, nitrocellulose, had several interesting properties. Cotton itself does not dissolve in water nor alcohol (ethanol), but nitrocellulose does. Solutions of nitrocellulose were used to turn it into fibrous or plastic form. Plastic form is called "celluloid." It is still used for some purposes. An interesting property of nitrocellulose is its explosiveness. It turned out that if all the three OH groups of glucose units in cellulose are nitrated, the resulting nitrocellulose is very explosive (see Chap. 8 for the explanation of explosive). But it is not explosive, though still quite flammable, if the degree of nitration is less than two. In 1884, nitrocellulose was made into fiber and sold as Chardonnet. Unfortunately, it occasionally burst into flame or even exploded.

Three British chemists discovered that cellulose could be solubilized when it was treated with sodium hydroxide and carbon disulfide. The product is called cellulose xanthate. The viscous solution was then extruded through a nozzle into an acidic solution, forming lustrous, silk-like fibers. They patented the process and commercialized the product in 1894. This is "rayon." It is chemically still cellulose, but its texture is different from that of cotton. It can be produced from not only from cotton, but also from any pulp.

Almost a century later, scientists at Courtaulds', a rayon manufacturer, chanced to discover that cellulose could be dissolved when heated in a solvent, N-methyl morpholine oxide. A new fiber (of cellulose) was produced from this solution. Its brand name is "Tencel." It has a luxurious look and feel and yet strong to be made into clothes like jeans.

Rayon used to be called "art silk," but the fibers mentioned above are not quite artificial. They are modified products of natural material. Another effort to produce silk-like fibers led to the invention of "nylon," a completely synthetic (artificial) fiber.

5.3.1   Nylon

In 1928, an organic chemist at Harvard, Wallace Carothers, was appointed to the director of organic research lab at the DuPont Chemical Company. He led a team to study polymers, the chemistry of which was just emerging. In 1938, he and his team invented "nylon 66." This is an entirely new, synthetic polymer, and the only resemblance (in chemical terms) to a natural fiber, silk, is that the two monomers are connected by a peptide bond as in proteins. The same type of bond, peptide, is called "amide" in synthetic polymer chemistry, and hence nylon 66 is an example of polyamides. Kevlar is another example of polyamide.

Nylon 66 is synthesized by the following condensation reaction between adipic acid and hexamethylenediamine:

$$HO-\overset{\overset{O}{\|}}{C}-CH_2CH_2CH_2CH_2-\overset{\overset{O}{\|}}{C}-OH \quad + \quad H_2N\text{-}CH_2CH_2CH_2CH_2CH_2CH_2\text{-}NH_2$$
$$\text{adipic acid} \qquad\qquad\qquad\qquad \text{hexamethylenediamine}$$

$$\xrightarrow{\text{- }H_2O} \quad \longrightarrow \quad (-\overset{\overset{O}{\|}}{C}-CH_2CH_2CH_2CH_2-\overset{\overset{O}{\|}}{C}-\underset{\underset{H}{|}}{N}\text{-}CH_2CH_2CH_2CH_2CH_2CH_2\text{-}\underset{\underset{H}{|}}{N}-)_n$$
$$\qquad\qquad\qquad\qquad \text{amide (bond)} \qquad\qquad \textbf{nylon 66}$$

The repeating unit in nylon 66 is the combination of adipic acid and hexamethylenediamine, both of which consist of six carbon atoms. Hence, the qualification "66." In each step as two molecules are connected, one molecule of water is removed. This process is a typical condensation polymerization. Because an amide group on one polymer chain can hydrogen-bond to that of the adjacent polymer chain as in silk (see Fig. 5.4), nylon forms a structure similar to the β-pleated sheet. The formation of nylon 66 can readily be demonstrated as some of you might have seen (Fig. 5.6). As you see, you can combine any compounds as long as one has amine

Fig. 5.6 Formation of nylon 66 [from Atkins P and Jones L, "Chemistry Principles, 3rd ed." (W. H. Freeman and Co., 2005)]

groups (NH_2) at both ends and the other carboxylic acid group (COOH). Indeed, many types of nylon have been synthesized, but only a few are still produced in quantities.

One of these is nylon 66, and the other is nylon 6, which can be synthesized by polymerizing a compound called caprolactam. This compound is first hydrolyzed by water to form amino caproic acid. It is a kind of amino acid. Whereas natural amino acids should be characterized as α-amino acid, as the amine group and the carboxylic acid group bind to the same carbon atom, the amino caproic acid is said to be ε-amino acid, because an amine group is situated at ε (5th) position from the carboxylic group. The amine and carboxylic groups in two molecules of ε-amino caproic acid can condense (with removal of water) to form a dimer. Its continuation will result in the formation of nylon 6 (see the reaction equation below).

Nylons are cheaper and much stronger than the natural fiber silk and yet give the feel and touch of silk. Nylon made an instant success, particularly as hosiery material.

5.3.2   Polyesters

Polyesters are produced by a chemically similar means; i.e., condensation polymerization. For example, methyl ester of terephthalic acid and ethylene glycol are polymerized by removing methanol (instead of water), and the poly(ethylene terephthalate) results. This polymer can be made into a unique fiber; crease-resistant fiber. As the hexagonal unit (benzene ring) is rather rigid, the fiber made of this

polyethylene terephthalate (PET)

polymer can resist shape change. Once it has been set in a certain shape, it can retain its shape, no matter how you treat it. Hence, it has been used for shirts that require no ironing after washing: a convenience for travelers.

Today, however, the polymer is more often used as the material to make PET bottle. In this case, the polymer is produced by heating terephthalic acid and ethylene glycol with removal of water molecule. The reaction is similar to that shown above, except that the CH_3 groups in methyl terephthalate should be replaced by Hs.

5.3.3   Polymers Obtained by Addition Polymerization

Carothers of DuPont company also invented a synthetic rubber, chloroprene. This polymer is produced by addition-polymerization unlike the polymers mentioned so far. Let us see how it is done with an example of polyethylene. Polyethylene is used in everyday life as a wrap, the thin plastic film. It is a polymer of ethylene, ethene in technical terms. Ethylene has the chemical formula of $CH_2=CH_2$. Addition polymerization can be conducted with a compound having one or two double bonds; such a compound is often called "olefin," Ethylene is the simplest olefin. One of the two bonds in a double bond is a σ-bond and the other is a π-bond, which is weaker and more readily splits than the σ-bond. Alternatively, the electron in the π-bond easily binds to an entity called a "free radical." A free radical has an unpaired (single) electron, which seeks another electron to pair up. The electrons in the π-bond provide such an electron. So what happens is this:

$$R^{\cdot}\left(\text{free radical,}\,^{\cdot\cdot}\,\text{indicates an electron}\right)+CH_2=CH_2 \rightarrow R\text{-}CH_2\text{-}CH_2^{\cdot}$$

In this reaction, the entity $R^{\cdot}$ is artificially produced or added and is called an "initiator" (of polymerization). The resulting entity $RCH_2\text{–}CH_2^{\cdot}$ now has an unpaired electron, which can react with another molecule of ethylene:

$$R\text{-}CH_2\text{-}CH_2^{\cdot}+CH_2=CH_2 \rightarrow R\text{-}CH_2\text{-}CH_2\text{-}CH_2\text{-}CH_2^{\cdot}$$

This process can go on and on, making a long chain of $–CH_2–CH_2–$ units; hence, the product of such a polymerization is called "polyethylene." Polyethylene is essentially the same as the molecules in gasoline or wax; only the chain length is enormous, often tens of thousands of ethylene units. [You might have noticed that you smell wax when you burn polyethylene plastic wrap]. In this polymerization process, you will note, no small molecule like water is removed as in condensation polymerization, and a monomer is simply added to the growing end of a polymer. Hence, polymerization of this type is called "addition" polymerization.

Derivatives of ethylene in which one of the hydrogen atoms is replaced by another entity are called "vinyls." For example, vinyl chloride is $H_2C=CHCl$, and vinyl acetate is $H_2C=CH(OCOCH_3)$. The polymer of vinyl chloride is poly(vinylchloride), PVC, which has been widely used for plastic cover and water pipe.

Other compounds of vinyl type that are used for polymers include acrylonitrile (vinyl cyanide), $H_2C=CH(CN)$, and styrene, $H_2C=CH(C_6H_5)$. Vinyl compounds are all polymerized by addition process. Tetrafluoroethylene, $F_2C=CF_2$, in which all the hydrogen atoms of ethylene are replaced by fluorine atoms, polymerizes similarly to form a polymer of unique character, teflon.

Another kind of monomer (repeating units) of interest is what is called "diene." Ethylene and vinyl compounds are so-called monoene, compounds that have in them only one double bond, as you saw. Diene molecules have two double bonds in them. Butadiene $CH_2=CH–CH=CH_2$, isoprene $CH_2=C(CH_3)–CH=CH_2$, and chloroprene $CH_2=CCl–CH=CH_2$ are important examples of diene. We have seen isoprene before; it is the repeating unit of natural rubber latex.

A monomer such as $XCH=CH_2$ can polymerize in different ways as illustrated in Fig. 5.7. In the first example, the groups Xs are arranged in the same direction when the monomers polymerize. In the second case, the directions of Xs alternate, whereas they are random in the third case. The resulting polymers of different specific structures (stereospecific) may have different properties. For example, the first two types of structure tend to crystallize well, as compared with the random one; hence, they are more suitable for use as a solid plastic. The modern process of polymerization can control the stereospecificity of polymer to a large extent.

In the case of a diene $CH_2=CX–CH=CH_2$ polymerization, there are two orientations of polymer chain across the central C=C double bond of each monomer unit, as seen in Fig. 5.8. In one, the polymer chain orients itself in the *cis*-direction (the same direction) across the C=C double bond; this is called *cis*-polymer. In the other, it orients in the *trans*-direction (the opposite direction); *trans*-polymer. The *cis*-polymer usually behaves like a rubber, while the *trans*-polymer is more like plastic.

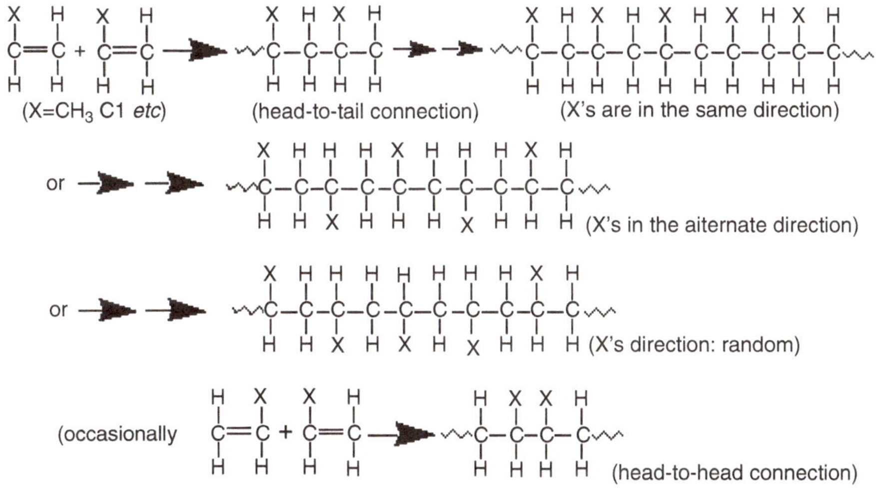

Fig. 5.7 Stereospecific polymerization of a vinyl compound

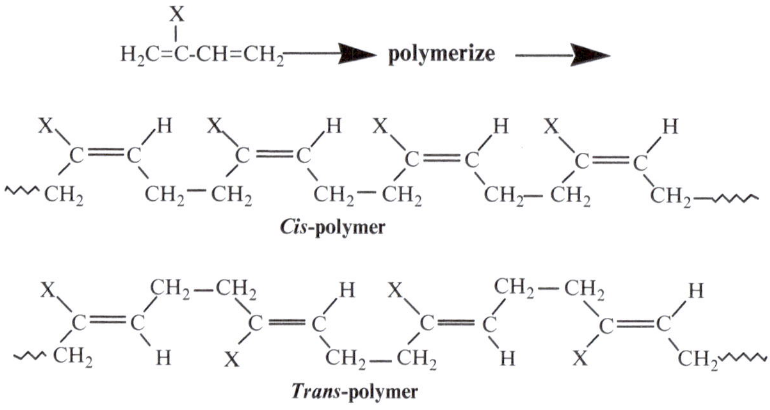

Fig. 5.8 *Cis*- or *trans*-polymerization

Addition polymerization can be accomplished not only through a free radical initiator as mentioned above, but also by some other means. The most important polymerization catalyst is of the type known as Ziegler–Natta catalyst. These two chemists discovered that a combination of chemicals titanium tetrachloride and tri-ethyl aluminum is an excellent catalyst for polymerizing a number of olefins. They were awarded Nobel Prize in 1963 for this discovery. Subsequent research by others found that similar combinations of chemicals: a transition element compound and triethyl aluminum or similar alkylating agent do catalyze polymerization of olefins. Specific combination of such chemicals allow formation of polymers of specific stereochemistry.

Ethylene is the simplest olefin. Propylene (propene) is the next simplest. The polypropylene produced by a Ziegler–Natta catalyst turned out to be of the structure of a special stereochemistry, and to form a fairly crystalline plastic. It was too hard for most of applications and people who developed this polymer had to invent some use for it. One of the uses they came up with was "hoola hoop." This is an example in which a product had been produced before they knew what they could use it for. Is this a rare exception? Would not people know the uses for a product before they decide to produce it? It is usually the case. However, there have been a number of cases in which a product had been produced first for no good reason or other pur-poses and only afterward they discovered its good application.

5.3.4   A Story of Vinyl Chloride

Let us talk about such an example. Paper industry needs a chemical called sodium hydroxide (caustic soda) to clean the pulp. The most convenient way to produce sodium hydroxide of a good quality is to use "electrolysis" of a brine solution con-taining sodium chloride. That is, if you put electric current through the solution, you can get a sodium metal which can easily be converted to sodium hydroxide by react-ing the metal with a steam. The other product of the electrolysis of the brine solution is "chlorine." Chlorine (Cl_2) is used for bleaching the pulp and to disinfect drinking water. It is also used for the production of chlorine-containing organic compounds such as described in Chap. 16, but the production of these compounds has declined for the reason of environmental concerns in recent decades. People did not know what to do with the excess chlorine. One of the uses they came up with was to produce vinyl chloride. Hydrogen chloride that can be produced from chlorine easily is made to react with acetylene; the resulting compound is vinyl chloride. Today, however, vinyl chloride is made from ethylene and chlorine.

Vinyl chloride is polymerized to produce polyvinyl chloride (PVC) as mentioned earlier. It is cheap, and fairly stable and sturdy, and hence has been used very widely for such purposes as plastic film, plastic cover, fake surface material (in any shade like walnut) for wall and furniture, and water pipe.

It has some problems. The plastic loses plasticity and becomes brittle as it is exposed to sunlight. This is due to a complicated series of chemical reactions initially caused by sunlight. Sunlight breaks one of the bonds, likely C–Cl, forming a Cl-free radical, which abstracts hydrogen atom on the nearby carbon atom, likely resulting in the formation of C=C double bond. The double bond then breaks, probably reacting with oxygen (O_2) in the air under the influence of sunlight. Otherwise, PVC is fairly stable and persists long in the environment.

Another problem is the formation of HCl when it is burned. It is argued that it even forms the toxic dioxin when it is burned at high temperatures in the incinerator. The biggest problem, however, seems to be that some toxic substances including dioxin are produced as by-products in the process of producing vinyl chloride itself. How to treat them and contain them is the problem.

Part II
Enhancing Human Health

Essential chemicals outlined in Part I are not sufficient to maintain health of living organisms. A number of the so-called inorganic elements (chemicals) are indeed required for proper workings of living systems; they are often called "mineral nutrition." Human bodies are often subject to health-disrupting causes, and remedying such ill conditions of health requires often chemicals (as medicine) in addition to a number of endogenous mechanisms to combat such disruptions present in the human body.

Mineral Nutrition

6

Men Do not Live by Bread Alone!

6.1   What Elements Are Necessary for Our Health?

Indeed! We need more than bread to live a healthy life. A balanced diet may consist of vegetables, meats and bread, or other cereals. Bread is mostly carbohydrates, which are made of elements such as carbon, hydrogen, and oxygen ($(C_6H_{10}O_5)_n$). Meat is a protein source; proteins are made of elements of carbon, hydrogen, oxygen, and nitrogen, plus a little of sulfur. Vegetables contain essential vitamins in addition to some carbohydrates and proteins. Vitamins are various organic compounds, made of carbon, hydrogen, nitrogen, and oxygen (and sulfur and phosphorus in some of them). The major portion of our body itself is made of "organic" compounds: proteins and DNAs, which are produced in our body from the material ingested (food). Our diets themselves come from other organisms, because all the organisms living on the Earth are, in large measure, similar in their biochemistry. Hence, all the organisms are connected through food chain. The major biochemical processes are talked about in Chap. 3. The biologically important organic compounds are made of four chemical elements: carbon, hydrogen, nitrogen, and oxygen. A few bio-organic compounds contain elements such as sulfur and/or phosphorus in addition to the four elements.

Milk, cow's as well as human milk, is close to a perfect diet, containing carbohydrates, fats, and proteins. In addition, it contains another important ingredient, calcium. Why is calcium important? For one thing, calcium is to make bones and teeth. Calcium is prescribed for women suffering from osteoporosis. So milk is good also for children as they need to build bones. Calcium, as it turned out, does a lot more essential things to the body than merely building bones and teeth. You have approximately 1 kg (2 pounds) of calcium in your body.

Women tend to lose blood more often than men and need to take more iron than men do to offset the loss. Iron is the essential ingredient of blood. In addition, iron

E. Ochiai, *Chemicals for Life and Living*,
DOI 10.1007/978-3-642-20273-5_6, © Springer-Verlag Berlin Heidelberg 2011

plays hundreds of different essential roles in all living creatures. Your iron content is about 5 g. Five grams may sound like a small quantity, but it amounts to a very large quantity of biologically important compounds that are dependent on iron.

Calcium and iron are two well-recognized "mineral" nutrients. In addition, altogether something like 30 elements have now been recognized to be essential to human and other organisms' health (Fig. 6.1). Let us explore here why they are necessary and what these elements are doing to our body and other organisms.

First let us take a look at Fig. 6.1. This diagram shows the contents (in terms of ppm) of elements in an average human body. "ppm" means "parts per million," and hence 100 ppm, for example, is one hundred grams in one million grams. For a man of 70 kg (155 lb), the quantity represented by 100 ppm would be $70 \, kg \times (100/1,000,000) = 70 \times 10^{-4} \, kg = 7$ g. Thirty-four elements are represented in Fig. 6.1. Other elements have also been detected in human body, but they are usually present at much lower levels than the indicated minimum in the diagram. Out of the 34 listed, 26 elements that are shown in boldface have been recognized to be essential to organisms. Not all of them have been shown to be essential to human beings. For example, boron ("B") is known to be vital to plants but not to animals. Strontium (Sr) is used as an outer skeleton for a group of oceanic plankton, but is not considered to be essential to humans, even though a significant amount of strontium is found in the human bones.

In order for our body to function properly (i.e., to be healthy), all the bodily physiological reactions should occur smoothly and at regulated paces. These are in

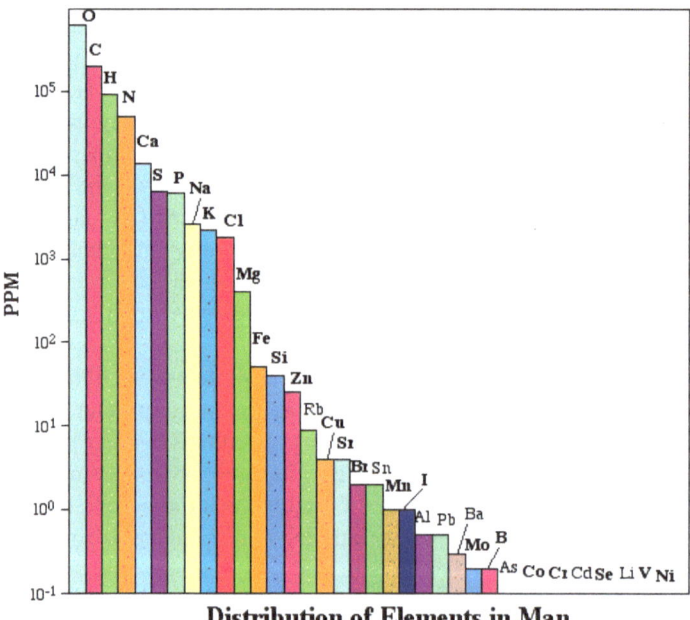

Distribution of Elements in Man

Fig. 6.1 How much of each element is in human body?

fact all chemical reactions, and the majority of biochemical reactions require catalysts, called enzymes. An enzyme performs two functions: one is to speed up the chemical reaction, and the other is to choose specifically the right compounds to work on. A cell may be likened to a soup pot, in which various ingredients are being cooked. For the cell to function properly, only certain kinds of reactions have to take place and at certain (usually fairly high) speeds. A catalyst, an enzyme, picks up only a specific ingredient and regulates its reaction. If it picks up a wrong ingredient to work on, the health of the cell may be jeopardized. It is not easy to be choosy, because there are hundreds of different compounds present in the soup (cell) and some of them could be very similar. The majority of those essential mineral elements assist the enzymes' functions. Let us look at some of the important mineral elements and how they work in our body.

6.2   Iron

Iron is ubiquitous and is required by all living organisms. It is used in a variety of ways in living organisms. There are three classes of iron-containing proteins (enzymes) according to the ways in which iron is bound in the protein (enzyme). The following is a brief discussion of some iron-containing enzymes and proteins of these different types. But first a brief description of chemistry of iron may be in order.

6.2.1   Chemistry of Iron

Iron, a familiar metal, tends to rust, as everybody has seen. What happens chemically when iron rusts? Iron atoms in the metallic iron carries no electric charge $Fe(0)$, in which 26 electrons (negatively charged) are orbiting around a nucleus that contains 26 protons (with positive charge) and 30 neutrons (with no electric charge). This applies to an isotope $_{26}Fe^{56}$, the most abundant isotope of element iron. But iron atom can lose its electrons. This process (loss of electrons) is called "oxidation." Iron becomes either $Fe(II)$ by losing two electrons or $Fe(III)$ by losing three electrons (under normal conditions), though it can take $Fe(I)$ (under special conditions). $Fe(0)$ is said to have been oxidized to $Fe(II)$ or $Fe(III)$. [$Fe(II)$ means an iron atom that carries two positive charges; this is so because there are now only 24 electrons (negative charges), but there are still 26 positive charges at the nucleus.] For this to happen you have to have a chemical entity that removes the electrons from the iron atom. Such an entity is called an oxidant or oxidizing agent. The iron atom is said to be a reductant or reducing agent in this process, for a chemical reaction in which a chemical entity (oxidant) gains electron(s) is called "reduction." Hence, oxidation and reduction reactions occur simultaneously and are like "head and tail" of a coin. Iron (Fe) reacts with oxygen in the air and is oxidized first to $Fe(II)$ and then $Fe(III)$ ending up with iron oxide Fe_2O_3. [$Fe(II)$ can also be expressed as Fe^{2+} or Fe^{II}, and such a state is called an "oxidation state"; in this case, the oxidation state of iron atom is +2 or II. Likewise, $Fe(III) = Fe^{III} = Fe^{3+}$. We will usually use Roman numerals to express the oxidation states in this book]. Here oxygen (O_2) in the air is the oxidizing agent. In the process, oxygen O_2 which is in "zero" oxidation state gains four electron and is reduced to two of O^{-II} (−2 oxidation state); therefore, the chemical reaction is $4Fe(0) + 3O_2 \rightarrow 2Fe^{III}_2 \left(O^{-II}\right)_3$. In this chemical reaction, 12 electrons are exchanged between four iron atoms and three oxygen molecules. When this is not purely oxide and contains hydroxide $Fe(OH)O$ or $Fe_2(OH)_2O_2$, it shows that rust color, brown. Pure oxide Fe_2O_3 forms an ore called "hematite," which is red. The red bed is found in many geological locations.

These descriptions suggest that iron, when forming chemical compounds, takes the form of Fe(II) or Fe(III). And it can go back and forth between Fe(II) and Fe(III) readily. Fe(II) gives off an electron to become Fe(III), and Fe(III) becomes Fe(II) when it accepts an electron. This kind of process is also called "electron transfer" reaction. Hence, iron (in the form of Fe(II) and Fe(III)) can readily undergo an "electron transfer" reaction or alternatively an "oxidation–reduction" reaction, because the process of Fe(II)s becoming Fe(III) is an oxidation and the reverse (Fe(III) → Fe(II)) is a reduction reaction.

Some of you might have experienced, as this author has, to have your toilet bowl and others stained brown by your well water. The water underground can contain (depending on the location and other conditions) iron compounds; the iron is in the form of Fe(II), which is dissolved in water and almost colorless. It remains as Fe(II) in the underground, because no oxidant such as oxygen in the air is available. However, once pumped out above ground and being exposed to the air, the iron soon turns into Fe(III) (through oxidation by oxygen). Fe(III) in water (neutral water, that is) is not stable, and soon reacts with water itself and forms iron hydroxide $Fe(OH)_3$, which is brown and precipitates. This is the brown stain. And the fact of the easy formation of iron stain suggests an easy affinity or reaction of Fe(II) with oxygen O_2. This is indeed the basis of the usefulness of iron in the biological systems and our health.

6.2.2   Heme Iron

6.2.2.1 Hemoglobin

The best-known example of iron-containing proteins is the one called "hemoglobin" that is the ingredient of red blood cells and carries oxygen throughout your body. A protein is a large molecule that consists of tens' or often hundreds' units of amino acids. The muscle or beef you eat is made of proteins. Many biological functional entities are proteins. Most enzymes (biological catalysts as we talked about above) are made of proteins. However, a substantial number of proteins and enzymes have been shown not to be able to function without added components, relatively small chemical entities. In the case of hemoglobin, this added component is a "heme group." The heme group is a beautifully shaped molecule with an iron atom sitting at the center (see Fig. 6.2). See also Fig. 21.10 for the entire molecule of myoglobin, which is a sort of one quarter of hemoglobin. Organic compounds including proteins, DNAs, and carbohydrates themselves apparently cannot perform the function of

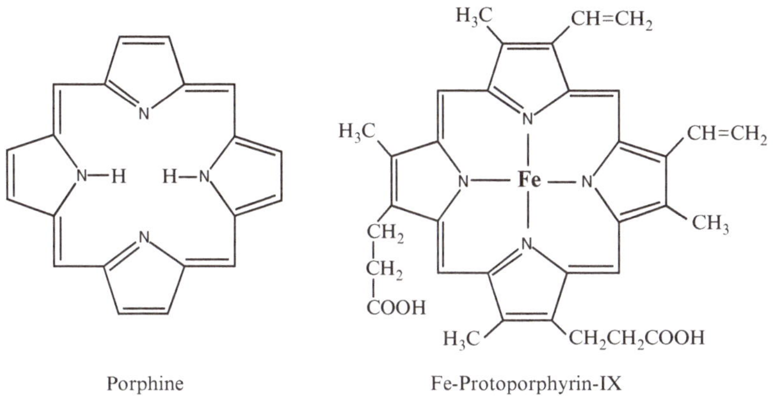

Porphine   Fe-Protoporphyrin-IX

Fig. 6.2 Heme group

binding oxygen (O_2) and carrying it through the circulatory system. That is the reason that hemoglobin uses an inorganic element iron to bind oxygen. Another inorganic element, copper is used for the same purpose in some marine organisms such as oyster and squid (blue blood).

Iron in the hemoglobin binds O_2. The chemical reaction equation for this binding is expressed as follows:

$$Fe^{II}\left(of\ haemoglobin=Hb\right)+O_2 \rightleftharpoons O_2^{-}-Fe^{III}\left(Hb\right)$$

An arrow in a chemical reaction indicates the direction of the reaction. In this case, the reaction can go in either direction. This situation is called "reversible." Fe(II) in hemoglobin binds oxygen in the lung (the forward reaction) and then carries the oxygen in blood vessel systems to a tissue where oxygen is needed. When red blood cell arrives at the target cell, the hemoglobin unloads O_2 (i.e., the reverse reaction). Myoglobin (Fig. 21.10) in muscle cells picks up this oxygen.

This reaction, binding of O_2, cannot occur when the iron is in the form of Fe(III). The hemoglobin with Fe(II) is red and becomes bright red when it binds O_2, but it becomes somewhat bluish when it becomes Fe(III) (without binding oxygen). Beef is red when it is fresh, but it turns to brown as you leave it in a refrigerator. This discoloration is due to the oxidation (by air) of Fe(II) in the blood to Fe(III). A disease called "cyanosis" is caused by an excess level of nitrite (NO_2^{-}) in drinking water. Nitrite oxidizes the Fe(II) of hemoglobin to Fe(III) and hence diminishes the oxygen-carrying capacity of the blood, exhibiting an bluish tinge on the skin (thus the name: cyanosis).

However, nitrite (NO_2^{-}) can be reduced to nitrogen oxide (NO) in the cells (if there is enough reducing material in the cells), and NO binds to Fe(II) of hemoglobin more tightly than oxygen (O_2) does. And the NO-bound hemoglobin is also bright red and resists oxidation by oxygen in the air. This trick (chemically using sodium nitrite) is used sometimes as a preservative of red color of the meat. A health problem of the use of nitrite is that nitrite could react with DNA and cause changes in its sequence and that these changes may lead to a cell abnormality and then cancer. Carbon monoxide CO also binds more tightly to the Fe(II) of hemoglobin than oxygen (O_2) does. The iron that is bound with CO cannot pick up oxygen; hence, poisoning with carbon monoxide leads to asphyxiation.

6.2.2.2 Other Heme Proteins and Enzymes

Besides hemoglobin there are many proteins and enzymes that contain the heme groups. Most of them are involved in the respiratory system. You ingest and digest foodstuff and then oxidize them ultimately using oxygen you inhale. This process of burning carbohydrates to extract energy (converting the energy to ATP; see Chap. 3) is "respiration." Carbohydrates are oxidized step by step and eventually the electrons extracted from the chemical compounds (metabolites) derived from the carbohydrates are transported to the oxygen. Oxygen (O_2) thus gains electron(s) and is reduced to water (H_2O). This process in which electron(s) have to be moved around is carried out by proteins containing irons. Some of these proteins contain heme groups, and the electrons are dealt with by the iron. For example, one heme group is now in the Fe(II) state. It can then give one electron off to an adjacent protein with Fe(III)-heme. The former turns to Fe(III) and the latter to Fe(II). Iron can do this sort of thing very easily. Hence, heme-containing proteins are heavily used in this kind of process. These proteins are called "cytochrome(s)"; it means pigment (chrome) of cell (cyto-), because they are major coloring substances in cells.

Hemoglobin binds oxygen but does nothing to change it. The heme iron in a few enzymes do bind oxygen and modify it in such a way that the oxygen becomes more reactive and does react with other chemicals. The heme iron then acts as a catalyst; so the proteins containing such heme groups are called heme-enzymes. A group of such enzymes includes a special type of heme called cytochrome P-450. Enzymes containing P-450 catalyze a unique reaction, called "monooxygenation." For example, a steroid hormone, progesterone, reacts with O_2 under the assistance of a P-450 enzyme to have one of the O-atoms of O_2 inserted into a C–H bond. That is, C–H turns into C–O–H, and this oxygen atom comes from the O_2 molecule. This kind of reactions is used widely in our body to metabolize certain compounds including many steroid hormones, foreign compounds such as drugs, and environmental polluting agents.

Other examples of heme-enzyme include catalase and peroxidases. When you get injured, you might apply an antiseptic solution containing hydrogen peroxide (H_2O_2) to the injured part in order to prevent it from being infected. You might have observed some bubbles forming when you did that. It is due to a chemical reaction of hydrogen peroxide to form oxygen gas catalyzed by the enzyme catalase contained in the blood seeping out from the injured skin. The chemical reaction is $2H_2O_2 \rightarrow 2H_2O + O_2$. By the way, this oxygen molecule formed is not quite the same as O_2 found in the air and is more active and can kill (oxidize) bacteria present in the injured spot.

In some of these catalytic reactions, Fe is said to take the Fe(IV) oxidation state as in FeO^{2+}. When iron is bound with a negatively charged N- or S-ligand(s) in cytochrome P-450 or peroxidases, the higher oxidation state such as Fe(IV) is believed to be stabilized and hence will be realized.

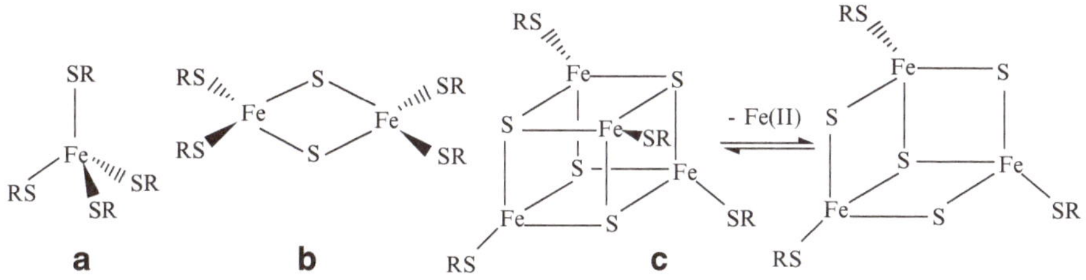

Fig. 6.3 Iron–sulfur proteins

6.2.3 Iron–Sulfur Proteins

Many proteins contain a peculiar unit that is made of iron atoms and sulfur (S) atoms. The most widely occurring such unit is a cube in which four iron atoms and four sulfur atoms are occupying alternate positions (see Fig. 6.3). An alternative is a unit that consists of two iron atoms and two sulfur atoms (Fig. 6.3). The proteins that contain such iron–sulfur unit(s) are called "iron–sulfur protein." Both of these units act as an electron-transfer unit, by changing its iron between Fe(II) and Fe(III) as in cytochromes. In particular, iron–sulfur proteins that are involved in photosynthesis are called "ferredoxins." Obviously, ferredoxins are not found in humans or other animals; they do not perform photosynthesis. However, there are a number of other proteins that contain "iron–sulfur" units even in human beings. Some of the important iron–sulfur proteins are involved in the electron-transport process in the respiration mentioned above.

The iron atom of an iron–sulfur unit is also used to catalyze a certain type of biochemical reaction, that is, adding water or removing water from a compound. A typical example is an enzyme called "aconitase," which is involved in a portion, called "TCA" cycle, of the energy-obtaining respiratory process. The iron in an iron–sulfur protein is sensitive to the presence of oxygen, because it can undergo a rapid oxidation–reduction reaction as seen above. The living cell takes advantage of this sensitivity to oxygen and uses such a protein as a sensor of oxygen.

6.2.4 Other Types of Iron-Containing Enzymes and Proteins

There are a large number of other enzymes and proteins that contain iron. They do not, however, have their irons in an organized manner as above and the irons are bound to certain specific amino acids. When you take in iron from foodstuff, the iron will be absorbed by the upper portion of intestine. Iron is then carried by blood to tissues and organs that need it. It does not move as a naked iron (either Fe^{II} or Fe^{III}), but iron is bound to a protein called transferrin, which literally means "iron-carrier." The iron in the form of Fe^{III} binds to several specific amino acids in the protein.

Another example is an enzyme called "pyrocatechase," which catalyzes the reaction between O_2 and catechol (1,2-dihydroxybenzene). This is rather a unique reaction and the enzyme is found in some bacteria (not in humans). The iron atom is bound to several specific amino acids in this protein.

Just one more example. When red blood cells age, they are scrapped and the iron recovered from them is then stored in a protein called "ferritin." By the way, the life of a red blood cell in human is about 120 days on average. Ferritin is an interesting protein in the sense that it does not literally bind iron atoms, but rather it sort of wraps up an aggregate of iron (iron hydroxide). This protein serves as storage of iron and participates in the control of iron level in our body.

6.3   Copper, Manganese, and Molybdenum

Iron, as mentioned above, is an "electron dealer"; it facilitates "electron transfer" or "oxidation reduction" reactions. Several other elements also function as "oxidation–reduction" catalysts. They include copper, manganese, and molybdenum. The basic common character is that they can readily change their oxidation states.

Copper can take oxidation states (I)(1+) and (II). So copper-containing proteins and enzymes can enhance "electron transfer" or "oxidation reduction" reactions. In this sense, copper behaves very much like iron. However, there is a rather subtle difference between them. This is due to a basic difference in the chemical characters between iron and copper. That is, copper(II) is more readily reduced to metal state (or (I) state) than Fe(III) is to the metal state in a medium of ordinary pH. Therefore, copper is found in nature often in the metallic state, whereas iron has rarely been found as native metal in the rocks. (It must be mentioned that the core of the Earth is essentially metallic iron.) You can see large specimens of such native copper in Natural History Museum in Washington, D.C. Hence, copper(II) in enzymes and proteins tend to work as stronger oxidizing agents than Fe(III), in the general sense. It must be pointed out, though, that the oxidizing power (reduction potential) of any metallic ion can be modified widely by the other entities bound with it. There are many copper enzymes and proteins in living organisms, though they are not so widely distributed as iron-containing ones. Copper is thus essential to our health. Lack of copper manifests, for example, in malformation of connective tissues.

Manganese is located beside iron in the periodic chart (Fig. 19.2). This fact suggests that manganese would behave chemically like iron. Indeed, there are a lot of similarities between them, and some manganese-containing enzymes play the same roles as iron-containing enzymes. However, there are also differences between iron and manganese. Manganese can take many different oxidation states; it can relatively easily take (II), (III), and (IV) oxidation states in enzymes (and (V–VII) in nonenzymatic compounds). As a result, manganese plays a very unique role in plants. Green plants synthesize carbohydrates (glucose and starch). Carbohydrates can be expressed in general as $(CH_2O)_n$ and are formed from simple compounds: water (H_2O) and carbon dioxide (CO_2). The reaction can be written schematically as $6CO_2 + 6H_2O \rightarrow C_6H_{12}O_6 + 6O_2$. The water molecule is decomposed into hydrogen and oxygen, and hydrogen is used to reduce carbon dioxide to form carbohydrate.

The photochemical decomposition of water molecule is carried out by an enzyme containing manganese. Manganese is hence uniquely crucial to green plants and algae, but also essential to all other organisms as well, as it constitutes an essential component of many other enzymes.

Molybdenum is another peculiar element. It constitutes the catalytic site for many oxidation–reduction enzymes. Such enzymes carry out reactions in which oxygen atom is given to or taken away from compounds. Xanthine oxidase, for example, attaches oxygen atom to xanthine, which is a derivative of one of the components of DNA. The enzyme is thus involved in the metabolism of DNA. The product of this enzymatic reaction is uric acid, which is to be found in our urine.

You might have noticed nodular attachments on, for example, the roots of clover. These nodules contain an enzyme called nitrogenase. It catalyzes the formation of ammonia from nitrogen in the air. The reaction is represented by $N_2 + 6H^+ + 6e \rightarrow 2NH_3$. That is, the nitrogen is reduced to ammonia. This reaction is called "nitrogen fixation." Ammonia is then used by the plants as fertilizer. Nitrogen is the essential ingredient of amino acids, proteins, DNA, and many other biologically important compounds. Ammonia is the only form of nitrogen that can be incorporated into these compounds. Nitrogen fixation is thus crucial in the ecology of the biosphere.

How nitrogenase works is under intensive study by many scientists. Besides, efforts are being made to incorporate a gene of nitrogenase into crop plants, in the hope of eliminating the necessity of artificial nitrogen fertilizer. Anyway, molybdenum is the essential element for this enzyme. In addition, the enzyme uses the iron–sulfur units mentioned earlier. It is thus dependent on iron, sulfur, as well as molybdenum and is one of the most complicated enzymes (see Fig. 21.12 for the entire structure).

It is to be noted, however, that other nitrogen compounds such as nitrate (NO_3^-) can be absorbed and utilized by organisms, both plants and animals. Nitrate needs to be reduced to ammonia level in order for it to be incorporated into bioorganic compounds. Some of the nitrate-reducing enzymes depend on molybdenum.

6.4   Zinc

Zinc is another very widely used element in all the organisms. Zinc is also used widely in everyday life. For example, one of the electrode, anode, of a dry battery (cell) is made of zinc metal. Today, most vitamin pills on the shelves of drug stores have labels indicating that they contain everything from A to Z. This "A" is vitamin A, and Z stands for "zinc (Zn)." Zinc is not a vitamin, though. You have in your body about 2–4 g of zinc. That does not sound much, but it is plenty. What does zinc do? Because of its very wide use as outlined below, Zn(II) often becomes short-supplied when you are ill or in a traumatic situation. Zn(II) is a necessary remedy for such a situation.

Let us look at the periodic chart (Figs. 19.2 and 19.5). Zinc is located at the end of the (first) transition metal series (Sc–Zn). It exists commonly as Zn(II) when it forms compounds. It does not change its oxidation state, i.e., would remain as Zn(II), unlike iron, copper, and others mentioned above. That is, it cannot transfer electrons and hence cannot be involved in oxidation–reduction reactions. That Zn is at the end of the series suggests that Zn(II) would be the smallest in size among the divalent (+II) transition metal ions. Besides, Zn(II) would be the highest in its effective positive charge among the same transition metal ions. Therefore, Zn(II) is one of the strongest (Lewis) acids, because of its high effective charge and small size among the divalent transition metal ions. Due to the other favorable effect called "ligand field stabilization energy," Cu(II), that is just next (left) to Zn(II), is actually the strongest acid, slightly stronger than Zn(II), among the divalent transition elements. ["Divalent" means "being in the oxidation state +2 or carrying +2 electric charge].

We talked about the chemical principles operating in biochemical reactions in Chap. 3. Most of the biochemical reactions can be either of acid–base type or of oxidation–reduction type. Many of the latter type reactions are catalyzed by enzymes that use transition elements such as iron, copper, manganese, and molybdenum, as mentioned already. The reactions of acid–base type are obviously catalyzed by enzymes of acid–base characters. A number of enzymes use their own resource, i.e., amino acid residues as acid–base catalysts. Such amino acids include serine, threonine, and tyrosine (the first two have slightly acidic OH group; tyrosine's OH is quite acidic), aspartic acid and glutamic acid (both have acidic carboxylic group), cysteine (slightly acidic SH group), and histidine with a basic N group. These amino acid groups act as the catalytic sites in many enzymes that work on reactions of acid–base type.

However, there are a number of situations where these amino acid residues alone are not adequate enough. Then "Nature" has tried to utilize acidic entities other than amino acid residues. Chosen were Zn(II) or other metallic ions such as magnesium (Mg(II)) and manganese (Mn(II)). In certain cases, even Fe(II) and Cu(II) are used for this purpose; aconitase mentioned earlier uses Fe(II) as an acid entity. Zn(II), however, is the most widely used metallic ion as Lewis acid in enzymes. The zinc-containing enzymes are found among all classes of enzymes. A few examples will suffice to illustrate the point.

An enzyme called "carboxypeptidase" splits a certain type of proteins. There are many enzymes in your body that split proteins (i.e., hydrolyzes peptide bonds in proteins). Proteins in meat need to be split, i.e., hydrolyzed in order to be digested. Pepsin and trypsin are two protein-hydrolyzing enzymes (a class called "proteinase (or protease) or peptidase") found in the stomach. These enzymes do not use Zn(II). As a matter of fact, there are as many non-Zn(II)-dependent proteinases as Zn(II)-proteinases. Zn(II) seems to be necessary for carboxypeptidase because the specific nature of the portion of a protein that is worked on by this enzyme. Other Zn(II)-dependent proteinases also need Zn(II) because of the specific needs of the specific proteins. However, the details of the specific needs are not very well understood.

Carbonic anhydrase, another Zn(II)-enzyme, catalyzes a simple reaction: $CO_2 + H_2O \rightleftharpoons H^+ + HCO_3$. [The reaction goes both ways]. Carbon dioxide is produced as a result of respiration (oxidation of carbohydrates) in cells. It has to be disposed of. CO_2 comes out of cells into the circulating system, and then it binds in the form of HCO_3^- to hemoglobin in red blood cells. It is then carried to the lung where it should be turned back into CO_2 and then exhaled. Hence that simple reaction above needs to take place very rapidly and requires a good catalyst. Carbonic anhydrase is that catalyst and its reaction rate is known to be one of the fastest of all the enzymatic reactions. Zn(II)-enzymes are also found among those that will hydrolyze phosphate ester bonds. However, we will omit the further details here.

Zinc(II) seems to be used also to maintain certain specific (tertiary) structures of proteins. A number of proteins that contain Zn(II) and that regulate the expression of DNA have been discovered in recent years. That is, this protein switches on and off the transcription of DNA, i.e., production of messenger RNA (see Chap. 4). Such a protein is called in general a "transcription factor," whether it contains Zn(II) or not. The transcription factors that contain Zn(II) are called often "zinc-finger" protein, because the Zn(II) here is used to maintain the specific structure of the protein (called "finger") so that it binds to a specific location of a DNA.

6.5   Calcium

All cellular organisms, that is, all the living organisms on the Earth, would not be able to live without calcium (Ca(II)). Calcium compounds play two very different types of role in living organisms.

First of all, calcium compound(s) is used as a solid substance to lend a mechanical strength to the body. The bone in our body is an example, which is essentially made of calcium phosphate $Ca_3(PO_4)_2$ (though the composition is close to mineral called hydroxyapatite $Ca_5(PO_4)_3(OH)$). The enamel of our teeth is also calcium phosphate. Many sea creatures including crams and many of coral-forming ones use calcium carbonate $CaCO_3$ as their shells. Another group of small marine creatures, called foraminifers also use calcium carbonate as their outer skeleton. Many terrestrial animals lay eggs outside their bodies, and the eggshells are made of calcium carbonate.

Why are calcium compounds used for these purposes? The answer is about the same as the reason for your choosing building material. That is, (1) the material serves the purpose and (2) it is readily available (in economic sense as well). Calcium (Ca(II)) forms insoluble compounds that can be mechanically strong enough to serve the purpose of providing cover and/or scaffold to the organisms; these are calcium carbonate and calcium phosphate. Calcium is one of the most abundant elements both on land and in seawater; that is, it is readily available.

Is there any other choice possible for the purpose? Well, magnesium and silicon come to mind as a candidate. Magnesium (Mg(II)) is more prevalent in rocks than calcium, and it forms solid compounds with carbonate and phosphate like Ca(II) does. However, magnesium compounds are much more soluble than the corresponding calcium compounds. This means that solid magnesium carbonate and magnesium phosphate are less stable and more readily dissolved than the calcium counterparts; perhaps then magnesium compounds are less suitable for the purpose.

Silicon forms many solid compounds; most rocks are indeed silicon compounds (Chap. 14). As a matter of fact, we use brick, rocks, and concrete for our shelter; these materials contain silicon. They are also amply available. Silicon, thus, satisfies the two criteria. But no organism except for a few has chosen silicon for their body covering or scaffold. Marine phytoplankton known as diatoms use silicon compound (silica gel, $SiO_2 \cdot$ x H_2O) as their covers and some sea squirts use silica (SiO_2) as internal scaffold. Some plants, grass and bamboo, use silica for the strength of their stem. Therefore, there must be some other reasons for the predominant use of calcium compounds as the building material in many organisms.

In addition to providing the mechanical strength through calcium carbonate and phosphate, Ca(II) ion is used widely in physiology of cells. When you want to move your right arm, for example, your brain starts sending that message as an electrical signal. The electrical signal itself is created by movement of sodium(I) and potassium(I) ion across the cell membrane of a neuronal cell. It will be transmitted along the cell (axon), and when it reaches the end (synapsis) of the cell, it causes emission of a neurotransmitter (a chemical such as acetylcholine). Acetylcholine travels a short distance to an adjacent cell and immediately bounds to a receptor and excites the cell; acetylcholine will then be decomposed by an enzyme. Eventually, the signal reaches a muscle cell in the arm. This signal then causes the increase of Ca(II) ion concentration, which then helps open suddenly the gate of endoplasmic reticulum which contains a lot of Ca(II). A sudden release of Ca(II) from the calcium sack (endoplasmic reticulum) follows; the calcium thus released then binds to a protein called TNC, a component of the muscle protein. This triggers the contraction of muscle. Ca(II) is also used in the process of acetylcholine emission at the synapsis. The effects of many hormones are also mediated by Ca(II). The clotting of blood consists of many steps of biochemical reactions that are dependent on Ca(II). Cells have to divide in order to grow. The process involves a number of biochemical reactions, many of which are dependent on Ca(II). This is only a partial list of physiological functions of Ca(II).

It is remarkable that hundreds of cell functions are dependent on Ca(II) and that leads us to the question of why: why Ca(II) works in such a variety of ways in cell physiology. Why have not the organisms chosen other chemicals for these purposes? What are the special properties of calcium that make it so suitable for these functions?

6.6   Other Elements

So far, we have mentioned six elements: iron, copper, manganese, molybdenum, zinc, and calcium. Iron, zinc, and calcium are the three most important minerals for living organisms, though others are also essential. We talk about here a few more examples.

6.6.1   Cobalt and Vitamin B$_{12}$

Vitamins are required in small quantities. All vitamins (except one) are organic compounds; that is, they are all made of carbon, hydrogen, oxygen, and nitrogen (and some of them contain sulfur and/or phosphorus in addition). Well, this exceptional one, vitamin B_{12} is also an organic compound, but it contains a mineral element: cobalt. Cobalt is not particularly abundant on the Earth, but vitamin B_{12} absolutely requires cobalt for its functions. Vitamin B_{12} is involved in manufacturing red blood, and hence we get pernicious anemia if we do not have enough of it. Humans cannot make vitamin B_{12} (neither all other vitamins for that matter), but we usually do not need to take it. The secret is that we harbor bacteria in our gut; and the predominant one, *Escherichia coli*, produces vitamin B_{12}, which we can use.

The physiologically active vitamin B_{12}, called B_{12} coenzyme or adenosyl cobalamin, contains a chemical bond between a carbon atom and the cobalt atom. Vitamin B_{12} coenzyme is thus an example of the so-called organometallic compounds. An organometallic compound contains a bond between a metallic atom and carbon atoms. A large number of organometallic compounds have been synthesized by chemists since 1950s. Not many organometallic compounds are known that occur naturally. Vitamin B_{12} is the first naturally occurring organometallic compound that has been discovered. The use of cobalt in the biological system seems to be limited to this, as the component of vitamin B_{12}. However, cobalt (II) has chemical properties similar to those of zinc(II), and some of cobalt(II)-substituted zinc-enzyme exhibit enzymatic activities. Enzymes requiring cobalt may yet to be discovered.

6.6.2   Selenium

Selenium (Se) is essential to most organisms including human beings. It is required in very small quantities. The recommended maximum daily intake is 450 microgram. Se becomes toxic when present in excess. The average human adult contains about 15 milligram of selenium. [milli = one thousandth; micro = one millionth].

Selenium is located just below sulfur (S) in the periodic chart (Fig. 19.2). This suggests that selenium behaves similarly to sulfur in chemical reactivity. It does so indeed, but there are a few significant differences. The problem is that organisms would not be able to distinguish Se and S very well, because they are similar enough. Sulfur is essential to all the organisms, and selenium may replace the essential sulfur in biocompounds when Se is present at high level. I would hasten to add that some sulfur compounds such as hydrogen sulfide as well as their selenium analogues are very toxic to most organisms. Most of the time, selenium-substituted enzymes and proteins cannot function as well as the proper ones containing sulfur, because of the subtle difference in their properties. For example, organisms have mechanisms to absorb the necessary sulfate SO_4^{2-}, and these mechanisms can also absorb selenate SeO_4^{2-}, that is not required. A selenium atom may substitute for sulfur in the essential amino acids, for example, cysteine. Such an amino acid, selenocysteine, may or may not function properly, though some special enzymes seem to require selenocysteine instead of cysteine.

An enzyme that requires selenium specifically is glutathione peroxidase (GPO). This enzyme decomposes very efficiently one of the so-called active oxygen species, i.e., hydroperoxide. Accumulation of hydroperoxides in cell membranes is believed to be one of the causes of "aging." GPO thus helps prolonging life.

It has been found that deficiency in selenium leads to infertility in male animals. Humans usually obtain a sufficient amount of selenium from what they eat, such as seafood, liver, lean red meat, and grains. Hence, humans are rarely deficient in selenium. However, scientists discovered decades ago that animals fed selenium-deficient diet often produced sperm that broke in the middle. A GPO (phospholipid GPO) turned out to be the major component of the capsule material of sperm, though this protein (GPO) did not show the enzymatic activity. Scientists (Ursini F and coworkers: *Science*, 285 (1999), 1393–1396) speculate that GPO acts as an enzyme early in sperm development, but that later it polymerizes into a protein mesh that contributes to the structural integrity of the mid piece of sperm.

Stories of Drug Developments

7

This chapter is not meant to provide a systematic discussion of drug developments, but merely gives a few interesting stories of how drugs were discovered and have been developed. A lot of complicated chemical formulas/structures are presented in this chapter, but you should not be worried about whether you understand them. It turns out that the drug function often depends not on the detailed structure but on the overall shape of the compound. Hence, you can look at it as a sort of picture with some shape, and that is sufficient to understand why drugs work and how it might be modified.

7.1 Penicillin and Similar Antibiotics: Human Battle with Bacteria

A Scottish physician, Alexander Fleming, was engaged in research at St. Mary's hospital in London. He was working with cultures of a disease-causing bacterium, *Staphylococcus*. Some culture plates (in Petri dish) were set aside and checked from time to time. One day in 1928, he noticed that one culture was contaminated by a blue–green mold and that the bacterial colonies had become transparent around the mold. It suggested to him that the bacteria there had died and dissolved away. His investigation led to a discovery that the broth in which this mold (*Penicillium notatum*) had grown indeed had an inhibitory effect on many pathogenic bacteria. Fleming named this antibacterial agent contained in the mold as "Penicillin." He could not isolate and identify the compound, but Howard Florey and Ernst Chain of Oxford University did purify in 1941 a product from the mold and called it "Penicillin G." Penicillin G had become widely available by the end of the World War II and saved many lives threatened by pneumonia, gonorrhea, and other infectious diseases. These scientists were awarded a Nobel Prize together in 1945. This was the first of the so-called antibiotics. Discovery of other antibiotics followed; some of the better known antibiotics that were obtained from microorganisms include streptomycin (from *Streptomyces griseus*), aureomycin (from *S. aureofaciens*), and cephalosporin C. These are different types of chemical compounds from penicillin. We would not pursue the details of these other antibiotics here.

E. Ochiai, *Chemicals for Life and Living*,
DOI 10.1007/978-3-642-20273-5_7, © Springer-Verlag Berlin Heidelberg 2011

What is "antibiotics"? It is a substance produced by a microorganism that inhibits growth of other organisms. Antibiotics were hailed as "miracle drugs" and revolutionized "chemotherapy," and have appeared to have saved mankind from most of the infectious bacterial diseases. This is the "mankind's" view of the antibiotics. Microorganisms do not intend to help mankind to control their diseases. Microorganisms have devised "antibiotics" for their own survival. Antibiotics are actually "chemical weapons" that a microorganism produces and releases into its surroundings in order to suppress the growth of other organisms. It is an agent of warfare among microorganisms. But would not penicillin also harm human beings then? It turned out that penicillin disrupts the formation of the bacterial cell wall. Animal cells including those of human beings do not have cell walls such as found on bacteria. Hence, penicillin would not have a significant physiological effect on human beings, but affect only the bacteria invading human body. [The allergic reaction to penicillin in some people is an entirely different issue.] These facts have a number of implications for its therapeutic uses. Two important ones are as follows (1) the naturally occurring antibiotics may not necessarily be suitable for human consumption; and (2) other organisms have had to and indeed did develop defense mechanisms for the weapon.

Let us start with implication (1). The natural antibiotics are to be taken up by other microorganisms. In the case of a human being, an antibiotic compound has to be taken up by some means and then travel via the internal circulating systems and be accepted by specific tissues and taken up into the cells of the tissues (or organs). Biochemistry and physiology involved in this whole sequence are quite different from the simple uptake through a single cell wall/membrane in the case of microorganisms. For example, penicillin G can readily be decomposed by acid. If penicillin G is taken through mouth, it has to go through stomach, which contains a high concentration of hydrochloric acid. Hence, penicillin G will be decomposed in stomach before it reaches its target. So what would you do? Modify it. It is a chemist's task to change a portion of it so that it becomes more resistant to acid. The chemists have developed over the years a number of derivatives of penicillin; some of them are shown in Fig. 7.1 below. Implication (2) is a more serious issue. Because a microorganism has devised an antibiotic, other organisms must have developed mechanisms to cope with it. Otherwise, those organisms would not have survived the evolution. Antibiotics are chemicals and they are subject to chemical changes. For example, an enzyme penicillinase has been developed to specifically decompose penicillin. This is a defense mechanism against penicillin. It turned out that bacteria often carry a group of genes on a special vehicle called "plasmid," which is separate from their main chromosome. Genes that would produce substances called "resistant factors" which counteract antibiotics (and other harmful agents) are carried on a plasmid. Not all bacterial cells carry such plasmids. But resistance factor genes carried on plasmids can readily be transmitted to other bacterial cells, and hence, other bacterial cells can also acquire an antibiotic resistance. That is what is happening now; many pathogenic bacteria have acquired antibiotic resistance. Now what can we do?

One solution is to restrict the use of antibiotics so that bacteria would not have much chance of becoming resistant. In other words, use of antibiotics should be

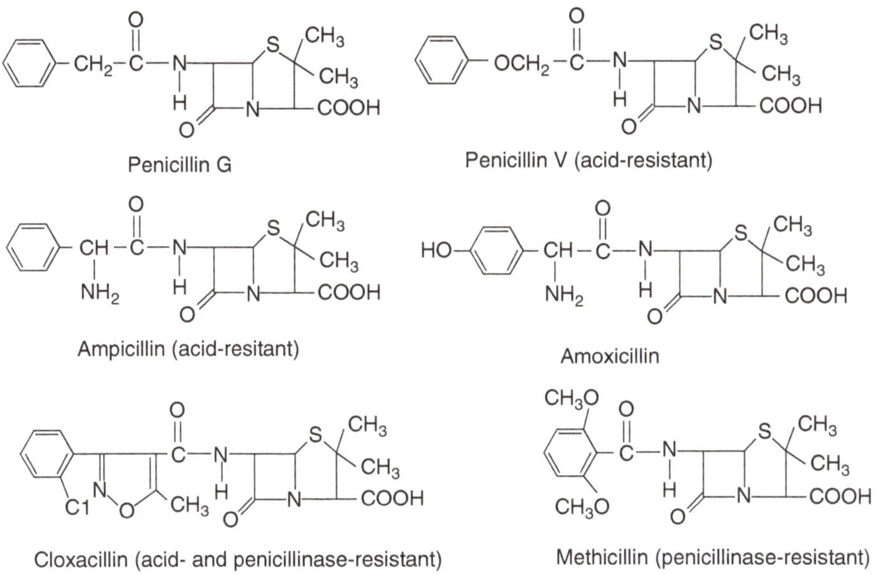

Penicillin G

Penicillin V (acid-resistant)

Ampicillin (acid-resitant)

Amoxicillin

Cloxacillin (acid- and penicillinase-resistant)

Methicillin (penicillinase-resistant)

Fig. 7.1   Penicillin G and some of its derivatives

strictly controlled. Another is to modify an antibiotic to defeat the specific resistant mechanism. This is what chemists are good at doing. Should we know how penicillinase works to decompose penicillin, we might be able to figure out how we might modify the penicillin to avoid the effect of penicillinase. Penicillinase hydrolyzes the β-lactam ring (the square ring in Fig. 7.1), but its activity is affected by the left-hand portion of the chemical formulas (of Fig. 7.1). Chloxacillin (see Fig. 7.1) and methicillin (dimethoxyphenyl penicillin) have been clinically found to be penicillinase-resistant.

Nature may as yet have in store some other mechanisms to defeat the effect of penicillin, which we are not aware of. In fact, many bacterial species have acquired in recent years resistance against many antibiotics (in addition to penicillin and other drugs). This threatens the well-being of mankind. Many once-forgotten diseases, such as tuberculosis, have returned, and returned with a vengeance, with resistance factors against many antibiotics. A boy in Madagascar was found in August 1997 to carry a pest bacterium (the one caused the "black death" in the medieval Europe) with multidrug resistance. Report came out also in August 1997 that strains of *Staphylococcus aureus* which were found in Japan and the United States were resistant to vancomycin, an antibiotic considered by many to be the last resort. Hence, we do not have effective cures for the diseases caused by these resistant bacteria.

However, there had been a general wishful notion that the resistant bacterial strains, because they are burdened with extra functionality, could not compete with the nonresistant strains once the drugs were removed, so that nonresistant bacteria would eventually come back and become dominant again. Therefore, the antibiotics may become useful again. We only have to restrain ourselves from using antibiotics for a while. This wishful notion has been shattered by recent discoveries that antibiotic-resistant strains do persist long after the drug went into disuse. The reason seems to be that those resistant strains do evolve farther to overcome the disadvantages.

So what more can we do? A very large number of antibiotics of various types have been isolated from microorganisms and others and have been modified chemically. Have we exhausted the natural sources? We may still look for more antibiotics in nature. We now have a fairly good knowledge about the physiology of pathological bacteria. Are there any other ways to disrupt their lives? Many researchers are focusing on ways to interfere with the normal physiology of bacteria that nature has not discovered to interfere with, so that bacteria may not be able to develop means to defeat. This could still be wishful thinking. Or we can try to defeat the nature's (bacterial) defense system so that we may be able to use the old antibiotics again, not by modifying antibiotics but by completely different means. We will discuss several such attempts here.

A new class of antibiotics is called oxazolidinone. An example linezolid has been developed by Pharmacia and Upjohn. The drug works by binding to one of the subunits of ribosome of a bacterium. This binding interferes with the protein (of resistance factor) synthesis. The drug has been shown to be active against gram-positive bacteria such as *S. aureus*. This drug was waiting approval from FDA in June, 2000. However, even before the approval, some bacteria seem to have developed some resistance against this type of antibiotic drug.

Some bacteria, yeast, and other pathogens have developed a mechanism to pump out tetracycline and other antibiotics. This is their way of defending against the antibiotics. Some drugs have been developed to block this pump, so that the antibiotics administered will do its work unimpeded. Recently, such a compound, 5'-methoxyhydrocarpin (5'-MHC), was isolated from barberry plants. When the antibiotics barberine was administered with 5'-MHC, it was found to inactivate the strain of *S. aureus* that was resistant to antibiotics such as norfloxacin.

Another attempt to defeat an antibiotic resistance mechanism is to try to trick bacteria to disrupt their synthesis of resistance factors (proteins). Altman and his colleagues at Yale University recently isolated an enzyme, ribonucleotidase P of *E. coli*. The enzyme chops down RNA, specifically at a certain sequence. They synthesized a short DNA piece called "external guide sequence" (EGS). This DNA can be transcribed into a short RNA, which can be linked to the messenger RNA of a drug resistance factor (protein). The ribonucleotidase P is then tricked into regarding this combined RNA as its target, and slices the messenger RNA part, disabling the bacterium to produce the resistance factor. The EGS portion emerges intact and binds with another molecule of the messenger RNA. This worked well in the lab, for chloramphenicol and ampicillin, the two drugs they attempted. The crucial step is finding ways to introduce effective EGSs into human cells, so that invading bacteria cells adapt EGS and proceed to disrupt their own machinery to produce antidrug factors. How can this be done?

Another, perhaps, potentially very effective way to combat the resistance to antibiotics is to eliminate the resistance factor-carrying plasmids altogether. Chemists at University of Illinois (Hergenrother P.J. and his graduate students) discovered that a small molecule named "apramycin," a kind of compound belonging to aminoglycoside, competes with plasmids for necessary RNAs for reproduction and wins the competition. Plasmids thus prevented from reproduction are eliminated from the cell. Hence, the bacterial cell will lose the source of resistance factor.

We seem to be in a perpetual war against disease-causing agents. These researches may appear to be far away from "ordinary chemistry," but all the players here, DNA, RNA, antibiotics, proteins and glycosides, and others, are chemicals, and understanding their chemical (as well as physiological) behaviors is essential for advancing our frontiers in medicine and pharmacology.

7.2   AIDS Drugs: AZT and Protease Inhibitors

Chemotherapy for AIDS is perhaps the most hotly pursued in pharmaceutical industry today. For quite a while, AZT, or its analogues, was the only usable drug.

7.2.1   AZT and Its Analogues

AZT is an acronym for azidothymidine and is an inhibitor for a pivotal enzyme, reverse transcriptase in HIV-1 virus. Well, let us look at the situation from the ground zero.

HIV-1 is a so-called retrovirus, and its gene is RNA. Our human gene is DNA, as in the majority of organisms. What is the difference between DNA and RNA? DNA is made of four so-called nucleotides, often abbreviated as A, G, C, and T. Each of these (nucleotides) is made of three chemical entities: base (this is the distinguishing factor – A, G, C, T), a sugar called deoxyribose, and phosphate. RNA on the other hand is made of four nucleotides A, G, C, and U. U (of RNA) and T (of DNA) work in similar fashion and are chemically not very much different; T has an extra methyl group (CH_3) than U. Another difference between RNA and DNA is the sugar part; RNA uses ribose instead of *deoxy*ribose. That is why RNA is *ribo*nucleic acid and DNA *deoxyribo*nucleic acid. This difference is structurally subtle (one less oxygen atom in deoxyribose than ribose), but makes the chemical properties of DNA and RNA quite different. DNA is much sturdier than RNA, so that the majority of organisms have adopted DNA as their genes [see Chap. 4 for more discussion of DNA as gene].

A retrovirus like HIV-1 uses RNA as its gene. But when they replicate themselves from their RNA gene, they make first the corresponding DNA. They have to do this, because they are going to usurp the machinery of the host organism that is DNA-based. The enzyme that carries out the synthesis of DNA from the RNA template is "reverse transcriptase." It is called so, because this is the reverse of the ordinary sequence: the DNA message transcribed into RNA. This enzyme, obviously, is not present in humans and other (DNA-dependent) organisms. The enzyme links one nucleotide after another and makes a long chain of nucleotides A, G, C, and T.

The enzyme binds one of A, G, C, and T (nucleotides) and then links it to the end of the growing end of a DNA chain.

What would happen if one adds a fake nucleoside to the reaction medium of reverse transcriptase? Note that we talked about "nucleotides" above, but we are here using word "nucleoside." This is not a typographical error. A compound that consists of base (A, G, C, and T or U) and ribose (without phosphate) is called "nucleoside." When this fake nucleoside is bound at the end of a growing DNA, the DNA cannot elongate further, because a phosphate group is necessary to connect two nucleosides. The fake nucleoside can also block the proper nucleotide binding site on the enzyme. Thus, it can inhibit the enzyme activity. This could stop production of DNA that is a copy of the gene RNA.

This is the basic idea. Is not this rather simple? One needs to choose a chemical that would act as a fake nucleoside. This may, hopefully, act on HIV alone, because the enzyme is present only in HIV. A number of choices are possible. First which of the four nucleotides A, G, C, and T would you choose to make a fake substitute for? A, G, or C? These are used also to produce RNA; and so if you make a fake for one of these, it would also stop the production of RNA, perhaps, unnecessarily. "T" is unique for DNA. So the best choice would be "T." AZT is a fake for "T" (thymidine). It has an essentially the same structure as that of thymidine with one minor difference (see Fig. 7.2). So the enzyme cannot distinguish AZT from the real "T" and binds it. Besides, the group called "azido" (N_3) on AZT has an ability to bind chemically to an important group on the enzyme. This means that AZT binds strongly to the enzyme, reverse transcriptase, and hence that it blocks its function. Now the virus cannot produce the necessary DNA and hence cannot reproduce itself.

AZT has an alternative name of zidovudine and is sold under the brand name of Retrovir. Other similar compounds that have been approved for AIDS therapy include ddI (2',3'-dideoxyinosine – didanosine, Videx) and DDC (2',3'-dideoxycytidine-zalcitabine, HIVID). They are fakes for G (guanoside) and C (cytosine), respectively.

AZT can only stop or retard the proliferation of HIV, but cannot cure AIDS. You can guess this from what we talked about it above. Besides, AZT has a number of side effects; the truth of matter is that no drug is without side effects. AZT is to block the action of an enzyme, reverse transcriptase. This enzyme is to polymerize

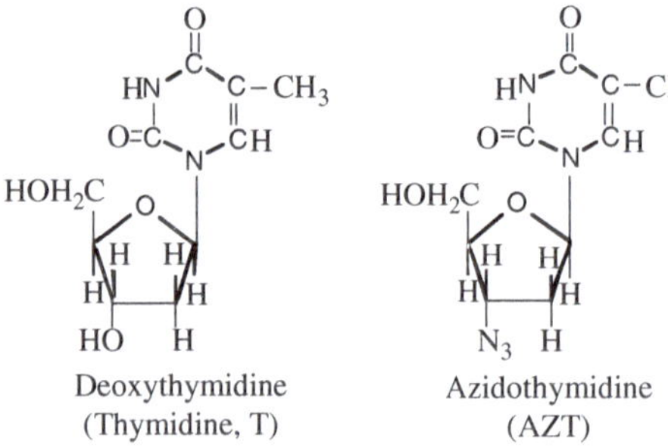

Fig. 7.2 The chemical structures of thymidine and azidothymidine (AZT)

Deoxythymidine
(Thymidine, T)

Azidothymidine
(AZT)

nucleotides to make a DNA. The host (human)'s regular cells also have enzymes that make DNAs. The mechanism of making DNA is similar in both enzymes reverse transcriptase and the regular DNA polymerase. This implies that AZT may be picked up not only by the virus' reverse transcriptase but also by the host cell's DNA polymerases. To what extent this happens is not very well known, and whether this is the only mechanism of the side effects is not known.

7.2.2   Protease Inhibitors

Another kind of drug is targeted at an enzyme called "protease" which is present in HIV-1 virus. The virus needs to produce some proteins and enzymes besides its RNA in order to replicate itself. Two genes called "gag" and "pol" (on the RNA) produces a multiprotein (let us call it "gag-pol protein") that contains two proteins fused together. The enzyme HIV-1 protease splits this fused protein into the individual active structural proteins. If this enzyme is made inactive, the essential structural proteins cannot be produced, and hence the virus becomes noninfectious. This suggests that a drug may be created that inhibits this enzyme.

Protease (also called proteinase) is an enzyme that splits the bond called "peptide bond" that connects amino acids in a protein. The reaction is a hydrolysis that means "splitting of a bond by adding water" and is expressed as follows:

$$\text{---CHR-CONH-CHR---} + H_2O \rightarrow \text{---CHR-COOH} + H_2\text{N-CHR---}$$

There are several types of proteases. HIV protease is an example of "aspartic acid proteases." The structure of HIV protease as determined by X-ray crystallography is shown in Fig. 21.11. How the peptide bond is cleaved is shown schematically in Fig. 7.3. The set of two aspartic acid residues shown catalyzes the addition of water molecule to gag-pol protein (a). The result is the formation of an intermediate shown as (b) in the figure. (b) is known to decompose into two separate entities as shown in (c). This completes the hydrolysis. The region in the gag-pol protein where this splitting occurs has an amino acid sequence of –Leu–Asn–Phe–Pro–Ile– (Asn = aparagine, Ile = isoleucine, Leu = leucine, Phe = phenyl alanine, Pro = proline).

Now, then how would you make the enzyme inactive? Compound (A) or (B) would bind to the enzyme with certain strength. If another compound binds to the same spot of the enzyme with a greater strength and yet cannot be decomposed, it will block the binding of the proper compound (A) or (B). And hence, (A) would not be able to bind and undergo hydrolysis. This is the idea of "inhibition" of an enzyme.

A number of drug companies have tried to develop AIDS drugs based on this principle. These drugs are hence called "HIV protease inhibitors." Four HIV protease inhibitors have been approved for use in treating AIDS. They are saquinavir (trade name = Inverase developed by Hoffman-La Roche), ritonavir (Norvir by Abbots Lab), indinavir sulfate (Crixivan by Merck), and nelfinavir mesylate (Viracept by Agouron Pharmaceuticals). All of these compounds try to mimic the structure of intermediate (B), but they are not decomposable.

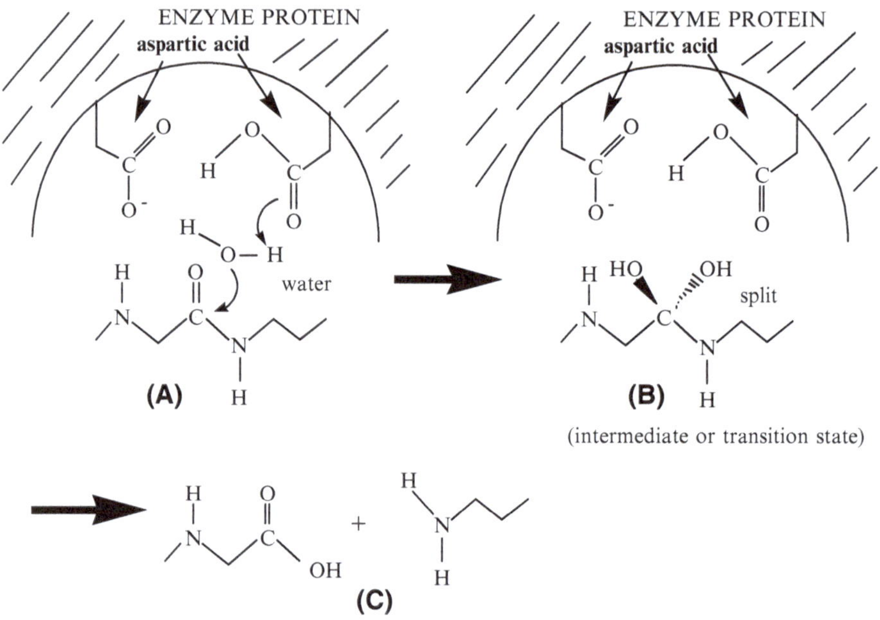

Fig. 7.3 The mechanism of HIV protease: This is only a schematic picture, not true structural representation. (A) represents only the crucial portion (in gag-pol protein) that is to be split (B) represents the intermediate. In (C), the first entity is the end portion of gag and the second is the beginning portion of pol protein

Well, (B) has a C–N bond (as in –C(OH)$_2$-NH–) which splits spontaneously once formed. An inhibitor molecule has, instead, a C–C bond such as –CH(OH)–C – which mimics structurally the C–N bond in (B) but cannot be cleaved in a similar manner. This is the principle in making such a compound (protease inhibitor) and is basically simple. However, the actual choice of a specific compound for a drug is not simple. In addition to the inhibitor character (i.e., binding there properly and strongly), an effective drug has to have a number of other characteristics. Among such important factors are the following: (1) The compound must work very specifically for HIV protease; otherwise, the drug may inhibit other similar proteases (there are many proteases in our body) and can cause adverse effects. (2) The compound should be well bioavailable, safe, and well tolerated.

Researchers at different companies used different reasoning in developing drugs. For example, chemists at Hoffman-La Roche imitated the hydrolysis site of the gag-pol protein, that is, Phe-Pro (see above). They modified the surroundings of the site and discovered that saquinavir is the most effective (see the structure in Fig. 7.4). In Abbot's Lab, the scientists made use of the symmetrical nature of the HIV-1 protease. Because most of other mammalian proteases are not symmetrical, a compound with a symmetrical structure about the mimicking site would inhibit the HIV-1 protease, but not other mammalian proteases, thus hopefully reducing side effects. With this notion in mind, they synthesized a number of compounds and discovered that ritonavir was most effective (see the structure in Fig. 7.4). As you see in Fig. 7.4, the

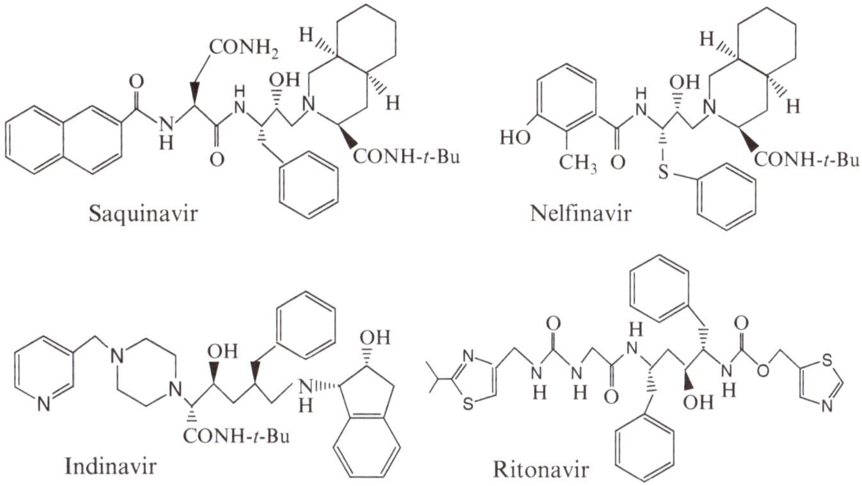

Fig. 7.4 Some of drugs, HIV protease inhibitors

other inhibitors, indinavir and nelfinavir, are modified from saquinavir. These drugs are now available commercially.

These protease inhibitors turned out to be effective when used in conjunction with AZT. Well, how about side effects? It has been discovered that they have side effects, including insulin resistance and type II diabetes. A cell biologist M. M. Mueckler, professor at Washington University, and his coworkers found that the protease inhibitors interfere with the ability of fat cells to store glucose. A glucose transport protein in fat and muscle cells, Glut4, brings glucose across the cell membrane. It seems that the protease inhibitors bind to Glut4 and hence block its transport ability of glucose.

7.3   Viagra and Others

Many people may remember the time when this drug was put on market in 1998. Every single mass medium seemed to report on the furors caused by the appearance of this wonder drug. It has become instantly a household name. It has been reported that 20–30 million American men suffer from some form of penile erectile dysfunction. More men young and old who were not physically disabled but felt insecure about their masculinity also welcomed this drug.

Viagra is a trade name marketed by Pfizer Pharmaceutical Company and is chemically sidenafil citrate. Researchers did not start creating a drug for impotence. The chemical compound was developed originally for angina (severe but temporary attack of cardiac pain) and is chemically an inhibitor for an enzyme phosphodiesterase (see the previous section, for example, for the idea of enzyme inhibitor). The drug was not successful in treating heart problems, but many of the test subjects reported their successes in bedroom activity. It turned out that the compound inhibits only a special kind of phosphodiesterase, called PDE5. PDE5 is not found much in the heart but is concentrated mostly in the penis.

How does it work? Well, a brief description of the physiology and chemistry of penile erection might be in order. The message sent by the brain or other stimulus causes production of a very simple chemical compound, nitric oxide "NO" (N is nitrogen and O oxygen atom; it is also called nitrogen monoxide). Actually, this simple compound is the chemical messenger. This compound diffuses into the smooth muscle cells of penis. Under normal conditions, the muscle is contracted. When NO arrives at the muscle cell, it binds to an enzyme called guanylate cyclase, and activates it. Guanylate cyclase then acts on GTP (guanosine triphosphate). GTP is similar to ATP (adenosine triphosphate) that has been talked about several places in this book, and shown in Fig. 7.5 below. GTP is converted into cGMP (cyclic guanosine monophosphate; see Fig. 7.5). cGMP is a sort of cellular messenger and causes relaxation of the muscle. This process itself is very complicated, but is not elaborated here. Blood can now flow into the arteries among the muscle cells, and hence, results in erection. cGMP cannot remain active for ever; it will be decomposed by a phosphodiesterase, PDE5, and the muscle relaxation (and hence erection as well) will be gone. This is where the drug interferes. Viagra binds to PDE5 and blocks the site where cGMP should bind. Hence, the decomposition of cGMP will be hampered and delayed. A result will be an erection maintained.

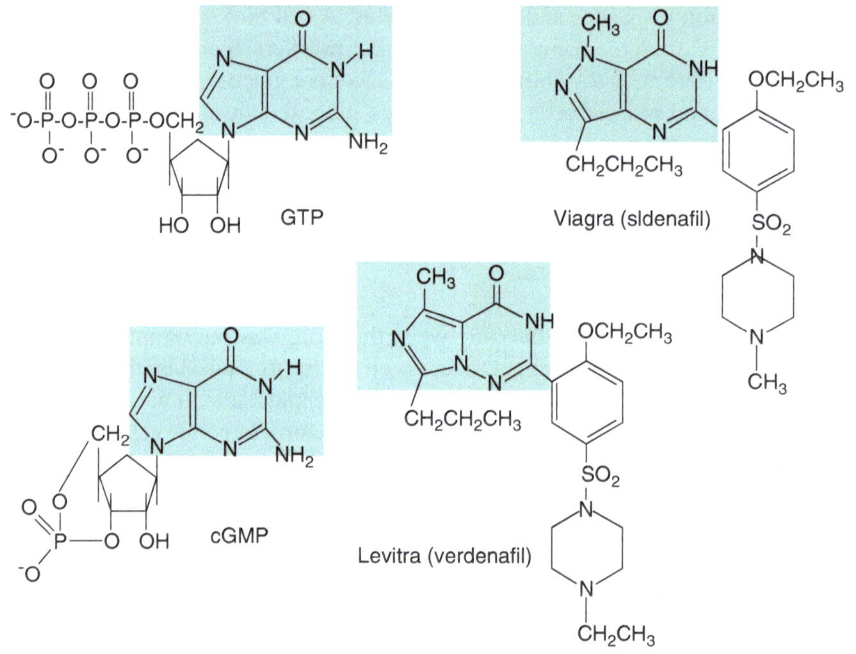

Fig. 7.5 Viagra and others

Why does Viagra work this way? Let us look at Fig. 7.5. Both cGMP and Viagra have similar structural features as indicated by the colored boxes. Because of this similarity, they bind to the same site of the enzyme. The other drugs put on market in recent years, Cialis and Levitra (see Fig. 7.5), are also specific inhibitors of PDE5.

Although Viagra binds very specifically to PDE5, a phosphodiesterase, it does bind to other kinds of phosphodiesterase (which work on cGMP or cAMP), though more weakly. One of the peculiar side effects reported is a temporary blue haze in the vision of the men using it or trouble distinguishing green from blue. It seems that Viagra also reacts with a phosphodiesterase (not PDE5) in the retina. It is conceivable that other kinds of phosphodiesterase may also be disrupted by Viagra. It might be important to note that cGMP (as well as cAMP) is used widely as a cellular messenger in other tissues and evokes a number of physiological processes.

A molecule, NO, was mentioned earlier. It is a messenger to cause relaxation of muscle cells (among other roles). Nitroglycerin is often prescribed for angina and heart attack. Nitroglycerin has a chemical entity called "nitro group, NO_2," which seems to be converted readily to NO in the human body. And that dilates the arteries. Drugs mentioned here (Viagra, etc.) carry a warning that they should not be used by persons who are on NO-producing medication. Perhaps, one can make a connection between this warning and the way these drugs work, as described above.

7.4   Taxol and Related Compounds: Anticancer Drugs

National Cancer Institute (NCI) sponsored between 1958 and 1980 a number of projects in which more than 35,000 plant species were tested for anticancer activities. Monroe E. Wall and M. C. Wani of Research Triangle Institute obtained a crude extract from the bark of Pacific yew tree (*Taxus brevifolia*) in 1963, which they demonstrated to be very effective against a wide range of cancers, particularly ovarian and breast cancers. They named the agent "Taxol." However, they did not apply for a patent for their discovery. It took several more years to figure out the structure of the compound (1971; see Fig. 7.6 for the structure).

Initially, the clinical trials to determine the efficacy of taxol were very slow. The reason was the limited supply of the compound. It was most effectively isolated from the bark of yew trees. You have to cut the trees to use the barks. Yet the content of taxol in the bark is not high and is somewhere between 70 and 400 ppm (parts per million). Suppose that the content was 200 ppm on average, then 1 ton of the bark (this requires felling as many as 100 yew trees) would contain about 200 g of taxol. In practice, however, 1 ton will produce something like only 80 g (because some will be lost in handling, isolation, and purification). It has been estimated that three large yew trees are required to provide enough taxol to treat one patient. NCI made a deal in 1991 with Bristol-Myers Squibb pharmaceutical company for the production of 25 kg of taxol for the purpose of clinical trials in exchange of a promise that FDA would give the company an "orphan drug" designation for the drug. "Orphan drug" designation means that the company will have 7 years' exclusive marketing rights for the drug, when it is approved by FDA. [How many trees have to be cut to produce 25 kg of taxol?].

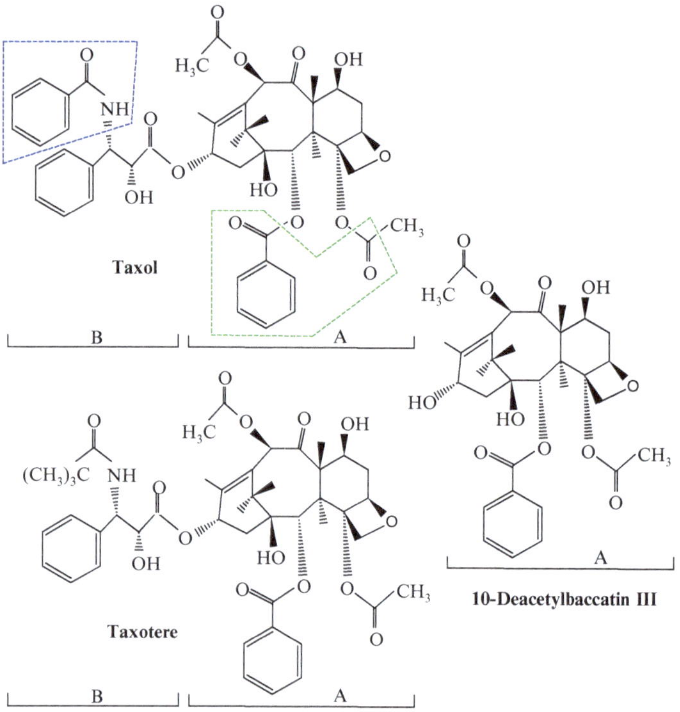

Fig. 7.6 Taxol and similar compounds

The Pacific yew trees are not particularly plentiful, and besides, they are rather slow-growing. Cutting trees for this purpose is wasteful and environmentally damaging. Pressures from cancer patients and these problems have caused people to seek alternative means of obtaining taxol or even alternatives for taxol. The needles and twigs from the yew trees are annually renewable and would be a better source; however, they contain even lower levels of taxol. How about the leaves of *Taxus brevifolia*? They also contain taxol, though at lower level, 20–70 ppm. However, the amount of leaves that can be collected is great, and hence the total amount of taxol obtainable from the same number of trees is about the same as that from their barks. Since you have to process more starting material (leaves), the product may be costlier.

It turned out that yew trees not only of the Pacific coast but of other species contain a compound called 10-deacetylbaccatin III (see Fig. 7.6); this is a precursor of taxol. You see in Fig. 7.6 that taxol consists of two portions A and B, and that this precursor provides the portion A. Now, to make another compound starting with a simpler compound is called "Chemical Synthesis," and *this is what chemists excel at*. Some people even say that this, chemical synthesis, is the only thing that is a "proper" chemistry. In this case, you have to add portion B on to the existing portion A. Robert A. Holton of Florida State University succeeded in synthesizing taxol from 10-deacetylbaccatin III.

Bristol-Myers Squibb Company started in 1993 to produce taxol by this semisynthetic method using needles and twigs as the source of the starting material. Meanwhile, the company filed a new drug application with FDA in July 1992, and taxol was approved in 1993. The term "Taxol" was designated in 1992 as "trademark," i.e., "brand name," though originally it was a chemical compound's name, i.e., nonproprietary. This decision seems to have been made due to a clever maneuvering by the manufacturer and is an interesting story by itself. An unfortunate circumstance was that a trade name "Taxol" had actually been used for a laxative long before the discovery of the anticancer drug taxol. Anyway, because of this, the nonproprietary name for taxol is now "paclitaxel," and the ending "-taxel" applies to all the derivatives of taxol. However, we continue to use "taxol" in our story.

Bristol-Myers Squibb Company sold $1.6 billion worth taxol in 2000. Phyton, Inc. in Ithaca, New York, is developing a technology to produce taxol by plant cell fermentation and is now collaborating with Bristol-Myers Squibb.

Meanwhile, on the other side of the Atlantic, French scientists were busy developing other derivatives of taxol. Scientists at the Institut de Chimie des Substances Naturelles, Université Joseph Fourier, and at Rhône-Poulenc (a pharmaceutical company) have synthesized 40 or so taxol-like compounds and tested their effectiveness and found that "taxotère" was the most potent. The structure of taxotère is shown in Fig. 7.6. It has a slightly different entity for portion B. This is chemically synthesized starting with 10-deacetylbaccatin III, which they obtain from the leaves of yew bush, *Taxus baccata*. This compound was patented by Rhône-Poulenc.

Another way of obtaining taxol is to synthesize it from scratch, that is, starting with a compound much simpler (than 10-deacetylbaccatin III) and readily available. This type of synthesis is called "Total Synthesis." A number of research groups attempted and competed for the challenge. A race was on! Early 1994, two groups simultaneously reached the goal, the total synthesis of taxol. They published their accomplishments in two different journals. The group led by R. Holton at Florida State University started with the off-the-shelf compound, camphor. It required 30 steps, and the overall yield was said to be 4–5%. The other group led by K. C. Nicolaou at the Scripps Research Institute started with a similar compound, but followed a quite different route. But here too, it required about 30 steps, and the yield was even worse than the competitor's. The former group boasted their superior yield, while the latter group countered by saying that the difference in yield occurred because of their different ways of evaluating the yield and that their own was not that bad as their published yield indicated, if reevaluated. These successes are academically quite significant, but the use of total chemical synthesis is not considered to be feasible for commercial production of the drug.

Taxol obviously is not made for human consumption. It is not made to fit human physiology. Taxol is not soluble in water, for one thing, and hence difficult to administer, and, though unusually well tolerated by humans, it does have some side effects and some resistance to it would develop. To improve the effectiveness of a drug and reduce the side effects, one tries to modify the drug. This part is the realm of chemistry (sometimes called "medicinal chemistry"). Two entirely different ways can be used to do this. One is "trial and error," and the other is a "more rational way." In the first, we synthesize a lot of different compounds derived from the drug and then test them (on experimental animals and then on human beings). This is a very expensive and wasteful venture, but may be the only way if we have no idea about how the drug works. In this manner, it has been demonstrated that some of the parts of taxol are essential for the anticancer activities. Those crucial parts include the oxetane ring (the square with one oxygen (O) atom inserted in the right lower portion of "A" in Fig. 7.6), C-13 side chain ("B" in Fig. 7.6), and C-2 benzoate (the left bottom portion of "A").

However, if we know how it works (mechanism of action) and how it behaves in human body (metabolism), we would have better ideas on how and what parts of the drug need to be modified. The more we know about the mechanism of anticancer action and metabolism, the more rationally we would be able to modify it. The elucidation of mechanism and metabolism is quite a basic science, but would contribute to enhancing the efforts to develop better drugs. It has been demonstrated that taxol affects the cell division. When a cell divides, the DNAs are duplicated and the resulting two sets of DNA need to be separated, as the cell divides and becomes two daughter cells. The process of separating of DNAs (chromosomes) is affected by a protein aggregate called "microtubule." It is a kind of rope that will pull the chromosomes apart from one another. For this to happen, microtubules themselves have to separate and reassemble. Apparently, taxol binds microtubules and stabilizes them, thus preventing them from separating. The detailed mechanism of this action is now under intensive study.

More recently, another compound, epothilone, was discovered to kill tumor cells just like taxol does by disrupting the microtubule's function in cell division. It was isolated from a soil bacterium from South Africa by German Gerhard Höfle and Hans Reichenbach and thus can be obtained relatively easily by growing the bacteria. Besides, it is water-soluble unlike taxol and hence easier to administer. Its chemical structure turned out to be quite dissimilar to and less complicated than taxol as shown in Fig. 7.7. The total syntheses of this compound from scratch were accomplished by three different research groups almost simultaneously, just like in the case of taxol. The first synthesis was accomplished in 1996 by Professor S. J. Danishefsky and his group at Columbia University and Sloan-Kettering Cancer Center. The other groups include one led by K. C. Nicolau at the Scripps Institute, who also succeeded in synthesizing taxol as mentioned earlier.

It turned out that taxol and epothilone A, despite their disparate structures, bind to a similar portion of β-tubulin. The details of binding mode are different, though.

On the other hand, a compound called cholchicine has been known for centuries to inhibit microtubule-dependent processes by inhibiting polymerization of tubulin protomers. Other compounds such as vinblastine and vincristine have also been found to inhibit microtubule formation.

All these compounds, as they affect microtubulin formation and/or depolymerization, can also affect noncancerous cells where microtubulin is involved in other cell functions such as cell movement. Hence, they are potentially cytotoxic. Well, can we find or make compounds that affect only the dividing cells? This is the direction

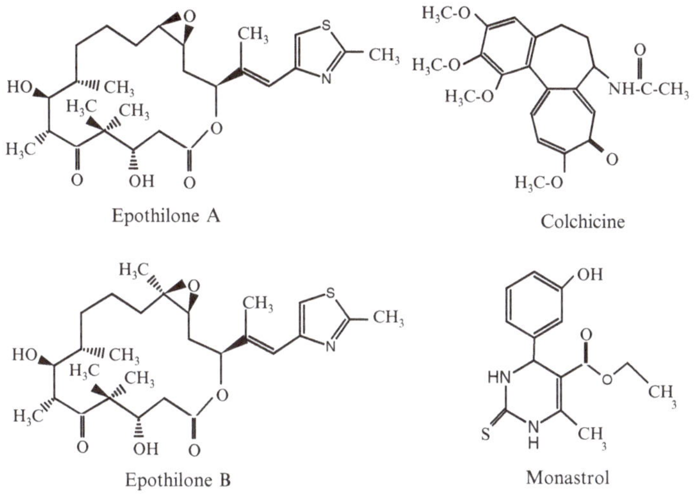

Fig. 7.7 Epothilone, etc

Thomas Mayer, Tarum Kapoor, Stuart Schreiber, and others (at Harvard) took. They combed through over 16,000 compounds and found several small compounds that affected mitosis in dividing cells but not in nondividing cells. One such compound, monastrol (see Fig. 7.7 for the structure), apparently affects the spindle formation in mitosis, but the details of mechanism are unknown.

7.5   Story of Cis-platin: A Unique Cancer Drug

In the 1960s, Barnett Rosenberg at Michigan State University was studying the effects of weak electric currents on the growth of bacteria, using platinum wires as the electrodes. One day he noticed that bacteria stopped dividing and grew longer. He guessed that something inhibited the cell division. A long search identified a platinum complex, cis-dichloro-diammineplatinum, cis-Pt(NH$_3$)$_2$Cl$_2$ (see Fig. 7.8), as the cause. Apparently, the platinum complex was produced from the platinum electrode, as a result of air oxidation. He reasoned that the compound might inhibit the uncontrolled divisions and growth of cancer cells, and so applied cis-Pt(NH$_3$)$_2$Cl$_2$ to tumors planted on rats, and observed a dramatic shrinkage of the tumors. This is one of the most interesting cases of serendipity. Today, platinum compounds derived from cis-Pt(NH$_3$)$_2$Cl$_2$ which is now called "cis-platin" are among the most widely used anticancer drugs. It has been useful in the treatment of testicular carcinomas, ovarian, head and neck, bladder, and lung cancers.

How does it work? Well the following is what the researchers believe to happen. Cis-platin is first transported into the cell via diffusion subsequent to an intravenous administration. The chloride ions are replaced by water yielding a positively charged complex.

Fig. 7.8 Cis-platin (the *broken lines* show the framework, not the bonds)

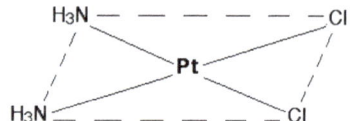

It takes about 2 h for the half of chlorides of cis-platin being displaced by water molecules at the body temperature, 37°C. This might be considered as an activation process. The Pt(II) complex now binds readily with DNA at a place called "N-7 nitrogen atom" of a guanine residue (G). [DNA is a chain of four different units designated as A, C, G, and T; see Chap. 4.] When Pt(II) binds G (and sometimes also with A), it disrupts the normal structure of DNA and hence its replication. It has recently been shown that cis-platin upon binding brings about a sharp kink in the double helix DNA, to which a protein called "high-mobility-group" (HMG) binds. In particular, a phenyl alanine residue on HMG inserts itself into the bent portion of DNA. How this will play out in the disruption of DNA duplication is not known.

So, as long as cis-platin enters exclusively into the tumor cells, it would work wonders. Because tumor cells are in rapid replication mode, they have more chance than ordinary cells to be affected by the platinum compound. And this is the reason for the effectiveness of cis-platin. However, it is likely that cis-platin enters other kinds of cells, too. This would lead to toxic side effects. Indeed, there are some severe side effects including diarrhea and nausea.

The recent efforts have been aimed at synthesizing cis-platin analogues that exhibit lower toxicity and have higher antitumor activity than cis-platin itself. Hence, various analogues of cis-platin such as cis-diammine(cyclobutane-1, 1-di-carboxylato)-platinum(II) or carbo-platin, 1,2-bis(aminomethyl)cyclobutane-platinum(II) lactate or loba-platin, and their derivatives have been studied in tumors that have exhibited cis-platin resistance. These are called "second generation platinum compounds." For example, carbo-platin that entered a clinical trial in the early 1980s has been shown to be effective against ovarian cancer and undergoes a reaction similar to that of cis-platin.

7.6   Curry, Another Anticancer Agent?

The bright yellow active ingredient of curry is a chemical called "curcumin." It has been discovered that curcumin slows the growth of new cancer. A Korean scientist, Ho Jeong Kwon, at Sejong University has shown that curcumin inhibits irreversibly an enzyme, aminopeptidase N (APN). APN breaks down proteins at the cell surface and helps cancer cells to invade the intercellular space. This is believed to lead to angiogenesis, that is, blood vessel growth. Hence, curcumin arrests angiogenesis, and hence cancer cells become unable to get blood to grow. It has been suggested that it might not have any significant side effects since curcumin is a regular ingredient of natural diet. How does curcumin do this? What is the mechanism of inhibition? This is yet to be studied.

7.7   PILLS: Controls of Reproductive Systems by Hormones and Their Analogues

The so-called "pill" freed women from worries about becoming pregnant, and hence revolutionized the man–woman relationship and then the overall human society at least in the Western countries. Human sexuality and pregnancy are governed by sex hormones. The representatives of sex hormones are estrogen and progesterone (these two are female sex hormones), and testosterone (male sex hormone). Estrogen is a collective name for estradiol and similar compounds including estrone and estriol. Estradiol is the most abundant and most effective estrogen. All these compounds belong to a group of compounds called "steroids" that also include cholesterol as well.

Is not everybody worried about cholesterol? And cholesterol is a steroid. Well then, are not steroids bad? What are steroids, anyway? Chemically, they belong to a group of compounds called lipid. Lipid is a fancy (collective) term for oil-soluble compounds and includes fats, phospholipids, and steroids. Fats are chemically long chain fatty acids and almost like wax. Phospholipids are combination of fatty acids and glycerine with some special groups on top and make up the membrane of cells (Fig. 1.5). By the way, cholesterol also makes up an essential part of cell membranes along with phospholipids. Cholesterol is indeed a necessity. Only excess presence of cholesterol is a health hazard.

Steroids are classified as lipids, only because they are soluble in oil, but they are quite distinct from the other lipids. Steroids all have a special skeletal chemical structure, as shown in Fig. 7.9, and all are chemically very similar. How closely they are similar can also be seen in their 3D structures. The 3D structures of estradiol and testosterone are shown in Fig. 7.10. They appear almost indistinguishable. A major chemical difference is that estradiol has one hexagonal group (the left-hand most) that is like benzene, i.e., three double bonds instead of one as in progesterone and testosterone, and that ring lacks a methyl group (CH_3). The other female sex hormone, progesterone, is in these respects similar to testosterone. The difference between them occurs at the other end: CH_3CO-group in progesterone vs. HO-group in testosterone. The whole point here is that the structural differences are rather small and subtle among the sex steroid hormones, and yet that they elicit very different physiological responses. [Sexual hormonal systems involve more than steroid hormones, such as follicle-stimulating hormone (FSH), luteinizing hormone (LH), and others].

Estrogens are normally secreted by the ovaries and induce syntheses of a series of proteins that manifest as feminine characteristic features in such organs and tissues as the female sex organ, the breast, and the fatty tissues. Progesterone instead is mostly involved in pregnancy. Menstruation is caused by the sudden reduction in both estrogen and progesterone at the end of the monthly ovarian cycle. Progesterone is secreted by the corpus luteum in nonpregnant women, but is formed in extreme quantities by placenta when a woman is pregnant.

Testosterone is mainly produced and secreted by the interstitial cells of Leydig in the seminiferous tubules of testes. It affects the development of the male sexual organ and has effects on body hair, muscle, bone calcium, and others.

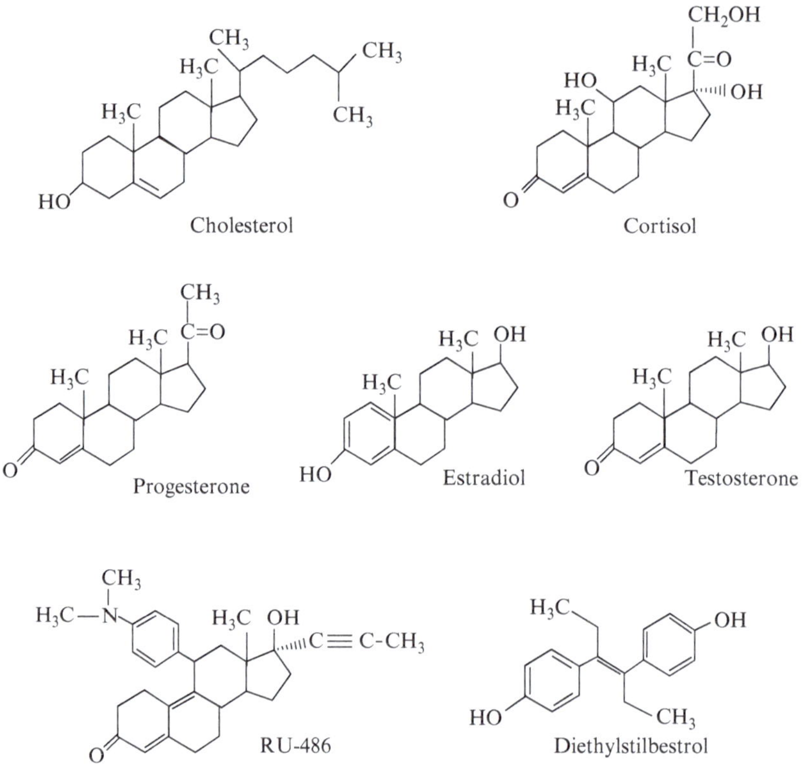

Cholesterol

Cortisol

Progesterone

Estradiol

Testosterone

RU-486

Diethylstilbestrol

Fig. 7.9 Steroids

Fig. 7.10 Estradiol (upper)
versus testosterone

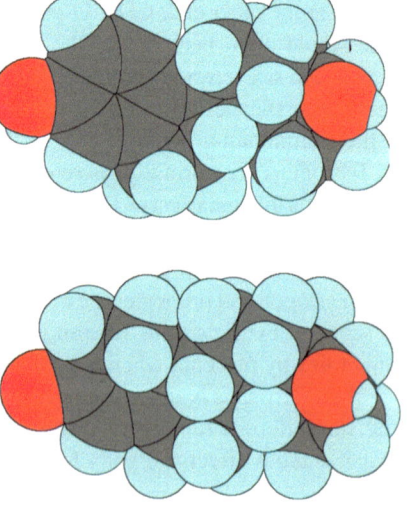

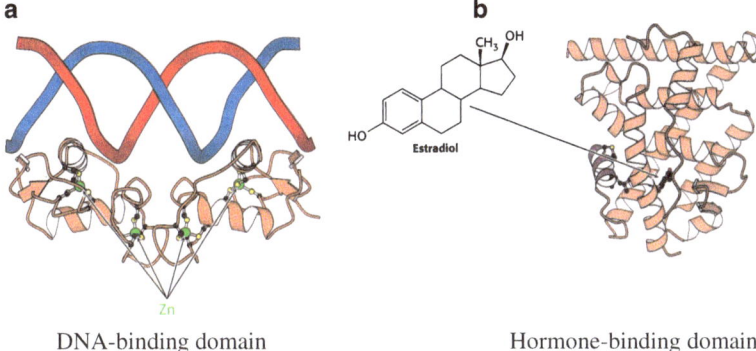

a — DNA-binding domain

b — Hormone-binding domain

Fig. 7.11 Hormone receptor; (**a**) zinc-finger portion binding to a DNA and (**b**) hormone-binding segment with estradiol bound [from J. Berg, L. J. Tymoczko, and L. Stryer, "Biochemistry, 5th ed" (W. H. Freeman, 2002)]

A steroid, being a lipid, has a high affinity to cell membranes, and hence it can enter a target cell readily, requiring no special mechanism. A steroid hormone binds with a receptor protein in the cell. The receptor–hormone complex is then transferred into the cell nucleus. The receptor protein binds to a specific place of DNA in the cell nucleus, with or without a steroid bound. However, when the receptor protein binds a steroid, it changes its structure (conformation) and becomes able to bind yet another factor (protein) called a "coactivator." The binding of a coactivator to the DNA–receptor–hormone complex acts as a signal to start transcribing a portion of the DNA into a messenger RNA (m-RNA). That is, the receptor–hormone (plus coactivator) complex acts as a "transcription factor." As with other common transcription factors, many of the steroid hormone receptors contain a portion of the so-called zinc-finger as mentioned earlier (Sect. 6.4). The cell machinery then synthesizes a new protein from the copy of the m-RNA, and hence, the manifestation of a physiological effect. An example of such complex, hormone–receptor–DNA, is shown in Fig. 7.11. The process of synthesizing proteins (DNA → m-RNA → protein) is discussed in Chap. 4.

7.7.1   The Pills

The idea of the pill was suggested by Margaret Sanger, a lifelong champion of women's rights and birth control. In 1950, she met a reproductive scientist Gregory Pincus and raised money to get Pincus started on a research on contraceptives. It had been known since 1930s that administration of excess hormones, estrogen and progesterone, prevents ovulation in rabbits. Pincus used progestins as contraceptive, which he tested on poor women in Puerto Rico. By the end of 1960s, a contraceptive pill became widely available in the United States and was welcomed by women. The earlier pills had high contents of hormones, both estrogens and progestins, and were recognized later to be quite an overdose. Today, the pills contain a much less quantity of estrogens and progestins, often progestins only.

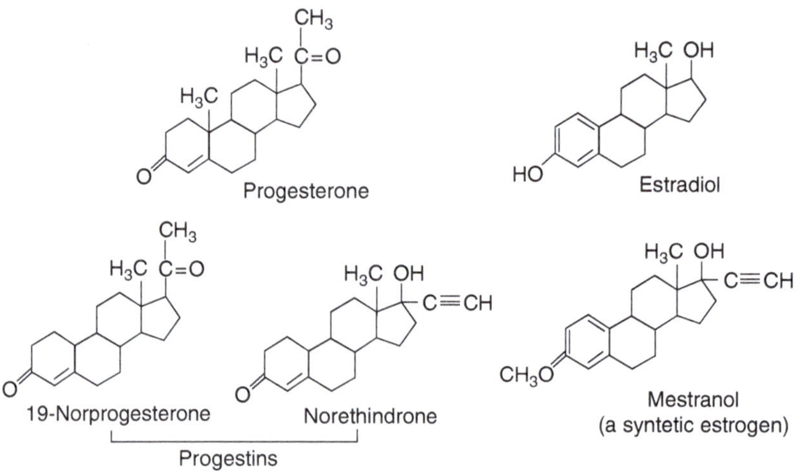

Fig. 7.12 The pills

The mechanism of contraceptive action is believed to be that the presence of an extraneous quantity of these hormones, particularly progestin, mimics the hormone condition of pregnancy. Hence, the body mistakes itself as being pregnant and suppresses ovulation. Recall that progesterone is secreted in a large quantity by placenta when a woman is pregnant.

What kind of progestin would work well? The indigenous hormone, progesterone, will work, but will the administration of progesterone itself be a good idea? Besides, production of progesterone may not be easy. These thoughts led chemists to an idea of using chemically synthesized analogues of progesterone and estrogen, particularly those that can be administered orally. The term progestins used above means any compound that has the basic progesterone skeletal structure. Carl Djerassi, a Stanford chemistry professor, and others managed to synthesize several such compounds. Some of such compounds are shown in Fig. 7.12. You can see their overall similarity to either of the natural hormones, progesterone and estrogen. Although the chemistry of synthesizing these compounds is quite difficult (and cannot be discussed here), their effectiveness can be gleaned from a look at their overall structures. In reality, the physiological effectiveness of each compound is determined by many factors in addition to the global structures. Such factors include how a compound is delivered to the target and how it is metabolized.

7.7.2 RU-486

RU-486 is a product of a French company, Roussel-Uclaf. Chemically, it should be named as 11-(4-dimethylaminophenyl)-17-hydroxy-(proy1ynyl)-estra-4,9-diene-3-one. When used in conjunction with a synthetic prostaglandin, RU-486 has been demonstrated to be effective in terminating pregnancy up until 9 weeks after the previous menstrual period. RU-486 has been shown to bind three times more strongly to the progesterone receptor than the endogenous progesterone itself.

Let us look at the chemical structure, as compared to progesterone (Fig. 7.9). The antagonist (RU-486) possesses a bulky dimethylaminophenyl group, which may prevent the binding of a coactivator that is required for activation of gene transcription. This antiprogesterone activity blocks development of the endometrium, which is necessary for implantation of the blastocyst, thus stopping the further progress of pregnancy. Six-hundred milligrams of RU-486 followed by administration of one or two synthetic prostaglandins 36–48 h later has been reported to be 96% effective in terminating pregnancy. RU-486 was approved by US FDA in September 2000.

Part III

For the Fun of Life

Chemicals are also involved in entertaining us. Our lives would be desolate without their efforts to entertain us. A few areas where we are entertained by chemicals are talked about in this part, including fireworks, light show by firefly and others, ceramic artworks, diamonds, and perfumes. One of the emphases here is the basis of the visual arts, i.e., the phenomenon of "color."

Fireworks: A Carnivals of Chemicals

8

8.1 Firework: A Chinese Invention

Boom! Bang! Millions of stars of red, yellow, and green. Oh! Wow! On July 4th the American sky is covered with hundreds of thousands of fireworks all over the nation. It is estimated that more than 100 million dollars are spent on this *spectacular display of chemicals* on this day. This is perhaps one of the most ancient uses (conscious uses) of chemicals for the purpose of enjoyment. The essential ingredient is gunpowder that detonates and helps display millions of colored particles. Gunpowder is that black stuff, and used for what else but gun; the same stuff.

Fireworks were actually invented by Chinese alchemists who sought the elixir of immortality, as recorded in a book (*Methods of the Various Schools of Magical Elixir Preparations*) published around 650 ad. What an irony! Chinese people love to use explosives for celebrations. As early as 200 bc as recorded in chronicles, they threw pieces of bamboo into fires; the bamboo would explode with a bang. By the way, the Chinese word for firecracker is still "explosive bamboo." So when the gunpowder became available, it is only natural that they utilized it for the purpose of fireworks.

E. Ochiai, *Chemicals for Life and Living*, 111
DOI 10.1007/978-3-642-20273-5_8, © Springer-Verlag Berlin Heidelberg 2011

A book published in 1275 ad describes: *"Inside the palace the firecrackers made a glorious noise... All boats on the lake were letting off fireworks and firecrackers... their banging and rumbling were like thunder..."*. The Chinese rapidly developed all sorts of fireworks.

8.2   Basic Ingredients and Principle of Firework

One of the essential ingredients of the black gunpowder is saltpeter, potassium nitrate in chemical terms, and was apparently plentiful in China. They used the gunpowder for flame-thrower (even often adding the toxic element such as arsenic – one of the earliest chemical weapons!), guns, bombs, and land mines. Anyway, by about 1300s when the gunpowder came to the attention of the West, the Chinese had perfected the gunpowder.

Let us see now how the fireworks work. A typical firework bomb consists of three components (see Fig. 8.1). The first is the black gunpowder, which propels the firework bomb up. Next comes the bang component, a mixture which explodes and gives the flash and the sound. The third is to give colored, red, blue, and green, stars. The black gunpowder's ingredients have remained about the same since its Chinese invention. It is a mixture of saltpeter (chemically, potassium nitrate, KNO_3), sulfur S, and charcoal C. A fuse ignites this. What happens? When the fire on the fuse reaches this portion (black powder), the temperature becomes high enough to start chemical reactions. A number of reactions actually take place, but the main ones are the "oxidation" of charcoal and sulfur by potassium nitrate. The main reaction can be represented by $4KNO_3^- + 5C \rightarrow 2N_2 + 5CO_2 + 2K_2O$. This is a simplified version; the actual reactions are much more complicated. Also oxidation of sulfur to sulfur dioxide SO_2 and sulfate (SO_4^{2-}) takes place ($4KNO_3 + 5S \rightarrow 2N_2 + 5SO_2 + 2K_2O$). Sulfur dioxide smells badly. You smell that when you go near the site of fireworks or after a gunshot. A lot of questions come to mind? What is the oxidation? Why does potassium nitrate oxidize something well?

Why does it explode? What is the explosion anyway? Chemistry can answer all these questions. Here, we shall consider the issue of "explosion," first.

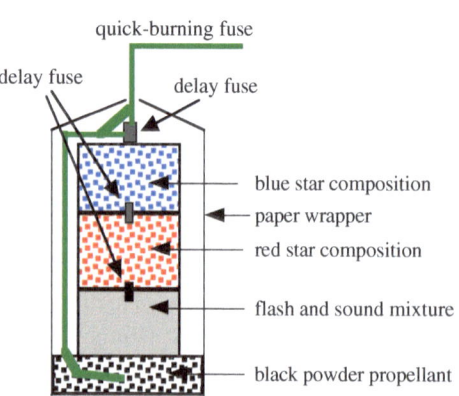

Fig. 8.1 A schematic diagram of a typical firework bomb

Two things usually need to happen for explosion. (1) The chemical reaction involved produces a lot of heat (energy), and (2) it also produces a lot of gas. Any chemical reaction would become faster at higher temperatures. Therefore, as a heat-producing reaction proceeds, the temperature goes up and hence the reaction becomes even faster, if the heat was not allowed to escape. Soon the reaction becomes so fast; it will complete the reaction in a flash; this is one component of explosion. The other factor is the production of gas. Gas is much more voluminous than the corresponding liquid or solid. Take an example of 10 g of water, which occupies about 10 cm^3 (cubic centimeter) or about one half of one cubic inch. When it becomes gas, i.e., steam at 100°C, the volume of this much of water would become 17 L (liter) or 17,000 cm^3; that is, the volume of water becomes 1,700 times as large when it becomes gas. What is more, gas will expand as temperature goes up. As we said in the last paragraph, the major gas products in the explosion of the black gunpowder are nitrogen (N_2), carbon dioxide (CO_2), and sulfur dioxide (SO_2). According to a study, 200 g (about half a pound) of gunpowder will produce about 43% of gas (i.e., about 86 g in this case) and 57% of solid products. From this, we infer that the volume of gas will be 200 L or so at 800°C. [By the way, you do not need to worry about these calculations, unless you really care]. Suppose that the firework ball is contained in, say, a 12″ long tube of 4″ diameter. The volume of the tube is about 2.4 L. 200-L gas suddenly produced in a 2.4 L container would exert a tremendous pressure (up to 100 atmospheric pressure), which propels whatever present above the gunpowder. This sudden expansion of gas creates a detonating sound (boom sound). And the gas produced will be ejected as a jet that propels the fireball.

The second stage is the formation of flash and bang sound, occurring in the second compartment containing aluminum (Al) and/or magnesium (Mg) powder, sulfur (S), and potassium chlorate ($KClO_3$) or perchlorate ($KClO_4$). Chemistry here is essentially "burning" (oxidation) of aluminum (and sulfur) by potassium chlorate or perchlorate. In this sense, this chemistry is essentially the same as that of the first stage. Only the oxidizing agent here is potassium chlorate or perchlorate, instead of potassium nitrate. Aluminum emits a lot of heat and light (white) when it is oxidized; i.e., it combines with oxygen (of chlorate and perchlorate). This heat (as well as the heat coming from the oxidation of sulfur) suddenly expands the surrounding air volume, creating the bang sound, and also emits sparkling light. The light is supposed to come from the product aluminum oxide or magnesium oxide, which is heated to very high temperature supposedly up to near 3,000°C. This is a phenomenon known as "black body radiation," and such a light is a continuous spectrum; that is, the light consists of lights of continuously varying wavelength. Such a light appears white at very high temperatures.

8.3   Color of Firework

The last stage is the formation of colored stars. Stars are made of compounds containing sodium, strontium, copper and/or barium compounds, and others. These compounds will give rise to colored light when heated. This is called "flame coloring." A strontium (Sr) compound, for example, when heated high, decomposes and forms an unstable compound such as strontium monochloride (SrCl). Sodium compounds would form sodium atom (Na), when heated high. As seen in Chap. 19, the electrons in an atom or molecule take discrete energy levels. When the atoms/molecules are at low temperatures, most of the electrons would be in the lowest energy level (called the ground state). Upon being brought to higher temperatures, some electrons (in atoms/molecules) would occupy higher energy levels because the energy supplied as heat can bring them to more energetic states. These energetic states (excited states) are not comfortable for an atom or a molecule to be in, for the electrons are high in energy and unstable, and they tend to go back to the lower or lowest energy level. This transition (from a higher energy state to a lower energy state) would emit a light of a specific color (wavelength or frequency in technical terms), determined by the energy difference between the two states (see Chap. 20). This is what happens in the firework's third stage. Strontium monochloride SrCl gives off "red" light, sodium (atom) "yellow," barium (barium monochloride BaCl) "green," and copper (copper monochloride CuCl) "blue" (see Fig. 8.2 for some of these colors). The other colors are made by mixing the compounds that contain these elements, and other ingredients.

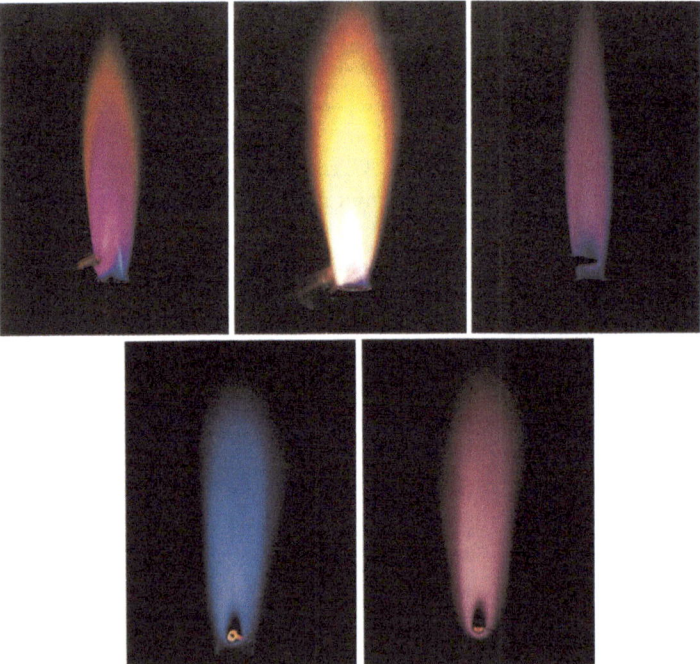

Fig. 8.2 Flame colors; they are lithium, sodium, potassium, cesium, and rubidium in the clockwise from the top *left* [from D. A. Mcquarrie and P. A. Rock, "Descriptive Chemistry" (W. Freeman and Co., 1985)]

8.4   Oxidation Reactions Involved in Firework

The last, perhaps most crucial and difficult, issue is why potassium "nitrate" or potassium "perchlorate" does what it does, i.e., oxidizes something else. As implied here, nitrate or perchlorate, not potassium, is the oxidizing agent. What is the "oxidation" anyway? Let us begin from the basics. As we argued (Chap. 19), some atoms hold their electrons more tightly than others do. This tendency is called "electronegativity." It is a measure of how strongly an atom (actually its nucleus) attracts electron(s) around it. Fluorine (F) has the highest electronegativity, and the next is oxygen (O). F has a strong tendency to become F^- by attracting one electron. Now let us look at nitrate NO_3^-, a combination of nitrogen and oxygen. Oxygen is more electronegative than nitrogen, and oxygen becomes more comfortable in a compound with two more electrons than its neutral atom has, i.e., O^{2-}. If this is so, then nitrogen in nitrate should carry formally +5 electric charge [i.e., $3 \times (-2)$ (from three O^{2-})$+(+5)=-1=$the overall charge]. The oxidation number of nitrogen in nitrate is said to be +5 in this case [and by the way, the oxidation number of oxygen in nitrate (and other compounds) is -2] (see also Sect. 3.2.1).

OK so far? Since the nitrogen carries a +5 electric charge in nitrate, you have to remove five electrons from a neutral nitrogen atom in order to create it. This requires a lot of energy, because an electron is attracted by the positive nucleus, and you have to pull them apart with force. "+5" state is called "+5 oxidation state," as removal of electron(s) from a chemical entity is called "oxidation" (we used "+V" instead of +5 in some other chapters). This implies that the +V (+5) oxidation state of nitrogen in nitrate is not very stable, because you have brought it to that state by expending a lot of energy. Anything that is in a rather unstable state wants to become more stable. This is one of the most basic rules of the physical world. In this case, the nitrogen (in V oxidation state) wants to gain electron(s) to become lower oxidation states, for example, the zero oxidation state, which is N_2 molecule. This can happen when the nitrate comes into contact with a compound which can provide that electrons. Charcoal and sulfur that are mixed with potassium nitrate are two such substances. Charcoal is essentially carbon; C can readily give up electrons to become nominally +4 (IV), if it can combine two of O^{2-} and forms carbon dioxide, CO_2. In this process, nitrate has gotten electrons from C and turns into N_2. The process in which a compound removes electron(s) from another is defined as "oxidation." You can say that nitrate oxidizes carbon and hence acts as an "oxidizing agent" or "oxidant." From the carbon's point of view, it has given electron(s) to nitrate. This process, giving electron(s) is called "reduction," and it can be said that carbon reduces nitrate and hence carbon in this case is a "reducing agent" or "reductant." As you see, oxidation and reduction always take place simultaneously; it is like a head and tail of a coin. Similar reactions occur between nitrate and sulfur; sulfur will become sulfur dioxide SO_2 and sulfate, SO_4^{2-}. It must be noted that nitrate is a very strong oxidizing agent, as the nitrogen in nitrate is in such a high oxidation state and hence has a very strong tendency to gain electron(s) from other compounds. You may have been confused by now. If so, read this paragraph and the last once more slowly and carefully.

What do you think happens if a compound contains in itself both an oxidizing part and a reducing part? You are right on! If you guessed that an oxidation–reduction reaction can happen within that compound; it can self-explode. That is exactly what happens. First let us take as an example NH_4NO_3 "ammonium nitrate," now a household term since the infamous bombing of the Oklahoma Federal Building. Ammonium nitrate consists of ammonium NH_4^+ and nitrate NO_3^- Plants can utilize both ammonium and nitrate for their nutrition. Hence, it is widely available as fertilizer. The oxidation state of nitrogen in ammonium is −3 (−III), which is very low. It wants to release electron(s) to become neutral nitrogen N_2. Nitrate, on the other hand, is eager to get those electrons. Ammonium nitrate is stable enough at ambient temperature. But at high temperatures, a reaction (that is, oxidation–reduction) can occur between the ammonium part and the nitrate part. In essence, electrons will move from ammonium to nitrate. This reaction gives off a lot of heat, and hence will become very fast once it has started. Besides, it produces a lot of gas; hence explosion! The chemical reaction is summarized as $2NH_4NO_3 \rightarrow 2N_2 + O_2 + 4H_2O$.

How about the other nitrogen fertilizer, ammonium sulfate $(NH_4)_2SO_4$. Sulfate is not a strong oxidant, though the sulfur in sulfate is in a high (+6, VI) oxidation state. [Think why? This is a difficult question but interesting one to think about. Anyway, ammonium sulfate will not explode].

What if nitrate is combined with a carbon compound? A source of nitrate is nitric acid HNO_3. If you let nitric acid react with, say, glycerin, a compound called "nitroglycerin" will form. This has an oxidizing part, i.e., "nitro" group in this case, which does essentially the same thing as nitrate, and a reducing part derived from glycerin. So, if the compound is given a shock or heat, a very fast oxidation–reduction reaction ensues; it detonates. This is, of course, the ingredient of the dynamite. It was invented by Swedish Alfred Nobel for the purpose of peaceful uses (construction, etc.). He made a fortune by this invention and others, but later regretted that his invention was used for weaponry, though he also thought that a very strong weapon may provide a deterrent against wars. This caused him to set up the "Nobel Prize." Nitro-groups can also be combined with other organic compounds. For example, tri-nitrotoluene (TNT) (there are three (tri) nitro groups in this compound) is a well-known powerful explosives.

As you might now imagine, chlorate ClO_3^- and perchlorate ClO_4^- will act as strong oxidizing agents such as nitrate, because their chlorine atoms are in high oxidation states (+V and +VII, respectively).

8.5   Nitrate and the World History

Now one last thing about this saga related to the fireworks. In many of these applications, the essential ingredient is nitrate. Where do you get it? It came from saltpeter ore, or soda nitre ore (sodium nitrate). The latter was the major source, which came mostly from the Northern arid region of Chile. Germany, because of her aggressive stance, was blocked by the Allied countries from importing soda nitre in the early 1900s. The German government encouraged their scientists to develop methods to make nitrates artificially. Chemists knew by that time that nitric acid is readily obtained by oxidizing ammonia NH_3. Ammonia is made of nitrogen and hydrogen as you see in this formula. Nitrogen can be obtained from air, as it contains 78% of nitrogen (N_2). Hydrogen is obtainable by decomposing water; this can be done by a number of ways. Once you have nitrogen N_2 and hydrogen H_2, all you have to do is let them react with each other; i.e. $N_2 + 3H_2 \rightarrow 2NH_3$. On the paper, it is a simple reaction. In practice, the reaction virtually would not take place. The reaction is too slow; it may take millions of years to get some ammonia made. You need to speed up the reaction. What would you need to do that? Find a "catalyst"! A catalyst is a substance that will be required in small quantities, but will increase a chemical reaction speed. Our body's function, for example, is maintained by thousands of chemical reactions, the majority of which are catalyzed by enzymes that are biological catalysts, as discussed in other places. Indeed, what the German scientists needed to do was to find appropriate catalysts for the formation of ammonia from nitrogen and hydrogen. Fritz Haber discovered some suitable catalysts for the process, and Carl Bosch developed an industrial process to make ammonia based on the catalyst. The first ammonia synthesis plant was built in 1911. By 1914 when Germany went into a war (WWI), they had an ample supply of nitric acid to produce explosives, from ammonia-producing factories.

The same process, of course, led to the production of chemical fertilizers, ammonium sulfate, ammonium nitrate, and urea, which benefited mankind. The green revolution (an efficient production of rice) was possible only with a massive use of chemical fertilizers. Now this has resulted in the deterioration of soils, and a further increase by this technology alone in agricultural productivity may not be very likely. This story tells us that use of technology has always both positive and negative aspects. We need to be very judicious in applying science to technology.

It should be noted, before we leave this topic, that some fungi and bacteria, including those contained in the nodules found on the roots of leguminous plants, convert the atmospheric nitrogen into ammonia. This is carried out by a biological catalyst (i.e., enzyme) called "nitrogenase." Scientists all over the world are eagerly trying to find out how these small organisms do it or how nitrogenase works. In addition, some of them are trying to incorporate the enzyme in other plants by genetic engineering. Such enzyme may theoretically reduce the dependence on artificial fertilizers of many plants including rice and cereal.

Light Stick, Firefly, and Color TV

9

9.1 Absorption of Light and Emission of Light

Light and color are perhaps the most important factors in visual arts and entertainment. Paintings cannot be done without color. This color is provided by pigments, chemicals that absorb light in the visible range. All natural colors, green leaves, and flowers etc., are based on the same principles of absorption of light by pigment molecules. The absorption of light has been talked about in several places in this book (Chaps. 10 and 20). The basis is that the energy of a molecule takes distinct values (quantized) and the molecule will absorb a light that has a frequency (or wavelength) that matches the energy gap between those distinct levels. This is the end of the story in short. And that is "physics." Physics is concerned with only how a phenomenon occurs.

Chemistry, however, goes beyond that. That is, chemistry is concerned with "what compound gives what color." It deals with individual compounds. This is what makes chemistry both difficult and interesting. Difficulty and interest both lie in figuring out what compound is appropriate and how to make it, and carrying out the synthesis of the compound, because there are a large number of possible compounds known and/or unknown. Colors due to absorption of light are the ones we are familiar with in everyday life as well as in some art forms. Colors are everywhere: colors of food, plates, and their decorations, our clothes, flowers, and paintings artistic or otherwise. We react to and live with colors: enjoy, be repulsed by, or be afraid of. Without colors, our lives would be rather dull and desolate. In the next chapter, we look at the coloring of ceramics that is due to absorption of light.

Now, there is another kind of colors; i.e., the colors of television screen, the greenish-yellow color of light produced by fireflies, and the several different colors of the so-called light sticks are examples. How are these colors different from those mentioned in the previous paragraph? They are all due to light emitted by chemical compounds rather than absorption of light. An emission of light is also responsible for the fireworks, which we talked about in Chap. 8. Again the physics of light emission is quite straightforward. It is the reverse of light absorption. A molecule takes several

E. Ochiai, *Chemicals for Life and Living*,
DOI 10.1007/978-3-642-20273-5_9, © Springer-Verlag Berlin Heidelberg 2011

distinct energy states (levels). Ordinarily the molecule is in the lowest, i.e., the most stable energy state, which is called "ground state." If the molecule is brought to a higher energy state (coined as "excited state") by some means, it will try to come down to the lower (more stable) state and eventually to the ground state. This can occur in two ways. In most of the cases, the extra energy (in the excited state) is lost as heat to the surroundings; this is called "nonradiative" transition. No extraordinary thing will be observed in this case. In some other cases, however, the extra energy may be released as a light, and the molecule will come down to a lower energy level. This is the emission of light, and the process is a "radiative" transition.

One common way to excite a molecule is by shining light on it. The molecule absorbs light and then is brought to an excited state. Some compounds do emit light, and there are two different ways to do it. In one way, emission of light comes right after absorption of light. In this case, the molecule emits light as long as it is shone by light and stops emitting light as soon as the light source is shut off. This type of light emission is technically called "fluorescence." In another kind, emission may come sometime after the light source is turned off. In this case, the energy absorbed is transferred to a sort of metastable state that does not immediately go down to a lower state. That is, the compound remains in an excited state for a while. After sometime, it comes down to a lower state by emitting light. So there is a delay in emission. This type of emission of light is called phosphorescence. The emission of light by a compound by either mechanism is called luminescence in general.

An emitted light may be colored or may not be. If the light has a wavelength within the so-called visible range (from about 800 to 320 nm), it is visible to human eyes. If the wavelength of the light is outside of this range, human eyes cannot detect it. There are millions of different colors, but all colors can be constructed with varied combinations of three basic colors: red, yellow, and cyan. How this occurs is understood in the realm of physics. The issue of how a light of a specific wavelength appears to us, i.e., its color, belongs to the realm of physiology/psychology. Chemistry deals with the issue: what kind of compounds will produce what kind of light (its wavelength), by emission or absorption of light.

9.2   Color Sticks

Many of you may be familiar with glow sticks or color sticks. Sticks of various sizes and colors are now available. A typical stick is a 6-in. long tube and is made of plastics. It contains a capsule of thin glass wall. The capsule contains a chemical called hydrogen peroxide. When you bend the stick and snap open the capsule, hydrogen peroxide mixes with the other chemicals present in the tube, and the tube starts to glow, i.e., emit light. It is a "cold" light. It is not caused by a heated material like flame or incandescent bulb. This kind of device was originally developed for military purpose, but now available for entertainment and other uses as well.

What happens? A chemical compound in the tube emits light. For a compound to emit light, the compound has to be brought to a high-energy state, an excited state. How is it done? A chemical reaction that takes place in the tube provides that

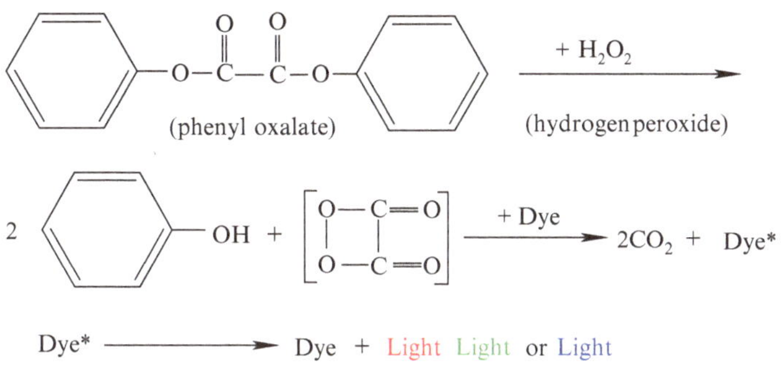

Fig. 9.1 Chemiluminescence

energy needed to bring the compound to an excited state. This kind of light-emitting process is called "chemiluminescence" (emission of light by a chemical means).

Intrigued by the cold light of fireflies, many people tried to recreate it chemically. In 1960s, Edwin A. Chandross, a chemist at Bell Labs was experimenting with chemiluminescence. He found that oxalyl chloride mixed with hydrogen peroxide and a fluorescent dye produced light. The efficiency of light production was not great, but it laid the foundation from which modern chemiluminescence developed. "Phenyl oxalate ester" used in today's light sticks is indeed a derivative of the original "oxalyl chloride" of Chandross. "Phenyl oxalate" was developed by chemists at American Cyanamid, and hence the trade name for some of chemical light products is "Cyalume."

Now, let us talk about chemistry. Phenyl oxalate reacts with hydrogen peroxide and produces phenol and an intermediate compound that is believed to have a formula C_2O_4. The reaction scheme is shown in Fig. 9.1. A square-shaped compound with such an O–O bond is often called oxetane, and is very unstable (that is, having a high-energy content), because the two oxygen atoms are bound in a very uncomfortable manner. It will be converted spontaneously to two molecules of a very stable compound, carbon dioxide, and simultaneously will transfer its energy to a fluorescent molecule like 9,10-diphenylanthracene. Now, the fluorescent molecule is brought to a high-energy state. The dye molecule in a high-energy state will come down to a lower state (ground state) by emitting light, a blue light in this case. Different colors are obtained by using different fluorescent dyes. The energizing mechanism, i.e., formation of a high-energy intermediate is common to all color sticks.

9.3 Fireflies and Other Bioluminescence

In a rather warm dump evening fireflies display their best light show. They are signaling to their mates. A school of squid illuminates the dark sea; squid fishermen look for that. Many jellyfish glow in the dark. Many living organisms emit light. These are called "bioluminescence." It is based on chemical reactions like those found in the light stick we talked about just before. Only it is a biochemical reaction that requires an enzyme.

The bioluminescence reaction involves a compound called luciferin, an enzyme luciferase and oxygen (O_2). [Lucifer is the morning star and is a light-bearer]. Luciferin and luciferase are collective names and differ from a species to another. Luciferin used in fireflies is luciferin (benzothiazole), but that in some fish, shrimp, squid, and some jellyfish is called coelenterazine and is a different compound from firefly luciferin. Some bacteria use an ordinary compound called flavin mononucleotide (FMN, which is used in all organisms for different purposes), and dinoflagellates (and some shrimp) use yet another compound. The luciferin used in crustacean (and some fish) is called vargulin. All these compounds, however, have similar reacting portions where the crucial reaction, oxetane formation, takes place. We will look at fireflies and ceolenterazine to illustrate how light is created.

The chemistry of these light-producing processes is summarized in Fig. 9.2. You do not have to worry about the details of chemistry. As usual, chemicals are described by

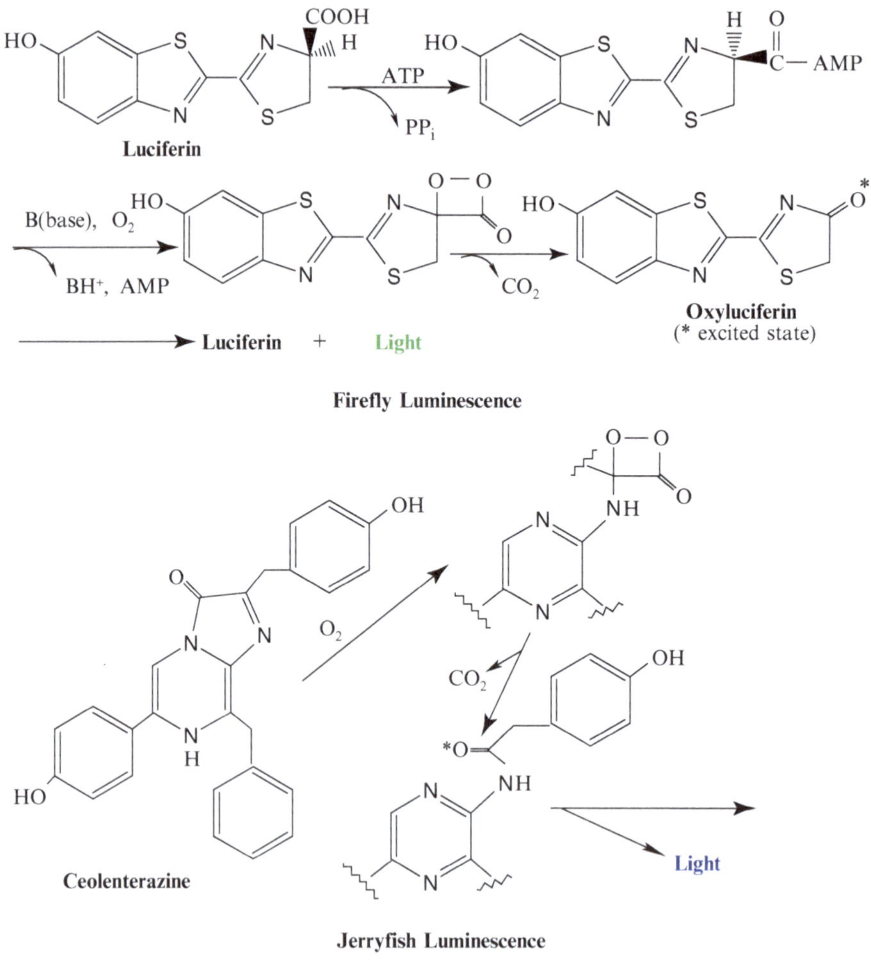

Firefly Luminescence

Jerryfish Luminescence

Fig. 9.2 Bioluminescence

chemical formula. Luciferase, the enzyme which fireflies produce (this is a very complicated chemical and is not shown), binds the chemical luciferin. However, in this case, luciferin itself is not sufficiently energetic and needs to bind AMP (adenosine monophosphate). You recall that ATP (adenosine triphosphate) is a universal energy carrier in the living organisms. What it does is to provide chemical energy for a chemical reaction that needs such an energy input. So upon binding a part of ATP in the form of AMP, luciferin is energized. This energized luciferin then reacts with oxygen (O_2 of the air) and forms the square-shaped unit with an O–O bond and a C=O bond (Fig. 9.2). This unit (oxetane) is very unstable (highly energetic) and spontaneously decomposes into carbon dioxide (CO_2) and the carbonyl unit (C=O). The energy present in the oxetane is left with the carbonyl unit; the star (*) attached to the C=O unit indicates that it is in a highly energetic (excited) state. This compound (called oxyluciferin) then emits a light of greenish yellow color and comes down to the most stable state (ground state). Oxyluciferin is then recycled chemically back to luciferin. This is what chemists think to happen, but they have not yet figured out the complete picture of the process.

The luminescence of firefly is very sensitive to the presence of the energy carrier ATP and is also sensitive to luciferase (the enzyme) in the presence of ATP and luciferin. So it provides a sensitive means to detect small quantities of ATP or luciferase. If firefly luciferase gene is coupled to a DNA containing some other genes and that region is expressed, luciferase is produced and instantly emits light (in the presence in the necessary chemicals). This light emission is an indicator of that portion of DNA being expressed and is conveniently used as a reporter for such events.

In the case of coelenterazine (of a jellyfish) as well, the crucial reaction intermediate is the formation of the square-shaped unit (oxetane) as seen in Fig. 9.2. This unstable intermediate decomposes spontaneously and forms an excited state of the carbonyl group; it then emits a light as it goes down to the ground state. This process requires the presence of calcium ion (Ca(II)), and hence this coelenterazine bioluminescence system is a sensitive detector of calcium.

There is another kind of protein that emits green light in jellyfish and others; it is called Green Fluorescent Protein (GFP). This protein, however, does not fluoresce on its own device. It needs to be excited by extraneous light. It is often used as a marker protein. The gene of GFP is incorporated along with that of another protein (which is the purpose of a genetic engineering) in a vector (e.g., a plasmid of *E. coli*). Whether the target protein is expressed or not can be judged by shining light on the cells. If the protein is expressed (produced), the associated GFP is also produced and the protein molecules expressed will show up as glowing green spots. The discovery of GFP and its application were rewarded with a Nobel prize in chemistry in 2008 (Osamu Shimomura, Martin Chalfie and Roger Y. Tsein).

9.4   Color TV and Color Monitor of Computer

The color TV or computer monitor (of now obsolete type) makes use of fluorescence, that is, an immediate light emission when a substance is excited or given an energy. In principle, its color is obtained in the same manner as those involved in fireworks. That is, "excite an atom or a compound and then it will emit light."

The only difference is the way of excitation. In fireworks, high temperature (heat) is the energy source of excitation, whereas in the case of the color TV screen, the energy source is an electron beam. Because of the difference in the setting and the energy source, the coloring material used for fireworks is not quite suitable for color TV screen.

All million different colors can be obtained by mixing three fundamental colors: blue, green, and red. The TV screen is covered with millions of spots, and each spot consists of blue, green, and red dots. An electron gun targets an electron beam at a red dot at one instant, and simultaneously the other two beams hit blue and green dots, respectively. The size of the dots is about 8–10 μm. We cannot see them as separate dots. [Micrometer is one millionth of a meter or one thousandth of a milli-meter]. The beam intensity determines the color intensity. The combination of these three colors of differing intensity brings about various colors.

Well then, what are those dots made of? They are made of inorganic compounds; they are typically resistant to repeated bombardment of electrons. They have to be. Blue is usually obtained by silver ion impregnated in zinc sulfide: Ag^{I}:ZnS. The electron in Ag^{I} is hit by an electron beam and excited to a higher level. Soon the electron comes back down to the ground level, emitting blue light. How soon deter-mines the so-called decay time. This one has a medium decay time. If the decay time is too short, the television screen flickers and makes viewing difficult, but on the other hand, if it is too long, the television image will be blurred, because an image is overlapped by the previous image. The green emission is from copper impregnated in zinc sulfide (Cu^{I}, (Al^{III}):ZnS). An exotic element, europium, provides the red color; Eu^{III}:Y_2O_2S (Eu = europium and Y = yttrium).

The requirements for computer monitors are slightly different. First, the monitor requires a higher resolution. Hence, the size of particles (dots) must be smaller, typi-cally 4–6 μm. Another difference is that it usually reproduces slower movements than televisions. This necessitates the longer decay time of the fluorescence. The materials used are similar to those used in the television screen, but they are modi-fied to make longer the decay time, by adding some other components.

Liquid crystal display (LCD) and plasma display in recent TVs and computer monitors are based on entirely different principles and material and are not discussed here.

Ceramics

10

Old and yet New staff

Pottery is probably one of the oldest craft mankind invented. It is believed that it may go back as long ago as 10,000 B.C. Yet it is still being used for practical as well as for artistic purposes, and generalized ceramics are now being pursued as high-tech material. That is, "ceramics is the oldest and yet the newest material."

Even those Paleolithic people who made jars and basins out of clay could not help but decorate them and also use the technique to create something pleasing and aesthetic without practical uses. Humankind, with its creative brain, cannot help but do something beyond absolute necessity. That has brought us "civilization" and "culture." Pots, the artifact, are actually often used to distinguish ancient civilizations, and how cultures were transmitted from one location to another. The reason is the long-lastingness of the ceramics. They are one of the sturdiest chemical materials.

E. Ochiai, *Chemicals for Life and Living*,
DOI 10.1007/978-3-642-20273-5_10, © Springer-Verlag Berlin Heidelberg 2011

They are essentially "rock," the inorganic material. The other artifacts made of plants or animal materials are organic materials, which are subject to chemical, physical, and biological decomposition. Particularly vulnerable are they to the attack by microorganisms, fungi, and bacteria.

10.1   Chemicals that Make Up Ceramics/Pots, and What Happens to Them When Fired in a Kiln?

The conventional ceramics, brick, earthenware, and porcelains, are made from clay. Clay is a complex material, but its main component is a mineral called "kaolinite." Pure kaolinite is a nice white powdery mineral. Such white clay is called "China clay." Clay is often colored though, because it contains small quantities of colored material such as iron oxide (Fe_2O_3) that gives a brownish color. Other iron compounds present give grayish colors. All iron compounds turn into iron oxide (Fe_2O_3) when fired, and hence the earthenware made from grayish clay usually ends up with a brownish color. Ordinary clays contain in addition other minerals such as quartz, soda mica (paragonite), and potash mica (muscovite).

One of the common minerals in granite rock is "feldspar," which has a chemical composition of $K_2O \cdot Al_2O_3 \cdot 6SiO_2$ [K is potassium, Al is aluminum, and Si is silicon]. Over the geological time, rocks containing feldspar are slowly weathered, and the mineral reacts with water and carbon dioxide in the air, and turns into kaolinite. The chemical reaction can be approximated as follows:

$$K_2O \cdot Al_2O_3 \cdot 6SiO_2 \, (\text{feldspar}) + 2H_2O \, (\text{water}) + CO_2 \, (\text{carbon dioxide}) \rightarrow$$

$$Al_2O_3 \cdot 2SiO_2 \cdot 2H_2O \, (\text{kaolinite}) + K_2CO_3 \, (\text{dissolved away}) + 4SiO_2 \, (\text{free silica})$$

The chemical composition of kaolinite can also be expressed as $Al_4Si_4O_{10}(OH)_8$; this formula represents better the crystal structure of the mineral as shown in Fig. 10.1 (as in a unit of the crystal (dashed line box), only two layers and two such units in each layer are shown). The crystal consists of layers of a sort of sheet which has the composition of $Al_4Si_4O_{10}(OH)_8$ (see more about the chemistry of rocks in Chap. 14).

One of the important properties of clay is its plasticity when it is wet. In other words, it can be molded in any form you want, and the shape will hold when you release your hands from it. Why is wet clay plastic? When water is added, water molecules actually bind to the surfaces of clay mineral (kaolinite) [this phenomenon is called "adsorption"]. What kind of chemical force binds them? This happens readily because of its chemical structure as shown in Fig. 10.1. Water acts then as glue holding clay particles together (due to surface tension). However, when you apply stresses (i.e., try to create a shape) sufficiently strong, water makes the flow (movements due to the stress) of particles easier. In other words, water can act like a lubricant as well. The clay particles are easy to flow because of its plate-like structure (Fig. 10.1). When the stress is removed, the flow stops and the effect of water of holding together the particles resumes; hence the shape will be maintained.

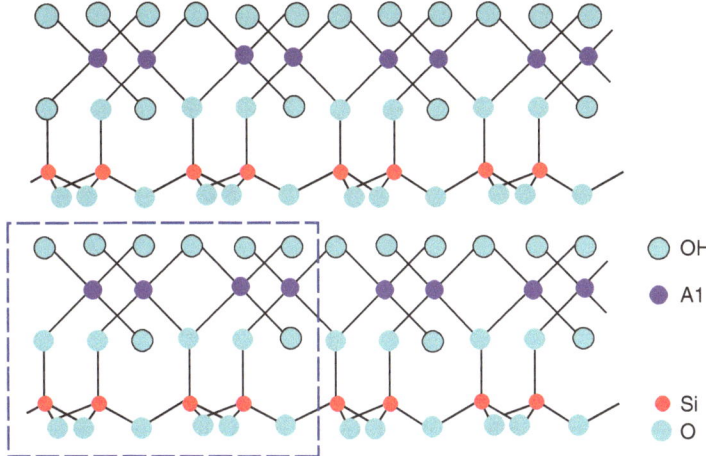

Fig. 10.1 Crystal structure of kaolinite; the portion in the dashed box is the basic unit

Further, when dried, the article becomes fairly rigid and strong, in the sense that it maintains its shape against a significant stress. This phenomenon is not very well understood. If you compact dry kaolinite powder, it will not very well be stuck together, and any ball formed will easily be crumbled. Therefore, we have to assume that water molecules that have been used have brought those small particles closer together and increased the close contacts among particles (so that the clay particles stick together better now) or water have formed some soluble cementing compounds.

In practice, you need more than just kaolinite to make a pot or a jar. The clay should contain the so-called flux and filling material. Flux is a mineral which melts and forms glass at temperatures lower than that at which kaolinite melts, so that this glassy matrix acts as a cementing agent between kaolinite mineral particles. The commonly used flux is feldspar (potassium (K^+)-containing one as seen earlier, or sodium (Na^+)-containing one) and its derivatives. The presence of either potassium or sodium in these minerals reduces the melting point (temperature), and, when cooled, they tend to become glassy, rather than forming crystals. The filling materials often used are silica (quartz, chemically SiO_2), alumina (Al_2O_3), and calcined bone (chemically apatite $Ca_5(PO_4)_3(OH)$ used for bone china). Most of these materials, except bone, are usually contained in the naturally occurring clay in different proportions.

The molded clay article is then fired in a kiln. The clay turns into ceramic. By the way, "ceramic" comes from the Greek word "Keramos," which means "burnt material." What happens in terms of chemistry? When you heat the clay, a number of chemical reactions take place. The clay first loses about 1.5% of its weight when heated up to about 150°C. This is the loss of moisture. At around 500–600°C, the weight loss is much larger, reaching about 14%. At this temperature, kaolinite changes to metakaolinite losing water from its crystal structure. That is,

$$Al_2O_3 \cdot 2SiO_2 \cdot 2H_2O \left(\text{kaolinite}\right) \rightarrow Al_2O_3 \cdot 2SiO_2 \left(\text{metakaolinite}\right) + 2H_2O.$$

Metakaolinite eventually changes to a mineral mullite around 1,000°C. Despite these crystal structural changes, the clay retains its shape over this temperature range. Beyond about 1,100°C, some of the minerals (particularly that of the flux) melt and become glassy. The melt then fills the pores (between clay grains), reducing the porosity down to close zero by about 1,200°C. [If you do not heat that high, the ware may be still quite porous; such porous wares were common in the ancient pots. Today's earthen wares are also somewhat porous.].

10.2   Glaze

Pots without glaze can be used and are used, but more often than not ceramics will be covered with glaze. It is a smooth, glassy layer usually about 100-μm thick. [Micrometer is one millionth of a meter, so 100 μm is one tenth of one millimeter]. It gives an attractive surface, glossy or nonglossy, that provides an impermeable layer. It is also decorative, especially if it is colored. In addition to the aesthetic effects, glaze can increase the mechanical strength of the ceramic article.

Glazes are basically similar to glasses and consist of silica (SiO_2) and other oxides. The melting point of pure silica is high, 1,700°C, and mixing with other oxides reduces its melting point significantly. Glazing mixtures are complicated, but usually contain sodium or potassium oxide (Na_2O or K_2O), calcium oxide (CaO), magnesium oxide (MgO), lead oxide (PbO), zinc oxide (ZnO), alumina (aluminum oxide, Al_2O_3), and/or borax (B_2O_3). These may be added as pure oxides or as complex oxide minerals. A glazing solution (slip) is a slurry of fine powders of these oxides. Inclusion of lead oxide is often desirable, as it has a high refractive index, giving smooth high-gloss surface. However, lead is toxic (see Chap. 15) and has to be converted into a nontoxic form. It is said that lead bisilicate $PbO·2SiO_2$ is nontoxic. To accomplish this conversion, the glaze components are preheated and fused to form a frit, and the frit is then ground to powder. A higher content of lead oxide reduces the melting point of the glaze mixture. Such a high-lead glaze is used for special art ware. Nonlead glazes require firing temperatures as high as 1,450°C.

10.3   Colors

The color is the most fascinating aspect of pottery. It is also the basis of all the visual arts (see Chaps. 8 and 9), aside from the shape. Potters from early on were very much involved in developing techniques to decorate pots by coloring them. Up until about a century ago, the selections and manufacturing of coloring material were made based on experience of trial and error and its accumulated knowledge of know-how. The chemistry of colors on ceramics is very complicated, particularly regarding various shades of colors. Some of the ancient beautiful ceramic colors have not yet been successfully reproduced, even though we have now a fair bit of understanding of the chemistry of color.

Fig. 10.2 Colors of aqueous
solutions of simple salts of
transition elements

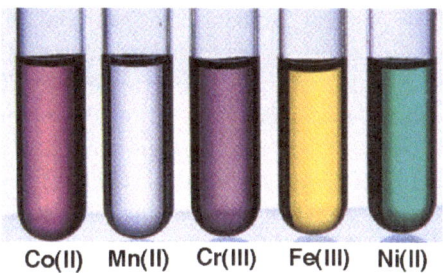

Co(II) Mn(II) Cr(III) Fe(III) Ni(II)

The basic coloring chemicals for ceramics are oxides of metallic elements, especially those of the so-called transition elements. Examples of such elements are chromium, iron, cobalt, and copper. These transition elements are located in the mid portion of the periodic table (see Fig. 19.2). The chemical compounds of these elements are usually colored, unlike most of the organic compounds (see Fig. 10.2 below). It must be pointed out, though, that some organic compounds are colored; the majority of dyes (colors on the fabrics) used today are organic compounds. These, i.e., compounds of transition metals and some types of organic compounds are the two kinds of chemicals that are colored. The organic compounds that are colored cannot be used for ceramics, because the organic dye compounds will be decomposed when the clay is fired to high temperatures. Oxides of metals, on the other hand, are stable even at high temperatures.

10.3.1 Red and Pink Colors

Let us look at some typical ceramic color stains. Red stains can be obtained with iron oxide, chromium–alumina combination, copper compound, or cadmium selenide. The red iron ore hematite is iron oxide (Fe_2O_3, di-iron trioxide). Iron oxide is perhaps one of the oldest red coloring materials for ceramics. The tint of the red color produced by iron oxide depends on how it is produced; particularly critical are the temperature of calcination (to be explained below) and the grain size. Larger grain sizes reduce the brightness of red color.

The iron oxide is typically produced by calcining iron sulfate, $FeSO_4 \cdot 7H_2O$. Heating a material without melting (fusion) is called "calcination." The chemical reaction that takes place upon heating up to 650°C is $2FeSO_4 \cdot 7H_2O \rightarrow Fe_2O_3 + SO_3 + SO_2 \left(+14H_2O\right)$. The addition of a small amount of zinc oxide (ZnO) brightens the red color. On the other hand, addition of boric acid and sodium chloride (common salt) will induce the formation of Fe_3O_4 (which is black) and hence make the red color darker. The iron oxide is also responsible for the color of the naturally occurring iron earth *ochres* and *siennas*.

When you boil chrome yellow (lead chromate, $PbCrO_4$) in a dilute caustic soda (sodium hydroxide, NaOH) solution, the color changes to bright red (called "coral red"). This is due to the formation of basic lead chromate $PbCrO_4 \cdot Pb(OH)_2$.

This pigment can be used as stain on the glaze, but can be applied only to art and studio pottery, as the toxic lead content is too high.

Mostly for this reason (toxicity of lead), coral red has been superseded by cadmium selenide (CdSe). Both cadmium and selenium themselves are chemically toxic (see Chap. 15), but their combination, cadmium selenide, is extremely insoluble and resistant to chemical attack, and hence would not be made to a form that can become toxic. In reality, this stain is a mixture of cadmium sulfide (CdS) and cadmium selenide. With a higher proportion of selenide, the color is bright red, while with an increasing proportion of cadmium sulfide color becomes orange.

Another red color is obtained with chromium in alumina; it is chemically chromium oxide plus alumina: Cr_2O_3–Al_2O_3. This is essentially the same as "ruby" in chemical terms, and the same is also used as the laser source used for reading bar codes in post office and at the checking counter of a grocery store. By adding tin oxide SnO_2 to this mixture, we can make "pink" color. Pink can also be obtained by using manganese compounds.

10.3.2 Blue Colors

The most important ingredient for blue colors is cobalt (Co) or vanadium (V). A typical formula for "cobalt blue" stains is one part of cobalt oxide (CoO), three parts of zinc oxide (ZnO), and nine parts of alumina (Al_2O_3). This formula will give matte blue. If we use silicate such as feldspar (see Chap. 14) and silica (SiO_2) instead of alumina, we get "royal blue" and "Mazarine blue."

Vanadium-zirconia is another blue stain used often. A mixture consists of zirconia ZrO_2, silica SiO_2, and vanadium pentoxide V_2O_5 (vanadium in +V oxidation state, which is colorless). The blue color is due to the formation of V_2O_4 in the calcination process. The V atom in V_2O_4 is in the +IV oxidation that has a characteristic blue color. The formation of V_2O_4 seems to be facilitated by the addition of a small quantity of sodium fluoride NaF.

10.3.3 Green Colors

The most common undergraze (emerald) green color is obtained with chromium oxide–zinc oxide. An addition of a calcium compound to the chromium gives a light green color called "Victorian green." On the other hand, the addition of a small amount of cobalt gives rise to blue–green. A mixture of nickel oxide and chromium oxide ($NiCr_2O_4$ – its structure is called "spinel") shows a moss green color.

A green color is also obtained by mixing three parts of vanadium–tin (oxides) yellow with the vanadium–zirconia blue mentioned above. Copper and vanadium combinations also produce stable clear green color. Copper itself gives also green color. Copper oxide (CuO) dissolved in boric glaze gives a transparent green color. Addition of alumina to this system gradually darkens the green color. An increase of alkalis in the glaze tends to change the copper color to more bluish.

10.4   Chemistry of Colors (of Inorganic Compounds)

Colors of ceramics are due to compounds containing the so-called transition (metallic) elements such as iron and copper, as seen above. Some of the colors of such compounds might be familiar to you. Iron oxide can be brown (rust) or red (hematite ore), and maybe you have seen a nice blue copper sulfate crystal. Compounds of other elements (nontransition elements) are rarely colored. What special is there about compounds of transition elements?

We need to discuss a little bit the different natures between transition elements and nontransition elements. Please refer to Chap. 19 for the definition of transition versus nontransition elements. First let us talk about the situation in compounds made of nontransition elements. Water is a typical such compound. It is made of two hydrogen atoms and an oxygen atom (H–O–H). An independent oxygen atom ($1s^2 2s^2 2sp^4$) has eight electrons altogether, but six in its valence shell (2s2p orbitals). [The two electrons in the core shell (1s orbital) are not involved in the bonding]. When an oxygen atom binds to two hydrogen atoms, it will be surrounded by eight electrons; i.e., six on the oxygen atom and one each contributed by two hydrogen atoms. As the valence shell consisting of 2s and 2p orbitals can accommodate up to eight electrons (two in 2s and six in three 2p orbitals), the oxygen in water molecule has a completed valence shell and all the electrons are paired up.

In methane, CH_4, the carbon atom has four electrons of its own plus four electrons contributed by the four hydrogen atoms. Therefore, eight altogether and hence the shell is complete, and all the electrons are paired up. How about an ordinary ionic compound, sodium chloride (NaCl, table salt)? It consists of Na^+ ion and Cl^- ion. Both of these entities have eight electrons (paired up) in the outermost shell (valence shell). You can make certain that this is indeed the case by referring to the periodic chart in Chap. 19. These compounds, i.e., compounds of the so-called nontransition elements (otherwise called typical elements) have the completed valence shell on each atom contained and all the electrons involved are paired up. You might note that all these compounds, water, methane, and sodium chloride, are colorless. As a matter of fact, the majority of compounds made of nontransition elements are colorless. Only colored compounds of nontransition elements are some halogen molecules (chlorine Cl_2 is yellow-green, bromine Br_2 is red, and iodine I_2 is purple), nitrogen dioxide (NO_2, brown), mercury sulfide (HgS, red) and similar compounds including cadmium selenide, and finally large molecules containing the so-called conjugated double bond system. Examples of the last category include carrotene, the pigment of carrot, and other organic pigment compounds that are used to dye your clothes. There are still a few other exceptions, but we cannot pursue this issue further here.

Transition elements have electrons in d-orbitals [again refer to Chap. 19 for a discussion of s, p and d orbitals]. A typical transition element is iron; it has 26 electrons. Its outermost shells consists of five 3d orbitals and one 4s orbital (plus three 4p orbitals), and the 3d orbitals would accommodate up to ten electrons (because each of five d-orbitals can pick up two electrons) and 4s up to two electrons. Iron atom (Fe) (in its free state) has six electrons in the d-orbitals and two electrons in the 4s orbital. When iron makes a compound, it usually loses two or three electrons, i.e., forms Fe(II) or Fe(III) ion. Fe(II) has six electrons in the d-orbitals and none in the 4s orbital, and Fe(III) has five d-electrons and no 4s electron [Note that the electrons were first lost from the 4s orbital in forming Fe(II) or Fe(III)]. Let us look at the case of Fe(III). Fe(III) in a free state has five electrons, one in each of the five d-orbitals, because all the five d-orbitals have the same energy in Fe(III) under a free condition. The d-shell in Fe(III) is incompletely filled, as the d-shell will be completely filled only when ten electrons occupy it. When Fe(III) binds with other entities to form a compound such as $FeCl_3$ (Fe(III)(Cl$^-$)$_3$), the d-orbitals change their energies a little bit, and the five electrons would occupy different d-orbitals depending on the compound. But still, the d-shell contains only five electrons, because most of Fe(III)-containing compounds are largely ionic.

Chapter 20 discusses how color will appear in a compound. It is due to the interaction between light and the compound. Light, with a certain energy, would move electrons around when it hits a compound. The movement of electron is due to a transition from an energy state to another. [Recall that the energy of a compound takes only certain distinct values (i.e., quantized)]. The energy required for moving an electron around should match the energy that light has, in order for this (transition) to take place. So, when a light hits a compound and causes the movement (transition) of an electron, the energy possessed by the light will be used (absorbed) by the compound. That is, that particular light will be absorbed by the compound. The light that will come out from the compound then will lack that particular component, and this (light with some component lost) is what we see.

The energy of light is proportional to its frequency. [That is, $E = h\nu$, where h is a number called Planck's constant and ν is the frequency]. The frequency (ν) and the wavelength (λ) are inversely proportional to each other in the form of $c = \nu\lambda$ (where c is the speed of light); a light of the shorter wavelength has the higher energy. The visible range (wavelength range of 320–800 nanometer (nm)) is relatively long in wavelength and hence relatively low in energy. A light in the ultraviolet range has a shorter wavelength (or higher frequency) and hence a higher energy than that in the visible range.

The energy that is required to move an electron from an energy state to another, i.e., the energy gap, is dependent on the type of orbitals that electrons are in. Most of the nontransition elements in ordinary compounds have a closed shell as mentioned in a couple of paragraphs before. Then, if the electron in that situation is to be moved (by light), it needs to be moved to another shell, because there is no room in the current shell. The energy gap between a closed (complete) shell and the next is usually large, and the light absorbed by such a compound occurs in the ultraviolet range. Our eyes cannot detect the ultraviolet light, and so such a compound would look colorless to us.

Now let us turn to the compounds of transition elements. Iron compounds such as iron oxide Fe_2O_3 have its iron in Fe(III) state. Fe(III) has five electrons in its valence shell (d-shell), as we mentioned earlier. The d-shell consists of five d-orbitals and can accommodate as many as ten electrons. Since there are only five electrons in the d-shell of Fe(III), the electron can be moved around within the d-shell by light.

The energy that is required to move an electron within the same d-shell is relatively small, and hence the wavelength of light that is absorbed by a Fe(III)-containing compound is relatively long (or the frequency is small) and happens to occur in the visible range. Therefore, the compounds containing transition elements of *incomplete* d-shell are usually colored. Examples are Cr(III), which has three electrons in the d-shell in water, is green, Co(II) with seven d-electrons is pink in water, and Cu(II) (how many d-electrons?) is blue in water. The colors of solutions of simple transition metal salts (such as metal chloride) are shown in Fig. 10.2. The other entities that combine with such a transition metal ion modify the energy states of the d-shell and hence its color. For example, all the following compounds contain Co(III), but $[Co(III)(NH_3)_6]$ is golden yellow, *trans*-$[Co(III)(NH_3)_4Cl_2]$ nice green, and *cis*-$[Co(III)(NH_3)_4Cl_2]$ purple. The last two compounds have exactly the same composition, but the modes of attachment of the entities (NH_3 and Cl) are different. The difference in structure is indicated by terms of *cis* and *trans*. This is an example of what is called "structural isomers."

The last of the transition metals in the periodic table is element zinc Zn. When zinc forms compounds, it takes the oxidation state of II, i.e., Zn(II). The compounds containing Zn(II) are usually colorless. Find out how many electrons are in the d-shell of Zn(II), and see why they are colorless. [In practice colorless compounds look white in solid (powder) state].

The principles of light absorption have thus been worked out well. Yet, the details are difficult to predict. That is, "what combination of chemical entities with what structure would produce what color" cannot be precisely predicted with current theories. Not that the theories are deficient, but the theoretical calculations necessary cannot be done very precisely for relatively complicated systems. Besides, the production of ceramics is in general difficult to control. That is, it is difficult to make a pot precisely the way one wants. [Though, today, industrial production of ceramics is fairly well controlled]. All these conditions make coloring of ceramics a kind of arts, rather than science. In order to obtain an exact tinge of color, one has to follow rigorously a prescribed procedure that has been laboriously developed by trial and error.

10.5 High-Tech Ceramics

The materials that are used for appliances, computers, and automobiles are mainly metals such as iron, copper, and aluminum. These are used for the purposes, because they have suitable properties. However, they are relatively expensive, as they require high energy-consuming processes to produce, and the resources are dwindling. Besides these metals have a number of limitations in mechanical properties, chemical reactivity, and heat-resisting properties. For example, metals such as iron and copper melt at relatively low temperature; iron melts at 1,535°C, and copper at 1,083°C. They cannot, therefore, be used for an application at high temperatures. The World Trade Center buildings collapsed after jet planes hit them on September 11, 2001. The major cause of the collapse is said to be softening (partial melting) of the iron beams due to the intense heat caused by burning of the jet fuel. (However, the true cause of the collapse is still being debated.) Of course, the use of iron metal for this purpose is all right under normal circumstances. But iron cannot be used for the nose of a spacecraft that returns back to the Earth. When the spacecraft enters the Earth's atmosphere, its nose heats up enormously, because of friction and somewhat air oxidation. Iron and other metals would melt. Only some ceramics withstand such a heat.

Ceramics are studied and used for such places that are subject to high temperatures, but many others have a variety of uses. Ceramics is defined as an inorganic, nonmetallic material processed or consolidated at high temperatures. Ceramics includes silicates, oxides, carbides, nitrides, sulfides, and borides of metal or metalloid. The traditional ceramics are mostly silicates as discussed earlier and used as a pot or similar purposes. But today ceramics are pursued as material for high-temperature, electric properties such as ferroelectricity, piezoelectricity, magnetic properties, high mechanical properties, and optical properties. In a word, they are pursued as HIGH-TECH material.

Some of the general properties of ceramics are (a) relatively light (less dense than metals), (b) resistant to high temperature (after all they are made at high temperature), and (c) brittle (though there are exceptions). Let us try to find why ceramics (in general) are so different from metals in these respects.

We have to remind ourselves of the structures of these substances (Fig. 10.3). A metal, say, copper, has in it an array of neutral copper atoms. Because each atom is electrically neutral, it would be relatively easy to change the relative positions of atoms; that is, to deform the structure of the solid metal requires a relatively

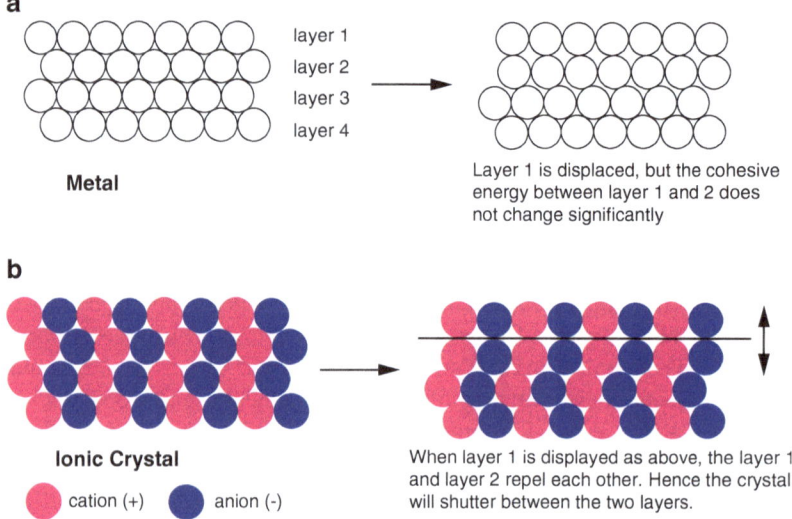

Fig. 10.3 Metal versus ionic crystal

small amount of energy. And even if relative positions have been changed, the cohesive energy among the neutral atoms would not change very much. Therefore, a metal is usually malleable; so that some of them can easily be made into a wire or a thin sheet.

What about a ceramic? Most of the ceramics are essentially ionic compounds. That is, they are made of positively charged ions and negatively charged ions. They are arranged in such a way that the lowest energy is attained. That means that attractive forces between particles of opposite electric charge are maximized, while the repulsive forces are minimized. If you change such an arrangement a little by an impact of hammer, repulsive forces will increase and attractive forces will simultaneously be reduced (see Fig. 10.3). The results may be that the repulsive forms overwhelm the attractive forces, shattering the crystal. This is the reason that ionic crystals such as table salts including many of the ceramics are rather brittle. [Exceptions include diamond (if you include it in ceramics), cubic boron nitride, and silicon carbide; these are not ionic crystals, but all the atoms in these materials are connected by covalent bonds throughout a solid. As a result, these materials are hard and difficult to shatter].

Ceramics are in general lighter, more precisely, less dense than metals. This can be attributable to two factors: the structures and the constituting elements. Firstly, most of the ceramics are ionic compounds, and the cations and the anions are arranged in such a way so that they are on average farther apart from each other than the case where the particles are electrically neutral (i.e., metals). This is to reduce mostly the repulsive forces. This increases the void space (between ionic particles) and hence makes such a crystalline solid less dense than otherwise.

Besides, many of the ceramics are made of elements lighter than typical metals. That is, typical ceramic-constituting elements such as oxygen, magnesium, aluminum, and silicon are relatively light; in the approximate molar atomic mass (g/mole), they are 16, 24, 27, and 28, respectively, as compared to iron (56), copper (63.5), and gold (197). The densities of some typical material are as follows (in units of g/cm^3): aluminum oxide (3.5–3.9), aluminum silicate ($3Al_2O_3 \cdot 2SiO_2$, 3.16), magnesium *ortho* silicate (Mg_2SiO_4, 3.2), iron (7.89), copper (8.92), and gold (19.3). Aluminum (2.7) and titanium (4.5) are two light metals that are used for airplanes, etc.

The gravest shortcoming of ceramics as an engineering material is therefore the brittleness. Besides, it is difficult to make advanced ceramic parts with uniform physical properties. The nonuniformity also degrades the mechanical and other (such as electric) properties. In order to improve the mechanical strength, ceramics are often combined with fibrous or whisker material; the resulting material is called "composite" material. The most often used fibrous or whisker material is silicon carbide (SiC). Silicon carbide fiber is made, for example, from polydimethylsilane ($-(-Si(CH_3)_2-)_n-$). It is converted to polycarbosilane ($-(-SiH(CH_3)-CH_2-)_n-$), which is then spun into fiber. The fiber is then heated in nitrogen at 1,250°C, turning into silicon carbide fiber. Whisker can be produced by heating rice hulls at 2,000°C. Whisker is a single crystal and stiffer and of a higher tensile strength than fibrous silicon carbide, which is polycrystalline. Boron nitride fiber, aluminum oxide fiber, and silicone nitride whisker are also used. Ceramic fibers (or whisker) improve fracture toughness of the associated ceramics by preventing tiny cracks from growing into large ones that may cause a stressed ceramic to shatter.

A single largest market for high-tech ceramics is based on their electric properties. Ceramics have interesting electrical properties including insulator, semiconductivity, super conductivity, ferroelectricity, and piezoelectricity. Ceramics have been traditionally used widely as insulator in large-scale applications such as electric transmitting towers and electrical sockets. The high-tech industry uses insulators such as aluminum oxide, beryllium oxide, and magnesium aluminate as substrates for integrated electric circuits.

Some ceramics are used for the material for capacitors, because they have high dielectric constants. Barium titanate ($BaTiO_3$) especially has a high dielectric constant, 100 hundred times greater than most other materials.

Diamond, Graphite, Graphene, Bucky Ball and Nanotube (Fun with Carbon)

11

11.1 Diamond and Graphite

"Diamonds are forever!" Are they really? We will learn later how true this statement is. Diamond is the hardest material, and that suggests sturdiness and long-lastingness. Yet amazingly, it is simply made of carbon atoms only. Coal is almost pure carbon. Graphite, an allotrope of carbon, is a better example in contrast to diamond because graphite is also pure carbon. Graphite is the material used in pencils among others. Diamond and graphite could not be more different. One is transparent, colorless and hard while the other is completely black and relatively soft. Diamond is much denser (density = 3.51 g/cm³) than graphite (2.25 g/cm³). Diamond is an electrical insulator, while graphite works like a metal, an electric conductor. Why are they so different, if made of the same carbon atoms and carbon atoms only?

To understand this difference we have to start with some basic ideas about the chemical bonds involving carbon atoms. Carbon, C, is the sixth element in the periodic chart, and has six electrons ($1s^2 2s^2 2p^2$ electronic configuration). Four of these electrons (in 2s and 2p orbitals) are the so-called valence electrons and are involved in bonding. [The two electrons in 1s orbital are so tightly held by the positively charged nucleus that it is hard to move them around; they are called "core electrons" and are not involved in bonding and chemical reactions]. As we discuss in the appendix (Chap. 19), a bonding between two atoms is made by a pair of electrons which is shared by those two atoms. Let us suppose that the type of bonding we are dealing with here forms when two atoms each contribute one electron to their bonding; this type of bonding is called "covalent bonding." Now carbon can bind as many as four atoms about it, as it has four valence electrons available.

One of the simplest such compounds is methane, CH_4, the major component of natural gas. The carbon atom is bound with four hydrogen atoms as shown below (Fig. 11.2). This shape is called "tetrahedral," because it forms a tetrahedron if adjacent hydrogen atoms are connected by imaginary lines (not "bond"). The connection, i.e., the bond between C and H (we simply express it as C–H), is made of two

E. Ochiai, *Chemicals for Life and Living*,
DOI 10.1007/978-3-642-20273-5_11, © Springer-Verlag Berlin Heidelberg 2011

Fig. 11.1 Graphite and diamond (from D. a. Mcqurrie and P. A. Rock, "Descriptive Chemistry" W.H. Freeman and Co., 1985), and an artificially synthesized diamond (called "HIME-diamond" http://www.47news.jp/CN/201012/CN2010121801000015.html)

electrons, one of which is contributed by hydrogen atom and the other by carbon. This type of bond between two atoms that is brought about by a pair of electrons is designated as a "single bond." Let us review. There are four bonds that require eight electrons altogether. Four of those electrons come from the carbon and the other four from the four hydrogen atoms, each of which has one electron.

Now replace one of the four hydrogen atoms with one carbon atom, which can bind, say, three hydrogen atoms. This will create $H_3C–CH_3$; this is ethane. You can replace one of the hydrogen atoms on the second carbon with another carbon. What you get will be $H_3C–CH_2–CH_3$: propane (these compounds are gases at ambient temperatures). You can continue this imaginary process, resulting in the real molecules. The compound with eight carbons is octane $H_3CCH_2CH_2CH_2CH_2CH_2CH_2CH_3$ (Fig. 11.1.). This compound is a major component of gasoline, and obviously a liquid at ambient temperatures. You can elongate the carbon chain further. Compounds with more than, say, 18 carbons become solid at ambient temperatures but melt at relatively low temperatures. These are "wax."

Okay, this time let us replace all the four hydrogen atoms of methane with carbon atoms; $CH_4 \rightarrow CC_4$. Since each carbon atom can bind four atoms, each of those four carbon atoms we put in can now bind three carbon atoms; that is, $CC_3(C\underline{C}_3)$. Each of the other three $\underline{C}$ atoms can bind with three C atoms. Of course there is no chemical distinction among C, C and $\underline{C}$. Figure 11.2 shows how it is done. You can continue this process over and over again; each and every carbon atom is tetrahedrally bound with four other carbon atoms throughout the whole space. This will lead to a structure that is quite rigid. And also it is transparent to light, as methane, gasoline and other similar compounds are. This is "diamond" in terms of chemical structure (Fig. 11.2). Why is diamond so "hard"? Well, what do you think? It is hard because it is hard (difficult) to break the bonds between carbon atoms and also distort that tetrahedrally bonded structure. Remember that each carbon atom is really very tiny. 1 Carat of diamond is about 0.2 g [which is 1.7×10^{-2} mol], and consists of approximately 10^{22} atoms of carbon. It is an enormous number. But the property of diamond can be understood, as expounded above, in terms of its structure, i.e., how each carbon is bound. This is the beauty of chemistry.

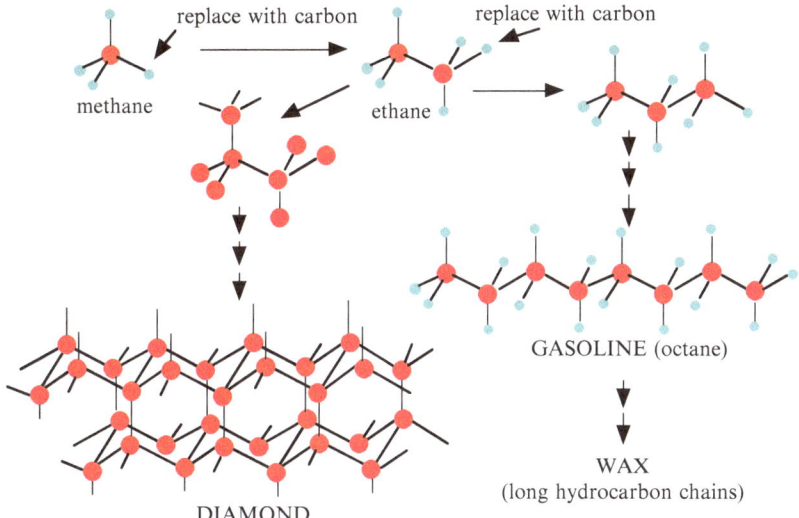

Fig. 11.2 How diamond might be constructed from methane – a schematic conceptual process, not the real process

Let us turn now to "graphite." This time we will start with ethane $H_3C–CH_3$. Suppose that we break one of C–H bonds on each carbon. The result will be $H_2C^·–^·CH_2$. Here the dot indicates an electron. So we have now an electron on each carbon atom. These two electrons can pair up and form a sort of bond between the two carbon atoms. This bond now formed is not exactly the same kind of bond already existing between carbons. The first bond, i.e., present in ethane, is called "σ"-bond (σ: sigma), whereas the newly formed bond is called "π"-bond. We can write it as $H_2C=CH_2$, with an understanding that one bond is of σ type and the other π, and we say that there is a double bond between the two carbons. This compound is called "ethene," but more commonly known as "ethylene." This second bond (π) is not as strong as the σ-bond (see Fig. 19.9).

Benzene is an interesting compound that contains three each of single and double bond. Its structure is shown in Fig. 11.3. It is hexagonal, and has three alternate double bonds and single bonds, as you see in (a). There is an alternative structure (b). The real benzene is neither (a) nor (b); it is a sort of mixture of (a) and (b) (see Chap. 19 for more details). You might picture that the six electrons involved (two in each of the π bond of the double bonds) are moving around above and below the hexagon. This suggests that these electrons are more readily movable than those involved in the σ-bonds. There are a great variety of compounds that stem from benzene. For example, you can fuse two benzene rings together as shown in Fig. 11.3c. In this imaginary process, you will remove four hydrogen atoms and two carbon atoms from the system. You can fuse another, and another, as you see in Fig. 11.3. The hydrogen atoms in periphery of the fused ring system are omitted in the figure. The ratio of Hs and Cs is 1.0 in the case of benzene, 0.8 in naphthalene

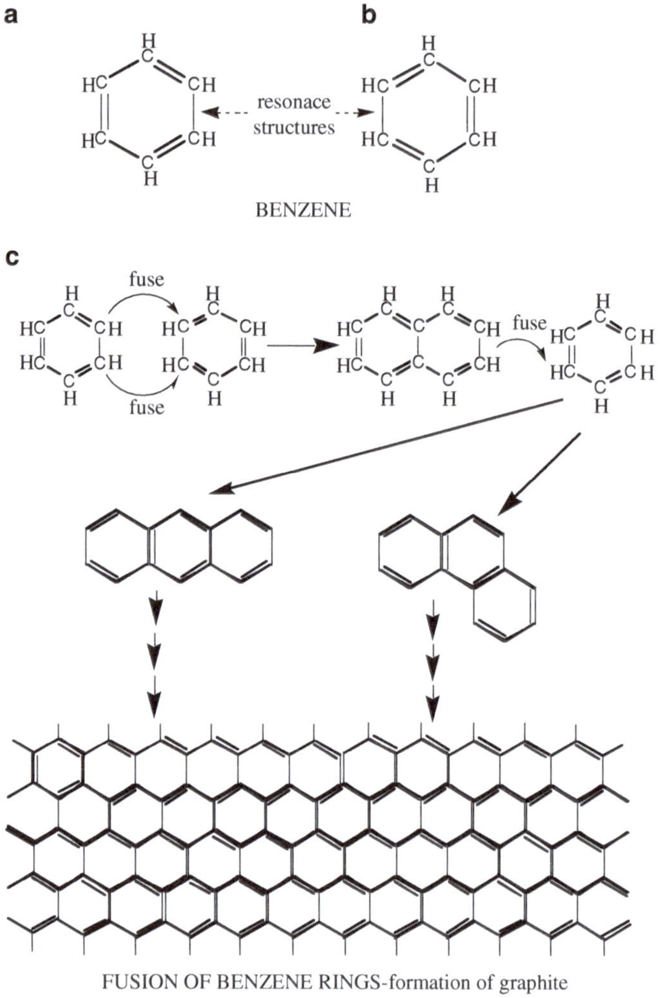

FUSION OF BENZENE RINGS-formation of graphite

Fig. 11.3 (**a**, **b**) Benzene molecule; (**c**) a conceptual picture of how graphite is formed as the fusion (condensation) of benzene rings

(two-ring), 0.714 (=10/14) in anthracene (three rings fused in a straight manner), and so on. As you increase the degree of fusedness (condensation) in this manner, this ratio (H/C) would become smaller and eventually it would become virtually zero, when you have had so many benzene rings fused together. It is noted, though, that hydrogen atoms are still attached to the periphery of the sheets of fused benzene ring. Because we are dealing with an almost infinite number (say, 10^{16} rings or so) of rings in a sheet, we can ignore a few hydrogen atoms on its periphery; that is, this can be regarded a pure carbon. [By the same token, the surface of diamond should have hydrogen atoms bound. It does. Only the number of hydrogen thus bound is negligible in comparison with the number of carbon atoms of the bulk of diamond.

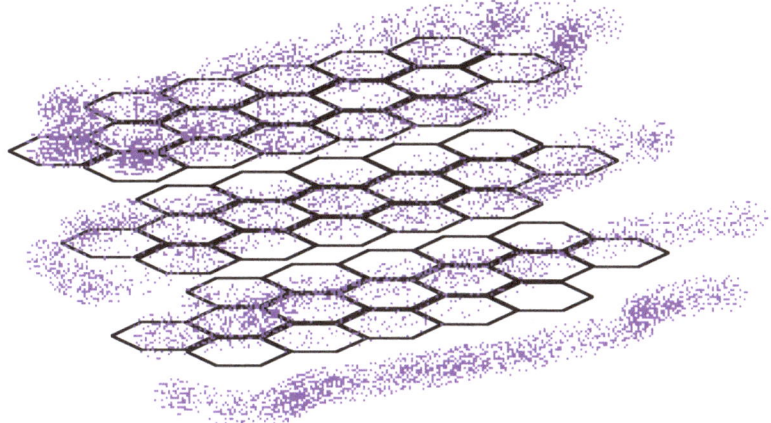

Fig. 11.4 Honey-comb layer structure of graphite and associated electron clouds

Fig. 11.5 An STM image of graphene (from E. Stolyarova et al. Proc. Natl. Acad. Sci. (USA), 104 (2007), 9209)

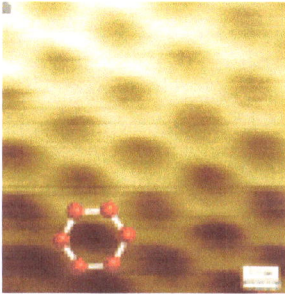

If you remove these hydrogen atoms, though, the property (friction coefficient, for example) of diamond surface changes significantly.]

So you would get a sheet of honeycomb mesh (Figs. 11.3c and 11.4). Indeed a recent STM image of graphite single layer (called "graphene," see Sect. 11.2) attests the honeycomb structure as shown in Fig. 11.5. And a cloud made of a large number of electrons (in π-bonds) hovers above and below the sheet. Then imagine that you have a stack of a very large number (say 10^8) of this sheet. And that is "graphite" (Fig. 11.4).

The nature of chemical bonding and electrons involved and the structure just described account for the chemical and physical properties of graphite. The bonds connecting carbons together within a sheet plane are of the regular kind (i.e., σ-bond) and very strong and hence difficult to break, the interaction between the sheets is not of the regular chemical bond and is relatively weak. Henceforth, it is easy to slide the sheets from each other. This is the reason that graphite is softer (than diamond) and feels slimy when you rub graphite powder between fingers. As the interaction between the layers is relatively weak, the interlayer distance is also relatively large

Fig. 11.6 Conversion of graphite to diamond. Colors are used to identify which atoms become which. The *small arrows* indicate movements of individual atoms' (relative) movements in the process

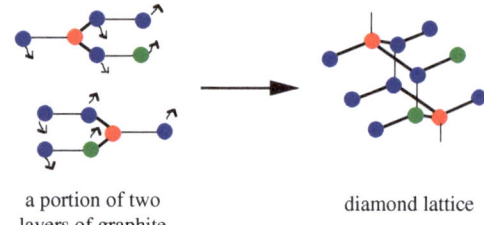

a portion of two layers of graphite

diamond lattice

compared to the regular chemical bond. This explains that graphite is less dense than diamond. In diamond, all carbon atoms are chemically bound (σ-bond) and fairly densely packed. As the electron cloud hovering on the condensed rings is fairly easy to move around, graphite can conduct electric current like a metal. On the other hand, all the electrons are tightly confined between carbon and carbon atoms in diamond and cannot be moved easily; hence diamond is not an electron conductor. The easily movable electrons also readily absorb light in visible range. Graphite absorbs all lights of visible range, and as a result, it looks "black." Whereas diamond cannot absorb light in the visible range and hence is transparent, as it lacks easily movable electrons. Thus, both diamond and graphite are pure carbon, and yet they are so different.

Can you convert graphite to diamond? Yes, you can, but it is not easy. You can appreciate the difficulty from their structural differences. As you see in Fig. 11.6, you have to distort almost all the carbon–carbon bonds in graphite and then recombine some carbon atoms in an entirely new arrangement (of diamond) in order to do so. Because the C–C distances, particularly the distance between the honeycomb layers, in graphite are larger than the C–C bond distance in diamond, this process to convert graphite to diamond would require a very high pressure to compress the graphite so that this rearrangement of bonds may be enhanced.

It turns out that graphite is the stable form under ambient conditions (temperature and pressure). This implies that diamond would spontaneously turn into graphite under the ambient condition. But do not worry. The process is so slow that "diamond is virtually forever," unless you keep it at high temperature, say 2,000°C. In that case, diamond may turn into graphite in tens of millions of years. To turn graphite into diamond, you have to apply a very high pressure, higher than 10,000 atmospheric pressures as we saw above. Yet it is very slow under such a high pressure and even at high temperatures. It took millions of years to form diamond under geological high pressures and temperatures deep in the crust. Then the diamond-containing layer had had to quickly rise up close to the surface. Otherwise, diamond may have turned back to graphite as it rose to less pressured places. Such geological conditions did not prevail widely. Hence only a few areas produce diamond, notably in South Africa, Congo, and Russia. The South African diamond is believed to have crystallized about 140 km deep in the crust at 900–1,200°C about three billion years ago. By the way, where do you think carbon originally came from? It has been shown that much of carbon of the diamond on the Earth originated from a pre-solar supernova.

General Electric Company invented a method to artificially produce diamond from ordinary carbon sources, organic compounds. It requires a high temperature like 2,000°C and a high pressure, 10^5 atmospheric pressures. Theoretically, they can produce diamonds of any size such as gemstone class, but it is too expensive. Instead they produce mostly small particles of diamond for industrial uses, such as abrasives. However, it turned out that diamond can also be produced under a much milder condition by a method called "chemical vapor deposition" (CVD). When methane CH_4 is decomposed, carbon atoms and hydrogen atoms are produced. Carbon atoms in the hydrogen atom stream then are deposited on a surface maintained at 600–900°C. Carbon atoms somehow bind each other in the diamond structure and form diamond film under this condition. It has been suggested that crystal growth of the diamond structure is faster than that of the graphite under this condition. Once diamond is formed, it is very slow (i.e., negligible) to turn into graphite even under this condition. [Remember that graphite should be the stable form under the condition]. Hence, a diamond film is formed. This method is now widely used to form a very rigid, strong coating on a number of articles.

In 2010, Japanese researchers at Ehime University together with Sumitomo Electric Co. developed another method to produce large diamond crystals (shown in the right hand side of Fig. 11.1) named as "HIME diamond."

11.2 Graphene

Graphite is made of a large number of honey-combed planes of carbon atoms, as described above. Then can we separate the planes, that is, can we make a separate sheet of single layer? The sheet of a single layer of carbon atoms is called "graphene." Theoretically it should exist, though it had been thought to be difficult to separate layers of graphite. In 2004, a group of scientists (Andre K. Geim and K. S. Novoselov of University of Manchester, England) succeeded to make such a sheet. They were awarded the Nobel physics prize in 2010. The method is surprisingly simple. A graphite sample was sliced mechanically as thin as 20 layers. The thin tiny sample (crystal) was then stuck between two adhesive tapes, and the tapes were pulled apart. This cleaved the thin crystal in two. This process was repeated several times, and finally the resulting slices were inspected. They found that some slices were indeed a single layer (one atomic thin) of graphene. Another group used transmission electron microscopy (see Chap. 21) to show that it is a single layer of carbon atoms arranged in the honey-combed manner, but it was not quite flat; it is slightly warped. A scanning tunneling microscopic image of graphene was obtained, as shown in Fig. 11.5. Such a single-layered graphene turned out to be surprisingly stable at room temperature.

Other methods to produce graphene in a large quantity have been hotly pursued. One method uses the same principle of making pancake. Graphite is highly oxidized chemically; this produces OH and other oxygen-containing groups attached to the carbon atoms of graphite. Then heating such graphite to a high temperature decomposes the oxygenated carbons producing carbon dioxide. CO_2 thus produced expands the space between layers, eventually separating layers. This is exactly what happens when a blob of flour mixed with baking soda (sodium hydrogen carbonate) is heated; the CO_2 produced from baking soda expands the pancake.

Graphene has a unique electronic character. As learned above, the carbon flat layer has loosely bound electron clouds above and below the plane. Those electrons move relatively freely, affording the electric conductivity to the graphite. Now, the same thing can be said with regard to graphene. A difference is that there is a restriction with the electron movement in graphite as electrons are confined between layers, whereas there is no such restriction with graphene. Hence, the electric conductance of graphene is expected to be much better than that of graphite. It has been shown that it is indeed the case.

11.3   Buckminsterfullerene and Nanotubes

11.3.1 Buckminsterfullerenes

Up until 1985, diamond and graphite were the only known elemental forms of carbon. In that year, Harry Kroto, an astrochemist, of Sussex University (UK) came over to Rice University in Houston, Texas and asked Richard Smalley, a physical chemist, to help him investigate experimentally how large cosmic compounds that had been discovered in the interstellar cloud might form (see Chap. 13). The Rice professor had developed lab techniques to produce carbon vapors (by lasers) in the stream of hydrogen or other gases. Smalley had been noticing from other investigators' works (notably those done at Exxon Research labs) that there are some unusual characters with carbon clusters. [A cluster is an aggregate of atoms, usually of the same kind, in a more or less ordered fashion]. They set out the experiments and indeed managed to produce the cosmic compounds in the lab, but in addition, they discovered an unusual cluster. The compounds obtained from the lab were analyzed by mass spectrometer, which determines the mass (in terms of atomic mass, see Chap. 19) of compounds. They found a prominent peak that corresponded to 60 carbon atoms C_{60}, that is, 720 g in molar mass. They suggested a structure shown in Fig. 11.7 for the cluster (1985). This structure was verified shortly afterward, not directly but using a derivative. Note that it is made of 12 pentagons (5-membered rings) as well as 20 hexagons (6-membered rings). The 6-membered ring is similar to the unit found in graphite. The cluster has been named as "Buckminsterfullerene," for the structure resembles the geodesic structure proposed and realized at 1967 Montreal Expo by Buckminster Fuller. It is also called "bucky ball" or "chemical soccer ball" for short. This discovery has created an enormous interest in carbon clusters. It had been noticed in the earlier works at Exxon Research that a large number of even-numbered carbon clusters with C_{40}–C_{200} do form in the laser vaporization of graphite. C_{70} is another prominent cluster among others that is now known to be as large as C_{400}. "Fullerens" are now used as the general names for C_{60} and other similar carbon clusters and their derivatives. In 1990, a German physicist together with University

Fig. 11.7 Buckmillerful-
erene (Bucky ball)

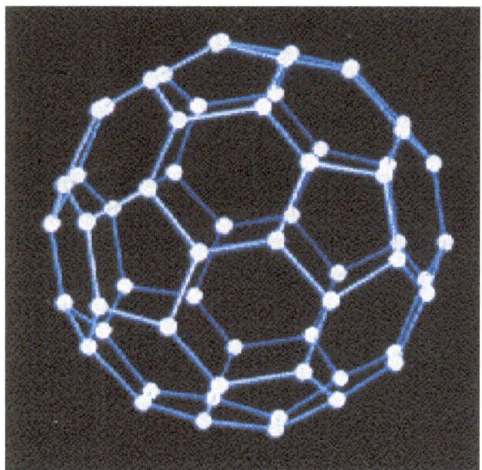

Fig. 11.8 Carbon basin
(T. Scott et al. J. Am. Chem.
Soc., 118 (1996), 8743)

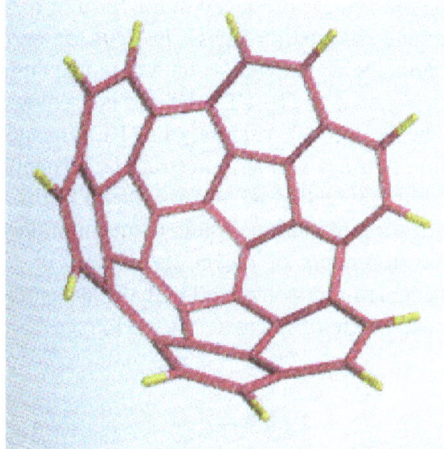

of Arizona physicists developed a relatively simple method to produce a large amount of C_{60} cluster. This procedure uses "arc" to vaporize carbon. As a result, C_{60} is now commercially available in grams-quantities. Kroto, Smalley, and Charles Curry shared a Nobel prize in 1995.

L. T. Scott, professor of Boston College and his coworkers reported in 1996 that they synthesized a stiff fullerene-like bowl of $C_{36}H_{12}$ (Fig. 11.8). These are made by heating decacyclene (a long-known flat aromatic compound) in vacuum. This contains more than half of carbon atoms of fullerene, and may be regarded to be a bridge between the old flat aromatic systems (as in graphite and simpler condensed aromatic compounds) and the new round aromatic systems of fullerenes.

11.3.2 Nanotubes

A Japanese physicist S. Iijima working at NEC Lab (Nippon Electric Company) discovered in 1991 a multi-layered carbon cluster tube among the graphite arc products. In 1993, he and a group at IBM research (headed by D. S. Bethune) independently found that they could make single-walled carbon cluster tubes by vaporizing carbon together with a transition metal such as iron or cobalt in an arc. As the diameters of these tubes are of the order of ten to several ten nanometers (nm), they are often referred as "nanotubes." "Nanometer" is one-billionth of a meter and the sizes of many molecules are of the order 1–100 nm. A nanotube may be regarded as a graphene cylindrically wrapped.

Smalley and his coworkers found in 1995 a more effective method to prepare single-wall nanotubes. It uses laser vaporization of a graphite-metal composite rod in a flow tube heated at $1,200°C$. The most effective metal is 50/50 mixture of cobalt (Co) and nickel (Ni). Apparently the nanotubes grow in the gas phase and are deposited on a cooled surface. This method produced in fairly good yield (70–90%) single-walled nanotubes of a remarkably uniform diameter ($13.8\,\text{Å} = 1.38$ nm). The nanotubes are produced in the form of ropes hundreds of micrometers long. A theoretical calculation (by G. E. Scuseria and C. Xu) suggests that the most favorable nanotube is one with a chain of ten hexagons around the circumference. This tube is designated as (10,10) tube and is predicted to have a diameter of $13.6\,\text{Å}$ (1.36 nm). The theoretical structure of a (10,10) nanotube is given in Fig. 11.9a. The visualization of the actual atomic structure of nanotube has been accomplished recently by a technique called STM. It is shown in Fig. 11.9b; compare them. This kind of visualization of atomic arrangements in molecules has become relatively easy, thanks to developments of clever techniques in recent decades, and has led scientists to become more convinced with the atomic/bonding theories and the atomic/molecular description of the material world (see Chap. 21).

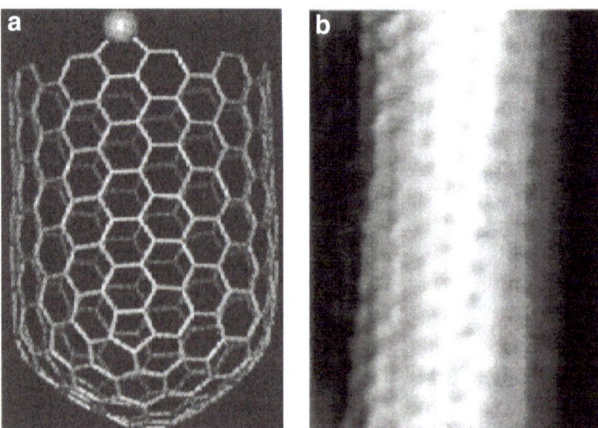

Fig. 11.9 Nanotube: (**a**) model; (**b**) STM (from L. C. Venema et al. Appl. Phys., A66 (1989), S513)

The electrical resistivity of the nanoropes was measured to be 3×10^{-5} ohm cm minimum, which is comparable to that of copper wire; that is, the carbon nanotube is a good electric conductor. You would have expected that judging from the structure, which consists of condensed aromatic rings as in graphite. Now a hope is up to make practically usable wires made of carbon nanotubes.

One of the basic problems for applications of carbon nanotubes is its high cost of production. In 2004, it cost about \$500 per gram; that is 30 times of gold. A Japanese group of scientists including S. Iijima, the originator of nanotube, found a way to produce a larger quantity of nanotubes of better quality. This finding is expected to reduce the cost of nanotubes. The nanotubes are produced by decomposing a hydrocarbon such as ethane on a metal catalyst, as talked about before. It tends to simultaneously produce amorphous carbon that is useless and covers the catalyst surface and reduce the catalytic activity. They found that if a weak oxidizing agent, water, is present, it oxidizes the amorphous carbon (burn it away), but leave the nanotubes intact. This process produced fairly long tubes of high quality relatively easily.

"Nano-something," e.g., "nanotechnology" is a collective prefix to indicate the dealing of molecular entities, not as an aggregate of molecules, but individual molecules. This notation applies now to science and technology of any molecules, not necessarily of carbon. Nanotechnology is now hotly pursued by the chemical industry to find new material for a variety of purposes from electric wires to cosmetic substances.

Perfumes

12

Odors can affect human psyche. A wonderful aroma from freshly brewed coffee will make us feel awakened and alive. A smell of a well-prepared dish waters our mouth. A perfume worn by a woman can even elicit sexual desire. On the other hand, putrid vegetables and meat make us sick.

Odors are brought about by chemicals. A chemical will waft through air and arrive at a receptor in our nose, and starts the sensation of smell. This suggests that odor-causing chemicals must be fairly volatile. That is, they have to have significant vapor pressures (or a significant number of molecules in the air). Therefore, the chemicals that cause this sensation are relatively small molecules.

How the binding of such a compound to an olfactory receptor causes the sensation of smell is physiology, and the whole physiological process itself is also dependent on chemistry. An odorant molecule binds to an olfactory receptor, which is similar to the receptors for hormones and neurotransmitters. It involves the so-called G-protein and cyclic ATP. There are about a thousand different olfactory receptors in human beings. They can distinguish remarkably similar compounds. For example, there are two compounds having the same structure: 1-methyl, 4-(2-propenyl) 6-on cyclohexene-1; they are merely mirror images to each other. Yet we smell different odors for them. However, the physiology and its mechanism are beyond the scope of this book (see for example: R. Axel, *Scientific America*, Oct. issue, 154–159 (1995); http://users.rcn.com/jkimball.ma.ultranet/BiologyPages/O/Olfaction.html), and we will be focusing on the chemical natures of odorous compounds.

12.1 What Kinds of Compound Give Odor?

Some compounds, though gaseous or very volatile, do not cause smell; that is, they are odorless. For example, oxygen (O_2), nitrogen (N_2) carbon dioxide (CO_2), and carbon monoxide (CO) gas are odorless. These compounds are small and rather non-polar and hence would not effectively interact with the olfactory receptor site.

E. Ochiai, *Chemicals for Life and Living*,
DOI 10.1007/978-3-642-20273-5_12, © Springer-Verlag Berlin Heidelberg 2011

It is rather unfortunate, though, that carbon monoxide is odorless, because it is a deadly gas. Other small inorganic gaseous molecules including ammonia (NH_3), nitrogen dioxide (NO_2), sulfur dioxide (SO_2), and hydrogen sulfide (H_2S) are polar and rather reactive and hence probably interact well with the olfactory receptor; thus they are all strongly smelly. Chlorine molecule (Cl_2) is itself non-polar but very reactive and hence gives a characteristic pungent odor. It is interesting to note that water molecule has properties to interact with many compounds, likely including the olfactory receptor molecules, and yet water is odorless (and tasteless). Could you think reasons why water is odorless?

Most inorganic crystalline compounds including salt (NaCl) and rocks (e.g., metal silicates) are odorless because they are solid at ordinary temperatures and do not give a significant amount of molecules in the gas phase; i.e., they are not volatile at all, and hence cannot come to our nose.

Most other compounds, particularly, organic compounds do smell. The simplest organic compounds are called "hydrocarbons," which are made of only carbon and hydrogen atoms. Hydrocarbons were also discussed in the previous chapter in connection with diamond and graphite. The smallest of hydrocarbon is methane (CH_4), which is odorless. The city gas in many areas consists mostly of methane. Usually it is tainted with a small quantity of foul-smelling chemical such as propane thiol in order for a leakage of the gas to be detected easily. The other small hydrocarbons include ethane (C_2H_6), propane (C_3H_8), butane (C_4H_{10}) and pentane (C_5H_{12}); they are odorless like methane. These are chemically non-polar and do not interact with anything else, let alone the olfactory receptor; hence odorless. Hydrocarbons larger than that, i.e., hexane (C_6H_{14}), heptane, octane, nonane, etc. are liquid at room temperature and do smell, though only faintly. (By the way, by now, you may have figured out the general chemical formula for these hydrocarbons called "alkanes"). A larger alkane hydrocarbon, though not polar, can interact with each other or another compound more effectively through London dispersion force as talked about in Chap. 19. As the carbon number increases, the compounds would become solid at ordinary temperatures; they are wax. The smell of wax is faint, as the compounds in wax are not very volatile. But the burning of wax often gives a strong odor, particularly the moment it is extinguished. The odor is very likely some incompletely burnt compounds such as aldehydes. The scent from decorative wax products is usually because of fragrant compounds added.

Another series of hydrocarbons exemplified by ethylene (ethene=C_2H_4, CH_2=CH_2) and propene (C_3H_6, CH_3–CH=CH_2) has a more reactive site, i.e., C=C double bond, so that they may significantly interact with a receptor site. They do indeed smell.

Benzene C_6H_6 has a shape of hexagon and has alternating single bonds and double bonds (see the previous chapter). The compounds derived from benzene (in the formal sense) are called "aromatic" compounds. The name itself implies that they have special odor.

Now let us replace the hydrogen atoms in hydrocarbons with other elements such as chlorine (Cl), oxygen (O), nitrogen (N), and sulfur (S). These elements are more electronegative (attracting electrons more strongly) than carbon atom. As a result, the electrons are not equally distributed between carbon and an atom X of one of these elements. That is, the electrons are more heavily distributed on X atom, and hence such a bond between carbon and X is polarized, and atom X is somewhat negatively charged (and the carbon atom carries a positive charge correspondingly). In addition, these atoms, Cl, O, N, and S have extra electron pairs (called "lone pair") on them when they constitute a part of carbon-containing compounds. Because of these two properties (i.e., polarity and lone pairs), these compounds interact fairly strongly with each other and other compounds, and hence the organic compounds containing these elements usually give strong odor.

Let us illustrate this point by a few examples. Chloroform $CHCl_3$ (i.e., three hydrogen atoms of methane are replaced by chlorine atoms) has a special scent, and was used as a general anesthetic. C_2H_5OH, ethanol, (one of the hydrogen atoms of ethane is replaced by OH group), is the main ingredient of alcoholic beverage. Everybody recognizes its smell. One can also easily recognize the smell of acetic acid, CH_3COOH, the main ingredient of vinegar. The compound in which the penultimate hydrogen atom of acetic acid is replaced by an ethyl group (C_2H_5) is called ethyl acetate, $CH_3COOC_2H_5$. It is used as nail polish remover, the strong odor of which many women might be familiar with. As a matter of fact, many of the fragrant substances used for perfumes are oxygen-containing organic compounds, as we shall see soon.

Nitrogen or sulfur-containing organic compounds have usually very strong, mostly unpleasant odors. You might have smelled the odor of ammonia (NH_3, this is not an organic compound, though), as it is one of the main ingredients of glass window-cleaning solutions. Similar (organic) compounds are called "amines" with NH_2 group(s); they all give unpleasant smell, provided that they are relatively volatile. Pyridine, which has a composition of C_5H_5N and a structure similar to benzene, has a characteristic malodor, which you cannot forget once you have smelled it.

Sulfur-containing compounds are usually worse in odor than nitrogen-containing compounds. The technical term for sulfur is "thio-." The most infamous malodorous material is the spray by skunk. The major components of the spray of the common skunk (*Mephitis mephitis*) are 2-butene-1-thiol ($CH_3CH=CHCH_2SH$) and 3-methyl-1-butanethiol (($CH_3)_2CHCH_2CH_2SH$). Most of the foul-smelling components of garlic are sulfur-containing compounds, diallyl thiosulphinate ($CH_2=CHCH_2-S(O)-S-CH_2CH=CH_2$) and its derivatives. Diallyl thiosulphinate is the immediate odor-causing compound when you grind a clove of garlic. It rapidly decomposes and the long-lasting odor is caused by its degraded substances including methyl mercaptan (methane thiol). Garlic also contains foul-smelling selenium compounds. Selenium is a relative of sulfur, and has properties similar to those of sulfur. Hydrogen sulfide H_2S mentioned earlier gives that odor of the rotten egg most of you are familiar with. This compound is not an organic compound, though.

Sulfur-containing compounds such as furylmethane thiol are found in roasted coffee. At very low concentrations, they give an odor reminiscent of freshly roasted coffee. During storage or further processing, the concentration of these chemicals increase and coffee loses its characteristic aroma. Coffee contains, in addition, a large number of volatile (odor-causing) organic compounds such as biacetyl and aldehydes. These compounds also contribute to the aroma of coffee.

12.2   Perfumes

We are all familiar with pleasing scents given off by some plants, flowers, and some animal tissues. These scents are due to chemicals used for communication among organisms. We, human, have several means of communication: auditory, visual, and others including language. Some animals use sounds and/or visual effects (like mating ritual) as means of communication. Living organisms also use chemical substances as means of communication. Social insects such as ants and bees use chemical signals to organize their activities. The chemicals used for communication between individuals of the same species are called "pheromones." Sex pheromones are most widespread. Male moths can detect females by smell at a range of many kilometers. Musks that are obtained from musk deer and civet cat can be classed as "pheromone" and are used in certain perfumes. Flowers send chemical signals of scent (or colors) to attract insects, because they need to be pollinated. Thus, these chemicals are technically called allelochemicals and benefit both parties. Jasmine is a typical fragrant flower, the extract of which is used widely in making perfumes. Rose, lily, geranium, and orange fruit are a few other examples, from which perfumery materials are obtained. It should be pointed out here that a similar term, "allelopathic chemicals" is used to indicate warfare chemical agents that are produced and emitted by insects, microorganisms, plants, and fungi. An allelopathic chemical may give some advantages for its emitting species and harms the organisms that receive it.

Several typical examples of naturally occurring fragrant compounds, that is, synthesized by plants and flowers and other living organisms, are shown in Fig. 12.1. Geraniol-nerol and citronellol (and their derivatives) are responsible for the rosy floral scents. Linalool causes the floral scent of lavender and bergamot. Linalyl acetate gives fruity odor. Methyl ionone is found in iris-violet. Menthol may be familiar to you; the odor of mint comes from it. Pine scent is caused by terpineol, borneol, and their derivatives.

It turns out that they are chemically related to each other, as you can imagine from the chemical structures shown in Fig. 12.1. They are called "terpenes" and "terpenoids." They can be regarded as derivatives from a five-carbon compound called isoprene (2-methyl butadiene). Two isoprene molecules combine to form "mono-terpene" (ten-carbon compound). The fragrant oils mentioned above are all the derivatives of mono-terpene. Terpenes are derived from a common metabolic intermediate of glucose, acetyl-CoA (coenzyme A). By the way, a tri-terpene (which three terpene molecules combine to form) called squalene leads to the formation of steroids, and if you connect a large number of isoprene in a linear fashion, you will get "natural rubber" (Chap. 5).

The fragrance of jasmine is due to several compounds including jasmone, methyl jasmonate, benzyl acetate, and indole. Jasmone and its relatives derive from an important lipid called "arachidonic acid." Arachidonic acid is the starting compound from which a number of physiologically important compounds including prostaglandins are obtained.

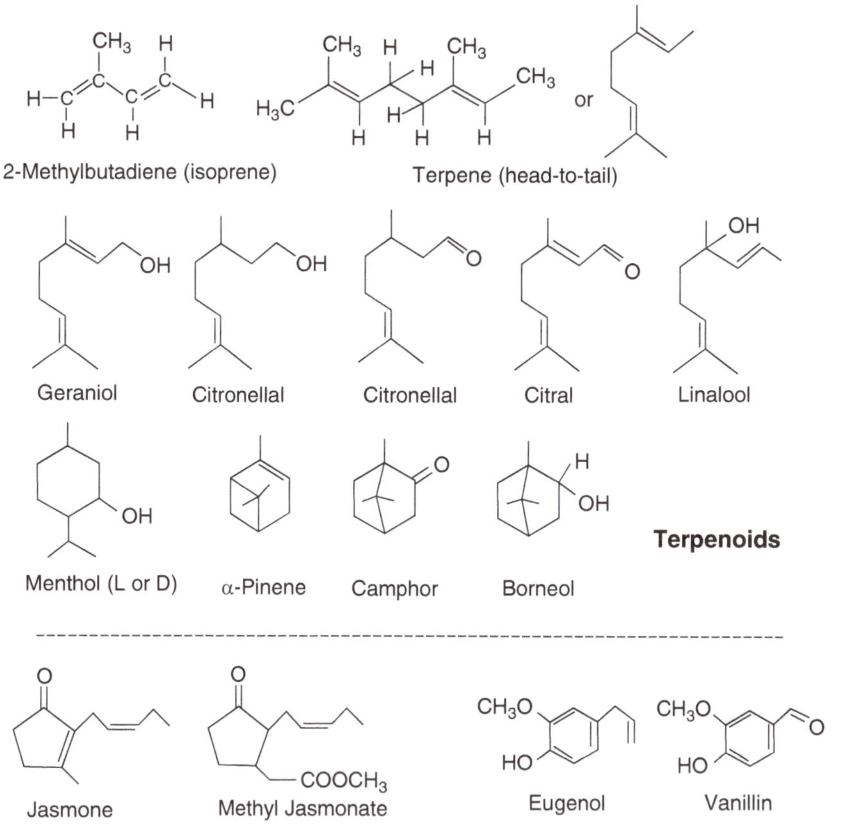

Fig. 12.1   Terpenes and others

Once the chemical natures of these natural compounds had been determined, chemists could start artificially making similar compounds and modifying the natural compounds. That is what synthetic organic chemists excel at. Perfumes used to be made from naturally occurring compounds such as those mentioned above. Today, however, artificially synthesized compounds are used in conjunction with natural ones. Some synthesized compounds have no resemblance in structures to natural compounds, but imitate the olfactory impressions of natural ones. For example, α-amylcinnamaldehyde reminds us of the jasmine odor, and 4-*tert*-butylcyclohexyl acetate imitates woody, violet odor.

The famous Chanel No.5 was the first and still is the commercially most successful perfume that used synthetic fragrant chemicals in addition to those from natural sources. Coco (Gabriel) Chanel asked her perfumer Ernst Beaux to produce several feminine perfumes. He presented to her ten products numbered, out of which Ms. Chanel picked No. 5, because "5" was her lucky number. Chanel No.5 was further popularized by Ms. Marilyn Monroe. It uses jasmine oil obtained from jasmine trees in South France, rose oil and oil from lily of the valley, vanilla, amber, musk, and some synthetic aldehydes.

Today, perfumes are made in analogy to, say a symphony. It consists of three movements; top note (first movement), middle or body note and bottom note or fixation. The top note consists of most volatile components and gives the initial sensation; this is what you smell upon opening a bottle of perfume. When you put a perfume on your skin, the bottom note lasts after a few hours. In between the middle note dominates and defines the character of fragrance. A typical perfume is made of 10–20% fragrance material dissolved in 90% or so alcohol (ethanol). An example of the composition of a rose-based perfume is: (a) top note (about 5% of the fragrance material) consisting of decanal and 2-undecylene aldehyde (in diethyl phthalate) and ω-denenol; (b) body (about 80%) consisting of phenethyl alcohol, citronellol, geraniol, phenethyl diethylacetate, nerol farnesol, etc.; (c) fixation (about 10%) consisting of phenethyl phenylacetate, cinnamic alcohol, and tricloromethylphenyl carbinyl acetate.

Part IV

What Are the Earth and the Universe Made of?

Lives and living systems on this earth are all dependent on the chemicals available on it. What are those chemicals and how do they behave? The earth is a small planet belonging to the solar system. The solar system is a minor part of the galaxy. Million different galaxies constitute the universe. What is the universe made of? Chemicals for sure. Some prominent features of chemicals found on this small earth and in the huge universe are very briefly talked about here.

Chemistry of the Universe: What Is It Made of?

13

13.1 Introduction

The simplest answer to the question in the title of this part is that the Universe and our Earth are all made of *chemicals*. We will talk about only the inanimate material world, though we ourselves, the living things, are also chemicals and a part of the universe. We talked about the chemistry of living things in other parts of this book. No attempt is made here, however, to give a comprehensive discussion of the issue. Besides, there is a lot to be learned yet. We try to give you some basic ideas for understanding the chemistry of this big universe and the tiny planet Earth.

13.2 Formation of the Elements

We are here together with all other organisms on this planet, Earth. We are using a lot of chemical material: proteins, carbohydrates, iron, oxygen, water, and so on. All of these chemical materials including our own bodies are made of atoms and molecules. Our body is made of about 2×10^{27} atoms. The earth itself is made of about 1.1×10^{50} atoms. So many atoms! But there are only a hundred or so different kinds, i.e., elements (see Chap. 19). Our bodies use about 30–40 of them (see Chap. 6). Well, where have all these elements (atoms) come from?

It is a long story, literally. We have to go back to the beginning of the universe. That is believed to be about 12–15 billion years ago. Our own Earth is believed to be about 4.6 billion years old. We human beings (*Homo sapiens*) are at most only about two million years old collectively. Fifteen billion years (the history of the universe) is then about seven thousands times as long as our *Homo sapiens'* history.

According to the current theory, i.e., the so-called Big Bang theory, there was no material at the beginning (the beginning of time). There was a kind of fireball, all energy. Somehow that fireball (at a very high temperature) all of a sudden exploded and expanded rapidly. As it expanded and cooled, the energy started to turn into material, atoms; "energy turned into material"; literally "materialized." You may

E. Ochiai, *Chemicals for Life and Living*,
DOI 10.1007/978-3-642-20273-5_13, © Springer-Verlag Berlin Heidelberg 2011

recall the famous Einstein's equation $E = mc^2$; this equation tells us that energy (E) and mass (m, i.e., material) are equivalent. [c is the speed of light]. The processes that created elements subsequently are all "nuclear reactions" (reactions at the level of nucleus) as opposed to "chemical reactions." Please review Chap. 19 for the fundamental difference between nuclear reactions and chemical reactions.

The first atom that formed from the fireball was the simplest one, hydrogen (H) or rather its nucleus. Hydrogen is still the most abundant element in the universe. Hydrogen atom (nucleus) is made of a single proton (p = $_1H^1$), an electrically positively charged nucleon. Another nucleon is neutron that is electrically neutral, but has about the same mass as proton. What were there before protons and neutrons? They were quarks; quarks combine to form neutron and proton. Chemistry can be understood at the level of neutron/proton. We do not need to look at quarks, as far as chemistry is concerned. A proton and a neutron combined to form the heavy hydrogen, deuterium (D or $_1H^2$, mass number is 2). A large cloud of hydrogen and deuterium atoms, if sufficiently large, starts to shrink under its own gravity force. This raises the temperature enough (as high as 10–20 million degrees °K) to allow the nuclear reactions between hydrogen atoms and deuterium atoms to take place. This leads to the formation of helium (He) atoms ($_2He^4$, positive electric charge = 2 (units), mass number = 4 (units)). This process, formation of helium from hydrogen and deuterium, is an example of nuclear fusion reactions, and is what is happening in a star like our own Sun. This process forms a very stable atom helium He. The stable state means a state that has a low energy content. Hydrogen atoms have higher energy content than the resulting helium atoms. Hence, when hydrogen atoms turn into helium, that process releases an enormous amount of energy. In this process, some mass of hydrogen and deuterium is lost as energy (that is, "mass turns into energy (or dematerialized)"). This energy is what we get daily from the Sun, the ultimate source of energy for all our activities on the Earth. Helium is the second most abundant element in the universe.

As a star becomes older, helium atoms accumulate in its core. If the star is large enough, the core becomes as dense as 100 kg/cm^3 and the core temperature as high as 100 millions degree (°K). This is a red giant star. Other fusion reactions can now take place in this star. In this case, helium atoms fuse together to form larger atoms, particularly carbon atoms and oxygen atoms. Three helium atoms combine to form a carbon atom ($_6C^{12}$). The reaction is $3 _2He^4 \rightarrow _6C^{12}$. In this process as well as all other nuclear reactions, the electric charge is conserved as $3 \times 2 = 6$ (with respect to the numbers on the lower left of the element symbol) and the atomic mass number is also conserved as $3 \times 4 = 12$. The mass itself may not be conserved, though. An oxygen atom (O) is made from four helium atoms [$4 _2He^4 \rightarrow _8O^{16}$]. Strangely, an atom that is a combination of two helium atoms (that is, an isotope of beryllium $_4Be^8$) is extremely unstable and would not survive long. It turned out that the nuclei of three elements, lithium (Li), beryllium (Be), and boron (B), are rather fragile, and as a result, they are present very little (relatively speaking) in the universe.

Further nuclear reactions take place in very massive stars. There, carbon atoms and oxygen atoms combine together to form heavier elements. Silicon (Si) forms from two oxygen atoms. Two carbon atoms combine to form neon (Ne) and

He $[2\,_6C^{12} \rightarrow\,_{10}Ne^{20} +\,_2He^4]$. Similar processes produce magnesium (Mg), sulfur (S), and other elements up to iron (Fe). Iron atom with 26 protons and 30 neutrons ($_{26}Fe^{56}$) is the most stable atom of all the elements. As a result, iron is one of the most abundant elements in the universe as well as on the Earth.

Elements heavier than iron are less stable than iron. So what would happen if the iron and similar elements accumulate in a massive star? The core would contract because of gravity, but no energy would be released even if nuclear fusion reactions occur. This will lead to an implosion of the core. It is then followed by an explosion of the star. This is considered to be what happens in a "supernova" phenomenon. One of the major processes that would happen in a supernova is the creation of a large amount of neutrons. Neutrons would combine with iron and other elements relatively easily as they are electrically neutral, and this neutron-capture process would yield heavier elements. Heavier elements are considered to be remnants of a supernova.

The demise of dinosaurs is now believed to have taken place as a result of sudden climatic change due to the impact of a massive asteroid on the Earth. And the asteroid is believed to be debris of a supernova. This was inferred from the high content of heavy elements such as iridium (Ir) in the layer between the Cretaceous and the Cenozoic levels.

Well, my body and your bodies are made of elements such as carbon and oxygen. They were made somewhere up in the space. Say one of my body's carbon atoms may have formed somewhere just outside of our solar system, several billion years ago. It came on a meteorite which hit the Earth on its early days, say 4.5 billion years ago. It reacted with other atoms that were present on the Earth, forming a part of an organic molecule. This particular molecule might have constituted a first living cell along with the other organic compounds. The organic compound then was decomposed and the carbon atom turned into carbon dioxide. It combined with calcium and formed calcium carbonate (limestone). It may have remained in that form for millions of years. Eventually, it was released back into the air as carbon dioxide. It then was incorporated into a plant through its photosynthesis. The plant was then eaten by an animal, which turned the organic compounds back to carbon dioxide. This was repeated millions, millions of times before finally it entered my body. This is a story of a single atom. But I have something like 10^{27} atoms in my body, all of which tell stories of their own. Think of it. How true is it that we are made of ash (debris) of the universe and then return to ash? How true is it that we are related to all other organisms and all nonliving material?

A diagram (Fig. 13.1) shows the distribution of elements in the universe. That is, what elements and how much. As we have already talked above, the most abundant elements are hydrogen and helium. Elements such as carbon, oxygen, silicon, and iron are relatively abundant, as you may have guessed from the arguments above. Another interesting point to be seen in the diagram is that above carbon the abundance alternates up and down, high at even atomic numbers and low at odd atomic numbers. This has a lot to do with how a nucleus forms with neutrons and protons, and what determines its stability, but its understanding is beyond this level of discourse.

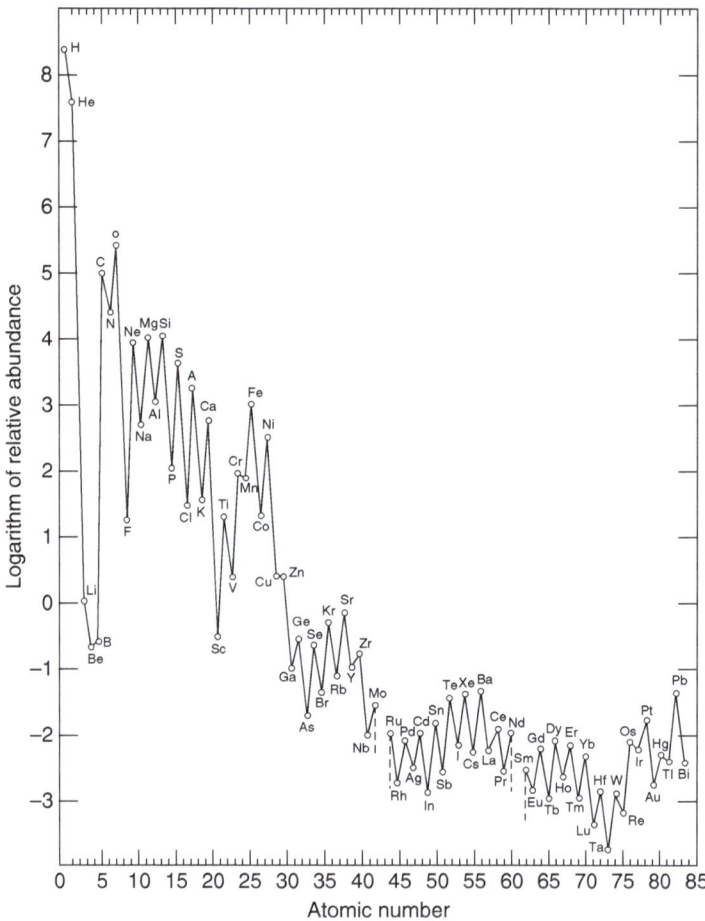

Fig. 13.1   The elemental composition of the universe [from B. Mason, "Principles of Geochemistry, 2nd, ed.", (Wiley and Sons, 1966)]

How has the diagram such as Fig. 13.1 been determined? How can you determine the relative quantities of elements in the universe? It would be very hard, virtually impossible, to determine the exact elemental composition of the Earth, let alone that of the whole universe. How then can we believe such a diagram as Fig. 13.1? Or how can they (those who put forward such a diagram) claim that the diagram is a meaningful estimate? These are only a few questions that can be raised for this kind of diagram. Think about it.

The universe is so vast, and we can reach only a tiny, tiny portion of it. How can we determine the quantities of material that are present far away from us? The determination can only be very indirect, and be made with a lot of assumptions. The diagram can represent at best a very rough estimate.

13.3   Formation of Molecules and Dusts in Interstellar Space

Our Sun is made mostly of hydrogen atoms, deuterium atoms, and helium atoms. Many other stars are also made mostly of these elements and other elements as mentioned above. Are those atoms (elements) present there just as atoms (in the atomic states)? Do not they form molecules or compounds?

On the Earth "independent atoms" are an "exception," not the "rule." That is, most elements on the Earth are combined in various manners to form molecular and ionic compounds. [Inert elements such as helium, neon, and argon do always exist in atomic states, because they are intrinsically very reluctant to form chemical bonds, i.e., molecules] Why do not they do so in stars or in the space? They do not do so in some stars, but they form some molecules in the interstellar space (space between stars). It all depends on temperature. Compounds, whether molecular or ionic, are made by connecting atoms by chemical bonds, as we learned earlier. The chemical bonds are not very strong and usually will be split easily at high temperatures above several thousand degrees (of Kelvin or Celsius). The temperature of the surface of the Sun is about 6,000°K and that of the inner core is estimated to be as high as 15 million degrees. There, any chemical bond would be broken and hence no molecules or chemical compounds would survive.

If the temperature of the interstellar space is low enough, formation of molecules becomes a possibility. The temperature of a typical interstellar medium is of the order of 100 K or below. Indeed, a number of compounds molecular or otherwise have been discovered and identified in the interstellar space, and on interstellar bodies such as comets. Our planet, Earth, is such an interstellar body. It is special because we are on it, and we will talk about it later.

The space between stars is not simply "empty" (vacuum), though there is a lot of vacuous space. There are gas (cloud) and dusts. These are made of many molecular as well as ionic compounds. How can we find them? We cannot collect those chemicals into a bag and bring it back to a laboratory and then analyze them, can we? Instead, we rely on the light each molecule is emitting. Molecules do not sit still. They are tumbling and vibrating, and undergoing other movements. And these movements do not occur arbitrarily; if you recall, they are quantized [see Chaps. 19 and 20]. For example, the tumbling (rotation) of a molecule occurs at a certain frequency, at the same frequency whether it is present on the Earth or the interstellar space. So does the vibrational motion. A molecule absorbs light with that frequency, and it also emits light of the same frequency. So those molecules in the interstellar space are emitting lights of their own frequencies. We can detect these lights with a radio telescope. The result of this observation is exhibited as a radio frequency spectrum. Figure 13.2 shows such a spectrum of the Orion molecular cloud (or Nebula) obtained at the Onsala Space Observatory. The Orion Nebula is shown in Fig. 13.3. The peaks seen in this spectrum are due to the light emitted by the rotational motions of molecules. This spectrum indicates the presence in the cloud of chemical species of HCN (hydrogen cyanide), SO (sulfur monoxide), SiO (silicon monoxide), and HCO_2CH_3 (methyl formate). [The height of a peak is dependent on two factors: how

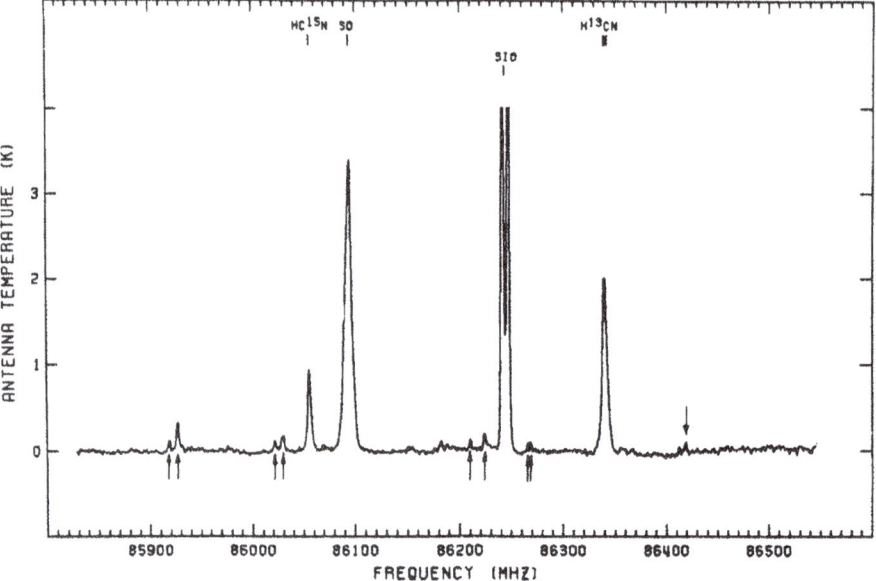

Fig. 13.2 Spectrum of the Orion molecular cloud [from W. M. Irvine and Hjalmarson, in "Cosmochemistry and the Origin of Life" (ed by C. Ponnamperuma, D. Reidel, 1983)]

Fig. 13.3 Orion Nebula

much of the chemical species is present in the view range of the telescope, and also how strongly a chemical species emits that light. The latter factor can be determined in a laboratory experiment, and hence the height (intensity) of a peak would give us an estimate of the abundance of a chemical species].

Table 13.1 Compounds
identified in the Orion
molecular cloud (Nebula)

H_2, OH, CH, H_2O, H_2S, NH_3
CO, CS, OCS, SiO, SO, SO_2, O_3 (?), CO^+
CN, HCN, H_2N-CN, $H_3C-C\equiv CH$, $HC\equiv C-CN$, $HC\equiv C-C\equiv C-CN$
$C\equiv CH$, $H_3C-C\equiv CH$, HN=C=O, HNC
$H_2C=O$, $H-C^+=O$, $H-C^+=S$, $H_2C=S$
$H_3C-O-CH=O$, H_3COH, $H_3C-O-CH_3$

A lot more chemical species (molecules) have been identified in the Orion Molecular Cloud. They are listed in Table 13.1 (above). There seems to be a preponderance of compounds that contain cyanide (CN) group: hydrogen cyanide HCN and cyano acetylenes consisting of C≡C (acetylenic group) and CN. Larger cyano acetylenes such as $HC\equiv C-C\equiv C-C\equiv C-C\equiv C-C\equiv N$ have been identified in other interstellar clouds such as that of Taurus. It is also worth noting that these compounds are made of the biologically important elements, that is, carbon, hydrogen, nitrogen, and oxygen. This suggests that life can be possible almost anywhere in the universe at least in the sense that all elements essential for organisms of the type found on the earth are found in the interstellar cloud.

How might have these compounds been produced in the interstellar medium? This question has not yet been adequately answered. It seems, however, that the presence of a kind of solid surface might be imperative, on which atoms and molecules bind and chemical reactions take place.

Are there such solids in the interstellar space? Yes, there are a lot of solids out there; they are called "dust" or "grain." What are they? A big problem with these kinds of material is again that we cannot hold them in our hands to study, as in the case of the molecular species in the space mentioned earlier. The study of the dust or grain in the space is more difficult than that of simpler molecular species. Sometimes, you may be able to obtain some spectra (in the infrared region) that give some indications as to the identities of compounds. In this way, some interstellar grains have been discovered to consist of silicates and frozen water. It is believed that such grains provide cores onto which other atoms and molecules are condensed. Those atoms and molecules condensed on the core react further under the influence of ultraviolet (and also other energetic particles available).

Another way to study this kind of situation is to recreate the situation in a lab. From a limited knowledge of the nature of grain such as above, we create such grains in a laboratory on this Earth and place it under such a condition as might prevail in the interstellar space. You need a lot of guessing, but science makes progresses essentially by guessing and testing. M. Greenberg of Leiden University tried such experiments. Simple compounds such as H_2O, NH_3, and CH_4 were condensed onto small (of 0.12 m diameter) silicate particles, and they were subjected to ultraviolet radiation. The researchers noted formation of complex organic molecules including such complex compounds as amino acids, in addition to some of the compounds listed in Table 13.1 above.

Chemistry of the Earth

14

14.1 How Was the Earth Formed?

The solar system formed about 4.6 billion years ago. The Earth and other terrestrial planets are believed to have formed by gathering together the so-called planetesimals. Planetesimals are formed by the coalescence of fine- or coarse-grained mineral matters, metals, and gases of various kinds (such as mentioned in the previous chapter). As planetesimals stuck together mostly by gravity and the body thus formed grew larger, it became a precursor of terrestrial planet. Some of these bodies were smashed by other bodies, and their fragments became meteorites. Hence, studies of meteorites would provide a lot of insight into the formation and the earlier state of the Earth.

The planet Earth was thus formed. Heat was created as the coalescence (of planetesimals) proceeded due to gravity, and heat also came from radioactivity of several radioactive elements such as aluminum-26. So the newly formed body was heated and the core was melted. As the material becomes liquid (as a result of melting), the materials contained in the liquid separate out according to their densities. The more dense material would sink closer to the bottom (core). Thus, the present layer structure of the Earth formed. The innermost core is a dense solid of about 1,200 km radius, whose density is about 12.6 g per cubic centimeter (12.6×10^6 kg/m^3). It is made of mostly iron metal and a small amount of nickel. By the way, the density of iron metal is only 7.8×10^6 kg/m^3 under the ordinary pressure. The next layer is the outer core (up to 3,500 km from the center of the Earth), which is liquid and has a density of $9.5-12 \times 10^6$ kg/m^3. The chemical composition seems to be about the same as that of the inner core. There is an abrupt change in density in the next layer, mantle. The width of mantle is about 2,900 km (3,500–6,380 km from the center). Its density ranges from 4 to 5.5×10^6 kg/m^3. The mantle is made of mostly magnesium–iron silicates (silicon oxides). The outermost layer is the thin crust of about 35–45 km on the land portion, and about 6 km under the ocean portion.

E. Ochiai, *Chemicals for Life and Living*,
DOI 10.1007/978-3-642-20273-5_14, © Springer-Verlag Berlin Heidelberg 2011

14.2   Chemistry of the Earth: Its Mantle and Upper Crust

The inner core of the earth is made of very much compressed metallic iron and a little bit of nickel; the outer core is not much different, only it is not as compressed as the inner core. These are essentially metals. However, the rest of the solid Earth is made of various minerals based on silicates.

The lower portion of the mantle (3,500–5,400 km from the center of the earth) is considered to be more or less homogeneous and made of magnesium–iron metasilicate (generally called "pyroxene") and magnesium–iron oxide (the mineral name is periclase). Pyroxene has the composition $(Mg,Fe)SiO_3$. The iron content is about 10–20% of the total of magnesium and iron. We will talk about the structures of silicates later on. The next 600 km is a sort of transition zone and the last 400 km or so is the upper mantle. The main minerals in the transition and upper mantle zone are olivine and amphibole-derived minerals. Olivine is basically magnesium silicate Mg_2SiO_4, and an example of amphiboles is calcium–magnesium silicate $(Ca_2(Mg,Fe)_3Si_8O_{22}(OH)_2)$.

The crust is the only layer we can have a direct access to. The upper portion of the crust consists of various constituents, granite igneous, metamorphosed, and sedimentary rocks, though the last two are secondary rocks. The granite igneous rocks come from the magmatic activities (volcanism). The lower portion of the crust is mainly basaltic and does also come from the molten material of the upper mantle. The igneous rock is made of quartz, feldspar, pyroxene, hornblende, biotite, titanium oxide, apatite, and others. The main components are the first three. Do not worry too much about these geochemical terms.

These rocks are weathered/eroded by air (air oxidation), water (rain/river), and temperature fluctuation, carried by river and wind, and deposited at the ocean bottom. Alternatively, dissolved substances may precipitate as the condition changes. These deposited substances eventually turn into sedimentary rocks.

Rocks can change their compositions and crystal form within the solid mass under the influence of the surrounding chemical environments (water and others), high pressure, and/or high temperature. This process is metamorphosis, and the resulting rocks are metamorphic rocks.

14.3   Igneous Rocks

14.3.1 Silicates

The rocks in the crust are predominantly (more than 90%) made of silicates (including silica SiO_2) and aluminosilicates. A question may be asked: why are silicates and aluminum derivatives the predominant components of rocks on the Earth? There are two reasons. One is that silicon and aluminum are among the most abundant elements on the Earth (and also in the universe). The second is that silicates form a variety of stable solid structures.

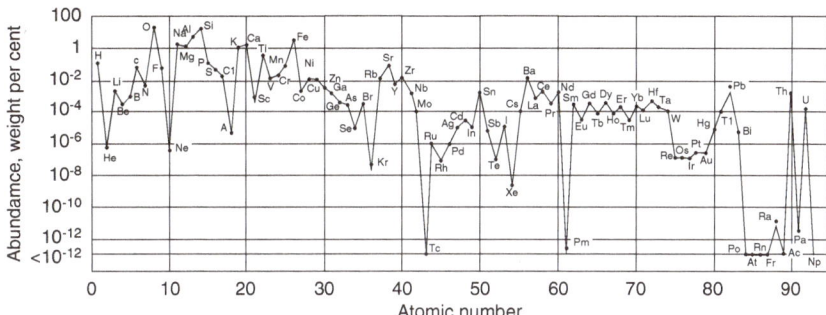

Fig. 14.1 The elemental distribution in the Earth's crust [from B. Mason, "Principles of Geochemistry, 2nd, ed.", (Wiley and Sons, 1966)]

Figure 14.1 shows the distribution of elements in the Earth's crust. You may note that it is significantly different from Fig. 13.1 (the cosmic element distribution) of the previous chapter. Lighter elements such as hydrogen, helium, carbon, and nitrogen are relatively less abundant on the Earth than in the Universe as a whole. The lighter elements seem to have been lost to the space when the Earth was forming, perhaps because the relatively small size of the Earth could not hold the lighter elements due to its relatively weak gravity force.

The most abundant element in the crust is oxygen (O), followed by silicon (Si), aluminum (Al), iron (Fe), and then calcium (Ca), magnesium (Mg), potassium (K), and sodium (Na). If silicon and aluminum, the two most abundant metallic elements, can do well in forming solid material, they would be the dominating elements in the crust. And, as it turns out, they are chemically very much suitable for making solid material, rocks.

First of all, they are situated in the middle of the periodic chart side by side, and they bind very strongly with oxygen. They form very stable oxides, SiO_2 and Al_2O_3. The former is quartz. Quartz is found in many different forms; transparent crystals, amethyst, smoky quartz, and agate (the colors in these minerals are due to impurities), and sand is mostly quartz. It consists of a structural unit of SiO_4, in which the silicon atom is nominally in the oxidation state of Si(IV) and the four oxygen atoms (O^{2-} each) are bound to the silicon in a tetrahedral manner. The tetrahedrons are bound to each other by sharing the corner (oxygen atoms) in a continuous three-dimensional manner. The most stable form of SiO_2 at room temperature and an ordinary pressure is α-quartz, the transparent and typical crystal familiar to many of you. In this crystal form, the tetrahedral units interlink three-dimensionally in a helical chain (Fig. 14.2). At 573°C, α-quartz changes to β-quartz crystal form. At higher temperatures, it will change to other forms, tridymite (at 867°C) and cristobalite (1,470°C), and it will melt at 1,713°C.

Many of the igneous rocks found in the Earth's crust are derivatives of silica, called metal silicates, in which the units SiO_4 bind in various manners. The unit SiO_4 has a tetrahedral shape and is nominally represented as $[SiO_4]^{4-}$; that is, each unit carries four negative charges. Among the simplest silicates is $M_2^{II}SiO_4$.

Fig. 14.2 The structure of
α-quartz

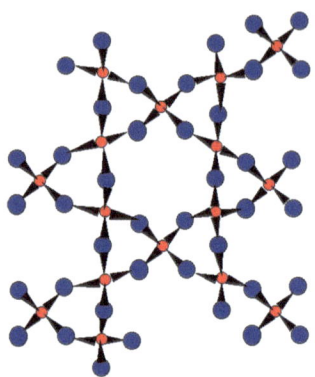

The most common M(II) is Mg(II), that is Mg_2SiO_4, and the mineral of this composition is called forsterite. Since Fe(II) has a size (radius = 92 pm) similar to Mg(II) (86 pm), the magnesium in forsterite can be replaced by Fe(II) in an arbitrary proportion. The resulting mineral is then represented by $(Mg,Fe)_2SiO_4$, and is known as olivine, one of the most common rock minerals. The structure consists of independent SiO_4 units interspersed by Mg(II)/Fe(II), which in turn is bound to six oxygen anions of SiO_4 units. Silicate of this type is called "orthosilicate."

The SiO_4 units can bind each other by sharing the oxygen atoms in various manners. They bind, for example, in a linear fashion sharing a single oxygen atom (see Fig. 14.3a); the repeating unit is then $[SiO_3]^{2-}$, which is called "metasilicate." The most common mineral with such a structure is pyroxene type $M^{II}SiO_3$. The pure $MgSiO_3$ is a mineral enstatite, and $(Mg,Fe)SiO_3$ is hypersthene. This is merely the simplest linear structure. There is a wide variety of ways to connect the $[SiO_4]$ units in chain fashion (as illustrated in Fig. 14.3). Two linear chains may form a double chain as shown in Fig. 14.3. This is the basic structure of amphiboles, where the repeating unit is $[Si_4O_{11}]^{6-}$. Tremolite with the composition of $Ca_2Mg_5(OH)_2(Si_4O_{11})_2$ is a typical amphibole. These are examples of long chains of the silicate units.

The linear structure of the basic units, single or double-stranded, suggests that some of these minerals under certain conditions may form fibrous material. Such material with flexibility, high tensile strength, and heat resistance is called "asbestos" in general. The most common asbestos is chrysotile, commonly known as "white asbestos," which has a composition of $Mg_3Si_2O_5(OH)_4$. It has the double-stranded structure similar to Fig. 14.3d, where Mg(II) ions bridge the two strands. A less common asbestos is the mineral crocidolite: $Na_3Fe_4Si_8O_{22}(OH)_2$, and is known as "blue asbestos." The repeating unit of this fibrous mineral has the structure shown in Fig. 14.3e. This asbestos' fiber is shorter than that of crysolite and is believed to be mostly responsible for the environmental hazard.

SiO_4 units can also condense themselves to form discrete polysilicate anions; examples are shown in Fig. 14.4. These possibilities suggest the wide variety in which silicates form minerals.

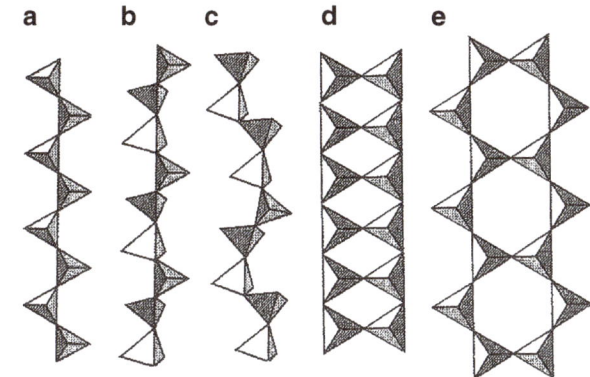

Fig. 14.3 The structures of some linear polysilicates; (**a**) Na_3SiO_3; (**b**) $Ca_3Si_3O_9$; (**c**) $(Mn, Ca)_5Si_5O_{15}$; (**d**) $[Si_2O_5^{2-}]$; (**e**) $[Si_4O_{11}^{6-}]$ [from N.N. Greenwood and A. Earnshaw, "Chemistry of the Elements, 2nd ed" (1997, Elsevier)]

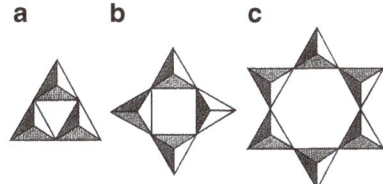

Fig. 14.4 The structure of discrete polysilicates: (**a**) $[Si_3O_9]^{6-}$; (**b**) $[Si_4O_{12}]^{8-}$; (**c**) $[Si_6O_{18}]^{12-}$ [from N.N. Greenwood and A. Earnshaw, "Chemistry of the Elements, 2nd ed" (1997, Elsevier)]

14.3.2 Aluminosilicates

The other interesting chemical character is that the silicon atom in $[SiO_4]^{4-}$ may be replaced by cations of similar sizes without significant modification of the structure. The most important one is aluminum Al(III). However, when this happens, the unit $[AlO_4]$ will now carry "5−" electric charge instead of "4−." That means that cations (M(I) or M(II)) must be added to compensate the electric charge difference when Si(IV) is substituted by Al(III), or that an O^{2-} unit is replaced by an OH^- unit. Al(III)-substitution occurs often in amphibole-type minerals. The amphiboles contain $[Si_4O_{11}]$ repeating units as mentioned earlier (Fig. 14.3), and Al can replace Si up to the extent of $[AlSi_3O_{11}]$. The extent of Al-substitution depends on the condition of formation of such minerals. Minerals of hornblende has the composition of $Ca_2Na_{0-1}(Mg,Fe,Al)_5[(Al, Si)_4O_{11}]_2(OH)_2$.

By far the most important rock minerals are feldspars, which constitute about 60% of the igneous rock. They are essentially made of three fundamental types of minerals: albite $NaAlSi_3O_8$, orthoclase $KAlSi_3O_8$, and anorthite $CaAl_2SiO_8$. Their structures are continuous three-dimensional network of $[SiO_4]$ and $[AlO_4]$ tetrahedrons, interspersed by positively charged sodium (Na(I)), potassium (K(I)), or calcium (Ca(II)). Albite and anorthite mix in arbitrary proportions; such mixed minerals are known as plagioclase feldspars. Albite and orthoclase also make mixtures; they are alkali feldspars.

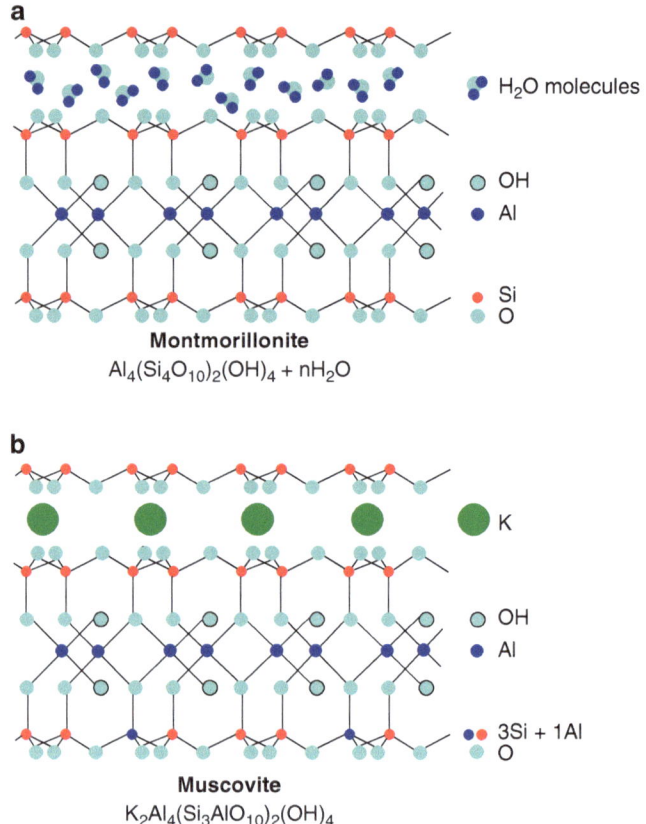

Fig. 14.5 Clay minerals: (**a**) Montmorillonite, (**b**) Muscovite

Another type of widely occurring aluminosilicates is classified as zeolite. Zeolites have much more open structures than feldspars, and, as a result, they can take up loosely bound water or other small molecules into their cavity. Now zeolites are artificially created, which have specific chemical characteristic and holes that accommodate only a specific substance.

Aluminosilicates of mica group have layer structures (Fig. 14.5). The $[SiO_4]$ units connect through three of the oxygen atoms and form a sheet structure. Representative minerals of mica group are muscovite $KAl_2(AlSi_3O_{10})(OH)_2$ and biotite $K(Mg,Fe)_3(AlSi_3O_{10})(OH)_2$. The sheet structure leaves one oxygen free. Two sheets made of tetrahedral $[SiO_4]$ are placed together with the free oxygen (vertex)'s pointing inward. These oxygen ions are then cross-linked by Al(III) in muscovite or by Mg(II) and Fe(II) in biotite. Then such sets of two sheets are held together by potassium ion (K(I)). This binding by potassium is relatively week, accounting for the characteristic basal cleavage of mica.

14.4   Clay and Soil

Igneous rocks that formed under high temperature and high pressure are unstable under the conditions prevailing on the surface of the Earth. Rocks are thus subject to various forces, physical, chemical, and biological, and will be slowly changed and eroded. This process is known as weathering. Only quartz is highly resistant to weathering processes, and all the other minerals tend to change.

The temperature on the Earth changes significantly by day and night, seasons, and over the long range of geological time. Temperature changes will expand and contract the rocks and eventually crack them. More significant is the effect of water. Water seeps into the cracks of rocks and it expands when it is frozen. This will facilitate the cracking. Rolling of hard rock upon softer rocks helps disintegrate the softer rocks. Wind and river then will carry smaller rock fragments away. These are merely physical effects.

Some components, for example, Fe(II), are oxidized by air to Fe(III). Water will dissolve the chemical substances in rocks. The presence of carbon dioxide usually helps dissolution process, because it makes water acidic. The fragmented and/or dissolved material may be redeposited on land, which forms soil. The fragmented and dissolved material may be carried to the ocean and eventually deposited and compacted at the bottom, which forms sedimentary rocks.

Soil consists of weathered material (clay) of the bedrock and other material. The main clay minerals are silicates and aluminosilicates derived from muscovite, biotite, olivine, and pyroxene. Iron oxide and aluminum hydroxide also constitute clay minerals. For example, olivine will be slowly eroded by the effect of air and water. The chemical reactions involved are summarized as follows:

$$2(Mg,Fe)\,SiO_4\,(olivine) + (1/2)\,O_2 + 5H_2O \rightarrow Fe_2O_3 \cdot 3H_2O + Mg_2SiO_4 + Si(OH)_4$$

The resulting Mg(II) ion (shown as Mg_2SiO_4) and $Si(OH)_4$ are relatively soluble in water and can be washed away. The iron oxide Fe_2O_3 colors the clay brown. Other silicates, pyroxene and others, are less soluble and more resistant to erosion, but will be weathered slowly. The only exception is quartz; it may be pulverized mechanically (to become sand), but is resistant to chemical weathering.

The major clay minerals are muscovite, montmorillonite, pyrophilite, and kaolinite. Kaolinite is the major ingredient of the clay used for making ceramics and is talked about in Chap. 10. These minerals all have layered structures (Fig. 14.5) like kaolinite (Fig. 10.1).

14.5   Sedimentary Rocks and Metamorphic Rocks

Sandstone, shale, and limestone are typical sedimentary rocks. The sedimentary rock that forms from finely fragmented mineral, mostly of silica and some silicates, is sandstone. It will be called "conglomerate" if it contains larger fragments and the texture is coarse. Mud (clay and others) leads to the formation of shale. Calcium and carbonate are combined to form calcium carbonate ($CaCO_3$) by many marine creatures, both plants and animals. Calcium carbonate can also deposit chemically without the help of organisms. Calcium carbonate thus created sinks to the bottom of the ocean, and eventually turns into limestone. The major mineral in limestone is calcite. Many marine organisms including diatoms form silica shell or use silica as skeleton. Silica thus concentrated and deposited at the bottom of ocean turns into chert.

Igneous and sedimentary rocks not exposed to the surface can also undergo physical and chemical changes in other ways. Such rocks are recognized as "metamorphic." The agents that effect such changes are high temperature and high pressure over long periods of time. A geochemical discussion of these different types of rocks and minerals is referred to more specialized literatures.

14.6   The Story of Iron

Iron is the most widely used material in today's human civilization. We obtain it from the Earth. Let us explore how we get it.

Iron is one of the most abundant elements on the Earth or in the Universe for that matter. The major isotope of element "iron" is $_{26}Fe^{56}$; the natural abundance of this isotope is 91.6%. The other isotopes are $_{26}Fe^{54}$ (5.8%), $_{26}Fe^{57}$ (2.2%), and $_{26}Fe^{58}$ (0.3%). It has been determined that the energy to bind nucleons, i.e., protons and neutrons, in the nucleus (called "nuclear binding energy" per nucleons) is the highest with $_{26}Fe^{56}$ among all the elements. It means that the nucleus of the major isotope of iron is the most stable of all the elements, and this is the basic reason that iron is among the most abundant elements in the universe. See the previous chapter for a more detailed discussion of element formation in the universe.

The core of the Earth is almost pure iron as mentioned earlier, but this is not available for human consumption. Iron, however, is also widely dispersed among the rocks constituting the Earth's crust. Originally, it existed as a part of silicate minerals such as pyroxene $((Mg,Fe)SiO_3)$ and olivine $((Mg,Fe)_2SiO_4)$. The iron in these rocks exists mostly in the form of iron(II) (Fe(II)), so that it substitutes partially Mg(II) of the magnesium silicates. Fe(II) ion is fairly soluble in water unlike Fe(III) ion. Hence, Fe(II) would dissolve in seawater or hot hydrothermal water.

Hence, the concentration of iron in seawater is supposed to be quite high. In fact, it is very low in the today's seawater. It is estimated to be about 10^{-5} mol/L on average. The problem is that the atmospheric composition on the today's Earth is quite different from that on the ancient Earth. Let us see why it is so. Iron, in the form of Fe(II), is soluble in neutral water, but Fe(III) is quite insoluble in water as it precipitates as the hydroxide $Fe(OH)_3$ or $FeO(OH)$.

Today's atmosphere contains about 21% of oxygen (O_2). Oxygen in the atmosphere is believed to have been produced mostly by photosynthetic bacteria at first and then algae and all plants (see Chap. 2). The first photosynthetic organisms, cyanobacteria, are believed to have emerged 2.5–3 billion years ago, though this issue has not been settled. Before that, the atmosphere is believed to have not contained much of free O_2. Therefore, it can be assumed that iron, in the form of Fe(II), was plenty in the ancient ocean before a sufficient oxygen accumulated in the atmosphere. The iron content in the ancient ocean can be estimated to be ten thousand times of the today's value. As the cyanobacteria and other photosynthetic microorganisms proliferated, more and more of oxygen O_2, a by-product of the photosynthesis, was produced. However, the atmospheric oxygen did not rise. It was because there was a vast amount of Fe(II) in the ocean. The oxygen produced was used to oxidize Fe(II) to Fe(III). The ocean was a huge oxygen sink.

The major iron ores are found in the so-called Banded Iron Formation (BIF). BIF is found all over the world, particularly in the USA (Minnesota) and Western Australia, and is believed to have deposited about 2.8 billion years ago through about 1.8 billion years ago. BIF, as the name suggests, consists of alternating layers of black iron-rich layer and whitish mostly cherty layer. The major ores in BIF are magnetite (Fe_3O_4) and hematite (Fe_2O_3). These are sedimentary minerals. It has been suggested that an iron-rich layer was formed during a summer season when the oxygen produced by cyanobacteria oxidized Fe(II) to Fe(III), which precipitated as hydroxide that turned into magnetite or hematite. In the lean season of low photosynthetic activity, silica continued to precipitate to form a cherty layer.

BIF ceased to form after about 1.8 billion years ago, though other types of iron oxide ores such as limonite and goethite (FeO(OH)) formed. Fe(III), when brought to a reducing environment where sulfate is reduced by sulfate-reducing bacteria to hydrogen sulfide (H_2S), reacts with H_2S, and forms pyrite FeS_2. Pyrite is often found in coal.

The iron ores commercially used come from mostly BIF, i.e., iron oxides, Fe_2O_3 or Fe_3O_4. How would you produce the iron metal from these ores? The iron in these ores is in oxidized states, either Fe(II) or Fe(III). Fe(II) is two electrons short of the iron metal Fe (Fe(0)), and Fe(III) is three electrons short. Therefore, in order to produce the iron metal from the ores, you have to add electrons to the iron oxides. Such a chemical reaction is called "reduction." In the modern industrial process, the reducing agent is coke, made from coal. It is essentially carbon, C. A carbon atom can gives four electrons to others when it turns to carbon dioxide. The chemical reactions involved are as follows:

$$2Fe_2O_3 + 3C \rightarrow 4Fe + 3CO_2$$

$$Fe_3O_4 + 2C \rightarrow 3Fe + 2CO_2$$

This is the overall reaction. In a blast furnace in which iron is produced, the real reducing agent is carbon monoxide CO which occurs from $C + CO_2 \rightarrow 2CO$. The temperature required is quite high, approximately 1,000°C or higher. This is because both the reactions require a lot of energy (Gibbs free energy) input to be effected. The basic reason for this is that iron Fe(0) can readily be oxidized (in thermodynamic sense: see below), indicating that the reverse reaction, reduction reactions such as shown above, are difficult with iron. The tendency of iron metal to get rusted is what every car owner knows from experience. The rust is an oxidized product of iron metal.

A pure iron monument known as "Ashoka's pillar" has been standing without much rusting near New Delhi, India, for at least several centuries. The old swords made by Japanese sword makers are known to resist rusting. Apparently, a very pure iron metal is slow to getting rusted, i.e., oxidized. How come? It was suggested above and is known widely that iron metal gets rusted easily. In Chap. 19, we discuss that there are two fundamental factors by which a chemical reaction is affected. One is the thermodynamic factor that determines whether or not the reaction should proceed; it says either the reaction should proceed in the direction designated or the reaction would not take place. Iron indeed should easily be oxidized in this thermodynamic sense. However, it does not say whether the reaction would in practice proceed at a reasonable speed. This is the second factor, i.e., a kinetic factor. The oxidation by oxygen of pure metal iron is indeed very slow. Why does then the ordinary iron metal get rusted fast? You might recall that you need to either raise the temperature or add a catalyst to speed up a chemical reaction. We are not concerned with the temperature issue here. Therefore, we have to look at the effect of a catalyst. In the older processes of making iron, the reducing agent was wood, whereas it is coke or coal in the modern process as talked about earlier. Coal typically contains sulfur (in the form of pyrite and others) as an impurity. Hence, the iron metal produced by modern methods tends to contain a minute amount of sulfur as an impurity; this sulfur seems to act as the catalyst for oxidation of iron. The mechanism of this catalysis is not yet well understood.

Let us now turn our attention to a completely different issue, though chemically related. Mankind had found the ways to use copper (as bronze) before they discovered the use of iron; that is, the "bronze age" preceded the "iron age." A Question is why? This is a chemical question in essence!

Part V

Chemicals that May Cause Problems: Poisons, Pollutants, and Others

We, human beings, are made of million different chemicals, and a chemical, when coming into contact with a human body, can potentially interact with many chemicals present in our body. Results of such interactions are varied; some can be beneficial and some may be harmful. The beneficial interactions are those we already talked about in Parts I–IV. Harmful chemicals are the subjects of this part.

Environmental Issues: Heavy Metal Pollutants and Others

15

Heavy metals are not well defined, but usually mean metals such as mercury, lead, cadmium, silver, and gold. Platinum, palladium, osmium, and other such metals are also definitely heavy (in terms of density) metals, but they are usually not included because there is little chance that they cause some environmental problems. We talk about mercury, lead, and cadmium here.

Arsenic is another commonly occurring pollutant and is included here. The element arsenic is often called a "metalloid," because its common form, gray arsenic, has a steely gray color and some other properties similar to metals. However, the properties of its compounds are typically those of nonmetallic elements.

15.1   Story of Mercury Pollution/Toxicity

One day in the fall of 1996, a professor of chemistry at Dartmouth College was transferring a liquid dimethyl mercury ($(CH_3)_2Hg$) with a pipette from a reagent bottle to an NMR sample tube (about 1/4 in. diameter glass test tube). She was well aware of the toxicity of the reagent and took every precaution including wearing a pair of latex gloves. She spilled several drops of the liquid on the glove. She did not think of anything particularly wrong at the time; she thought that the glove would protect her. But she became ill toward the end of that year; her symptoms were typical of mercury poisoning, and she died in February 1997.

This is a dramatic example of the severity of mercury poisoning. The toxicity of mercury had been well known. There is even an expression: "mad hatter," for example. Professional hat makers used to use mercury nitrate in making felt for hats, and some showed a sign of mental degeneration; hence "mad hatter." This practice lasted until about 1920s. It has long been known that the cause is mercury. Some mercury compounds have been used as fungicides and pesticides. The basis of these effects is, of course, the toxicity of mercury compound to fungi or bacteria, i.e., living organisms.

E. Ochiai, *Chemicals for Life and Living*,
DOI 10.1007/978-3-642-20273-5_15, © Springer-Verlag Berlin Heidelberg 2011

Toward the end of 1950s, people started to notice that a number of cats went crazy in villages along the Ariake Bay area in Kyushu, the southern most main island of Japan. Some of the cats danced wildly and others ran to their death. A few years later, human beings themselves started to get sick of unknown causes. Their eyesight deteriorated, became narrow (tunnel vision), and eventually was lost completely. The symptoms were accompanied by a number of other systemic as well as localized malfunctions, particularly loss of coordination of body movements, and many died. The major city in the area is "Minamata," and hence this malady became known as "Minamata disease." People suspected that a waste product that was released from a chemical plant into the bay was the cause, but the company strongly denied it. A lengthy battle between the company and the people of the area erupted. First the people were unorganized, but eventually the patients organized and the supporters gathered. As the scientific evidence mounted that the cause was indeed the mercury compounds released from the plant, the battles were waged in the courtroom as well. The settlement bankrupted the company.

15.1.1 Anthropogenic Sources of Mercury in the Environment

Meanwhile, similar diseases had been discovered at a number of locations around the world. A friend of this author visited a number of these places and found that they were indeed "Minamata disease." In most of those places, the sources of mercury appeared to be chemical plants, particularly associated with paper/pulp mills. The chemical plant in Minamata was not involved in the paper/pulp production; instead they were producing acetaldehyde.

Today, acetaldehyde (CH_3CHO) is produced from ethylene (C_2H_4), a petroleum product. In earlier days, though, it was produced by adding water to acetylene (C_2H_2); (the reaction is $C_2H_2 + H_2O \rightarrow CH_3CHO$). This reaction requires a catalyst, mercury sulfate ($HgSO_4$). The waste liquid containing mercury compounds was released into the Minamata bay.

Paper/pulp mills often produce sodium hydroxide they need in small factories attached. A simplest way to produce sodium hydroxide (NaOH) of high quality is to electrolyze a brine solution (NaCl). That is, $2NaCl$ –electrolysis– $>2Na$(metal)$+Cl_2$. When steam (H_2O) is added to the sodium metal, sodium hydroxide is obtained: $2Na + 2H_2O \rightarrow 2NaOH + H_2$. Because electric current is used to accomplish this reaction, electrodes are required. The electrodes used are carbon rod (for anode) and mercury (liquid) metal for the cathode, because mercury binds many metallic elements including sodium. Lax procedures at small factories often allow some of the mercury to escape into the environment. This is the major source of mercury in the environments.

By the way, note that chlorine (Cl_2) is a by-product of this process. Chlorine-containing compounds have caused a number of environmental problems as talked about in Chap. 16, and the chlorine source for these compounds was essentially this process, i.e., electrolysis of the brine to produce sodium hydroxide. Because of the environmental concerns, this process called "chloroalkali" process is now often conducted without using mercury as the electrode. In that case, the cathode produces OH⁻ by reducing water; i.e., $2H_2O + 2e$ (e = electron, from cathode) → $2OH^- + H_2$ (gas).

Mercury is also used to extract gold from the gold ore. Mercury has a peculiar property. It is liquid at ordinary temperatures, and binds (dissolve) many metals such as silver, gold, cadmium, and sodium, as mentioned above. These are a kind of alloy, and called "amalgam." Anyway, liquid mercury is a convenient medium to dissolve gold metal that is finely dispersed in crushed ores. As mercury has a relatively low boiling point, you can heat the amalgam liquid to remove mercury, and you will find gold left behind when mercury is gone. This procedure is still being practiced in many parts of the world, and can cause a serious environmental problem, as mercury enters both workers and the environment.

Another anthropogenic source is power-generating plants, where coal is used. Coal often contains a little of mercury, and as coal is burned mercuric substances turn to metallic mercury, which is volatile. Hence, mercury will be spewed out along with the smoke.

The mercury that is discharged into the environment from these sources is in the form of either metal mercury (Hg) or some inorganic mercury compounds such as mercury sulfate ($HgSO_4$ that consists of Hg(II) ion and SO_4^{2-} ions). Mercury metal itself is, chemically speaking, not toxic, but mercury (Hg(II)) ion is. A problem is that mercury metal can be oxidized to mercury(II) (Hg(II)) in the environment or living organisms and hence becomes toxic. The so-called inorganic mercury, e.g., mercury sulfate, is not very potent in inflicting damage to human beings or other animals. The reason is that mercury(II) may not be absorbed well into the bloodstream and is mostly removed through the intestinal system, though it can cause a severe diarrhea. It is obvious that the small quantity that does enter the bloodstream, of course, causes a problem.

A major form of mercury found in the environment is not metal nor inorganic, but often monomethyl mercury chloride (CH_3)HgCl or dimethyl mercury (CH_3)$_2$Hg. Monomethyl mercury turned out to be the most toxic form of mercury. If you recall, The professor of Dartmouth was handling dimethyl mercury. Dimethylmercury itself is chemically inert, but it can be readily converted into monomethyl mercury and that was the cause of death. Two major questions are as follows (1) Why is mercury, i.e., mercury(II) in general and monomethyl mercury in particular, so toxic? (2) How is the inorganic mercury in the environment converted into dimethyl or monomethyl mercury?

15.1.2 Toxicity of Mercury

Mercury ion (Hg(II)) has a unique character. It binds strongly with many chemicals, especially those containing the so-called thiol group. The thiol group is "S–H," that is, sulfur (S) bound with hydrogen. Some of the commonly occurring compounds with S–H are hydrogen sulfide (H–S–H which smells of a rotten egg, because a rotten egg actually produces this compound) and cysteine (HS–CH_2(NH$_2$)COOH), one of the amino acids which comprises proteins. Mercury(II) displaces the hydrogen of S–H group and forms an unusually strong Hg–S bond as in Hg–S–CH$_2$(NH$_2$)COOH. By the way, HgS (mercury sulfide) is the form of mercury ore (which is called cinnabar) found in the nature. Cinnabar is bright red and is used to be employed as a red pigment. Studies have shown that the mercury administered to experimental animals is found to be bound often to a cysteine residue in proteins. Many physiologically important organic compounds as well as enzymes/proteins contain such S–H groups, which play the vital role in their functions. When mercury(II) binds to these S–H groups, the normal physiological functions will be blocked. This is the basis of mercury toxicity.

Mercury(II) obviously has to reach the target to exhibit its toxicity. When mercury is ingested through the mouth, it has to be absorbed into the bloodstream to reach a target. Most of the inorganic elements are absorbed by the intestine. The intestine wall is covered with membranes that are not very permeable to cations (positively charged ionic species). For cell membranes are essentially made of long-chain hydrocarbons, i.e., oil. The oily material does not have affinity toward water or water-soluble ionic species. On the other hand, an electrically neutral, oily material can readily bind and go through cell membranes. Hence, inorganic mercury ion such as Hg(II)/SO$_4^{2-}$ would not readily be absorbed by the intestine. On the contrary, the dimethyl mercury, neutral and with an organic oily entity (CH$_3$ – methyl group), can readily go through the membranes and would show up in the bloodstream. The bloodstream will carry the mercury compound throughout the body and will deposit it where it is bound strongly. It turned out that mercury tends to bind strongly to the brain, and particularly the "optical site." As mentioned before, dimethyl mercury ((CH$_3$)$_2$Hg) itself cannot bind these sites, i.e., the S–H entity. That is because both the binding sites of Hg are already occupied by the bound methyl groups. Hence, the mercury in dimethyl mercury cannot bind another entity. As dimethyl mercury circulates in the body, it degrades to monomethyl mercury CH$_3$Hg$^+$. Now the mercury in this entity can bind to whatever it chooses to bind. Hence, it binds very strongly the "S–H" of important proteins and enzymes, resulting in dysfunctional proteins and enzymes. Monomethyl mercury compounds are soluble in water and hence can be readily carried by bloodstream, and can also readily enter cells if electrically neutral as in CH$_3$Hg(OOCCH$_3$) (monomethyl mercury acetate) or CH$_3$HgCl. These are the reasons that dimethyl mercury or a monomethyl mercury compound is most toxic. The unfortunate Dartmouth professor had dimethyl mercury droplets on her latex glove. Dimethyl mercury quickly seeped through the glove and her skin (membrane), as dimethyl mercury has a high affinity toward both the glove material and the skin, and went into her circulation system. It was gradually decomposed to monomethyl mercury, which exerted the toxic effects.

15.1.3 Mercury in the Natural Environment and the Biological Defense Against It

Mercury exists in the natural environment and has done so throughout the whole history of the Earth. Some organisms might have been exposed to mercury quite often.

For example, some microorganisms might have happened to be near the mercury-containing ores. They must have devised some means to reduce the toxic effects of mercury; otherwise, they would not have survived. We human beings have been exposed to it only rarely under the natural conditions. Consequently, we do not have very effective means to deal with it. However, now we encounter it more often because of our own activities as outlined above, and that is the cause of our problem.

So first let us explore how some microorganisms might cope with mercury. As mentioned above, the toxic form of mercury is Hg(II) which has at least one binding site left open. Let us make this chemistry a little clearer. The mercury ion (Hg(II)) can bind two entities under ordinary circumstances. So if it is bound strongly with two entities, it cannot bind another and hence cannot exert the toxic effects. Therefore, dimethyl mercury $(CH_3)_2Hg$ where two methyl groups are bound strongly with Hg(II) is chemically nontoxic, and the common toxic forms are free Hg(II) ion and monomethyl mercury ion CH_3Hg^+ or similar entities such as $C_2H_5Hg^+$. It is also to be remembered that mercury metal (the shiny liquid used in thermometer for example) itself is not chemically toxic. It shows some toxicity, only because it can be converted to Hg(II) by some means in your body.

Nature, i.e., microorganisms, takes advantage of the chemistry described in the previous paragraph to devise means to detoxify or reduce the toxic effects of mercury. There are four major modes: (a) to produce compounds or proteins containing S–H groups in order to grab Hg(II) fully and strongly; (b) to bind mercury in the form of HgSe; (c) to convert Hg(II) to metallic mercury Hg(0); and (d) to convert Hg(II) to dimethyl mercury.

15.1.3.1 Metallothionein, A Heavy Metal-Binding Protein

Metallothionein is a relatively small protein that contains an unusually large number of cysteine residues. Cysteine is a S–H containing amino acid: $HSCH_2CH(NH_2)$ COOH. This protein is produced only at a very low level under an ordinary condition. However, when an excess level of cadmium (Cd(II)), zinc (Zn(II)), copper (Cu(I)), or mercury (Hg(II)) is present in a cell, the protein synthesizing machinery of the cell starts to produce metallothionein. In other words, this protein is induced by these metallic ions. With its cysteine residues' S–H groups, a metallothionein molecule binds as many as 7–12 metallic ions. As a result, the toxic cations become unavailable for the unwanted binding, and hence the toxicity of mercury (or other metals mentioned above) is reduced. Metallothionein and its analogues have been found widely distributed among all kinds of organisms from bacteria, fungi, and plants to mammals including humans. However, there is a limit to the quantity of metallothionein produced by a cell. So mercury cannot be detoxified if present in a higher level than a threshold value.

15.1.3.2 HgSe

Selenium Se is below sulfur S in the periodic chart of elements (Figs. 19.2 and 19.5). This suggests that Se would have properties similar to those of sulfur. Indeed, Hg(II) binds Se^{2-} as well as or even more strongly than S^{2-}. So, once Hg(II) is bound with Se^{2-}, it would not be able to exert its toxic effects. This is chemistry. In the

United States, "mercury pollution" has been associated with some special fish. Some scientists discovered in 1960s that, for example, tuna fish, fresh as well as canned, contained abnormally high levels of mercury. First, they attributed the cause of the high level of mercury in tuna to the high industrial activities in the recent decades. However, it turned out that even very old specimens of tuna found in museums contained high levels of mercury. That is, the high mercury level in these fish does not seem to have been caused by recent industrialization. Besides, these fish show no sign of mercury poisoning.

The unusually high level of mercury in such fish as tuna and bonito is believed to be due to the fact that they are very active fish. A high activity requires a large amount of oxygen that comes from seawater; they have to keep moving in order to be alive. So these fish have to filter a large amount of seawater, and the mercury level in their bodies becomes high even if the mercury level of seawater is quite low, because mercury cannot easily be excreted once it gets ingested. A mercury compound added in the feed to experimental animals causes "mercury poisoning" symptoms in them, but addition of tuna meat to the feed was demonstrated not to increase the extent of mercury poisoning. Studies have shown that tuna fish meat contained a high level of selenium (Se) as well, and that the ratios Se/Hg were close to 1:1, suggesting the formation of HgSe. This mercury, being bound tightly with selenium ions, cannot bind any other entity and hence cannot exert the toxic effects. Selenium is another very toxic element, though a very low level of it is required for a normal health. These fish somehow have developed means to incorporate this toxic element to combat the toxic effects of mercury. It has also been shown that the people who worked in a mercury-mine (in Yugoslavia) and yet did not show the symptoms of mercury poisoning did indeed also have selenium in their body to the extent of the Se/Hg ratio of close to 1:1.

15.1.3.3 Reducing Hg(II) to Metallic Mercury

Some bacteria including the common *E. coli* have on their plasmid (extranuclear gene) Hg-resistant factor. This gene, when faced with the presence of an excess level of mercury, produces an enzyme called "mercury reductase." It reduces Hg(II) to metallic mercury Hg(0). Metallic mercury is not toxic and is somewhat volatile [remember that it is a liquid at room temperature]. Hence, the reduced mercury can diffuse out of the bacterial cell in the form of mercury vapor, and hence mercury is removed.

15.1.3.4 Converting Hg(II) to Dimethyl Mercury

Many microorganisms living in the sediment in stagnant waters can produce methane CH_4 from other organic compounds. You might have seen a gas bubbling up when you disturbed the bottom of a stagnant pond or bog. That is methane gas. An important factor in producing methane is a methyl derivative of vitamin B_{12}. Yes, vitamin B_{12} (Sect. 6.6.1). Vitamin B_{12} itself has a structure similar to the hem shown in Fig. 6.2, is called also as "(cyano) cobalamin," and is a beautiful red crystalline compound. It contains cobalt (Co) at the center of the molecule (instead of Fe as in heme). Vitamin B_{12} in our body is a slightly modified version, called adenosyl cobalamin. However, the cobalt in vitamin B_{12} can also bind methyl group (CH_3). This is methyl cobalamin. It has been shown that methyl cobalamin reacts readily with mercury ion Hg(II) and transfers its methyl group to Hg(II). That is:

$$CH_3 - Co(cobalamin) + Hg (II) + H_2O \rightarrow H_2O - Co(cobalamin) + CH_3Hg^+$$

If the monomethyl mercury CH_3Hg^+ gets another methyl group, it will end up with dimethyl mercury $(CH_3)_2Hg$. Dimethyl mercury, being relatively volatile and electrically neutral, freely goes through the membrane and the cell wall of a bacterium and hence will escape into the surroundings. The bacterium has now accomplished a task of getting rid of the toxic mercury. But the result is that dimethyl mercury has been released to the environment. It is slowly decomposed to monomethyl mercury in the environment or in the bodies of other organisms. Thus, it causes a serious pollution problem.

15.1.3.5 General Tolerance Mechanisms

Organisms have some general tolerance mechanisms in addition to the specific ones (a)–(d). Mercury, Hg(II), binds to cysteine residues as mentioned above. Hair is made of a protein called keratin and its associated proteins, which also contain many cysteine residues, and hence would bind Hg(II). It does, and it also binds other metal ions such as Zn(II), Cd(II), and Cu(II) (and also arsenic As). As hair is not physiologically critical, Hg(II) there would not be much harmful. Another entity, mother's milk, has affinity toward oily mercury, i.e., dimethyl mercury and monomethyl mercury. For mother, this can act as a mechanism of removing mercury into a nonessential locale. Minamata disease is known to have been transmitted from mothers to their infants through milk (and also through the placenta). Such a baby showed typical symptoms of mercury poisoning, though the mother did not show an appreciable symptom of the disease.

It needs to be emphasized that these defense and tolerance mechanisms have only limited capacities and that mercury present beyond the threshold levels determined by these limitations would exert toxic effects. Once mercury has been ingested, there is no known effective means to remove it from the body. Therefore, it is imperative that industry and others that utilize mercury make sure to contain it at the source and not to allow it escape into the environments. This principle applies to all waste material; once they are dispersed into the environment, it is almost impossible to recover them. Industries have also been trying to change their processes to reduce the use of mercury as far as possible. The other important measure would be to restrict the use of mercury compounds as fungicides, bactericides, and for other purposes.

15.2   Story of Lead Toxicity and Pollution

Occupational diseases due to lead have been known among miners and smelters, foundry and storage battery workers, potters, and paint sprayers. Lead-containing pigments were widely used in paints: white lead (lead oxide/carbonate), chrome green, and chrome yellow (lead dichromate). During the prohibition era, some illegal distillers of whisky used lead containers, resulting in illness and death.

The decline of the ancient Roman civilization is believed to be partially attributable to their widespread use of lead in their water piping systems, goblins, and wine sweetener (lead acetate). Famous painters Goya and Van Gogh, it is now believed, had gone mad and died from lead poisoning, because the paints they used contained lead-based pigments, and so they were constantly exposed to lead. In 1952, the scientists of Johns Hopkins University hospital discovered that chronic lead poisoning was quite common in children aged between one and three, and that it was frequently fatal. Lead poisoning is caused in many children by their habit of chewing foreign substances such as flaking paint. Older paints contained lead compounds.

Two sisters aged two and three died in convulsions in England in June 1967. At the same time, several domestic animals also died similarly in their neighborhood. Officials found that several more children and adults in the community showed similar symptoms. It turned out that plastic cases of lead batteries had been burned in their neighborhood and that some people apparently inhaled the lead fume that came from the residual lead adhered to the plastic cases.

The experience of a physician family in Illinois illustrates the typical toxicity symptoms and some common source of lead. Five of the six members of the family experienced fatigue, poor appetite, pains in the stomach, and vomiting. (One exception was the infant who was on a formula). They became increasingly irritable. The blood test disclosed severe anemia. An exhaustive search identified an earthenware pitcher as the source of lead. The pitcher was glazed with lead-containing compounds and it was insufficiently fired. It was used for orange juice container, and the acidic juice leached out the lead.

Gasoline in the United States and many other countries used to contain "lead." The actual name of the lead compound used in gasoline is tetraethyl lead. The chemical formula is $(C_2H_5)_4Pb$, where Pb is the chemical symbol of lead and "C_2H_5" is called ethyl group. It was added as an antiknocking agent. The lead content was as high as 4 g per one gallon of gasoline. As we drove cars extensively, we had spread lead widely. This has caused an enormous increase of lead level in the environment. Tetraethyl lead itself exhibits an acute toxicity, particularly attacking brains. For example, several workers fell unconscious when they were cleaning a large container of tetraethyl lead that was carried by an oil tanker, and they died soon after. An increased awareness of lead poisoning led to a ban of "lead" as an ingredient of gasoline in many countries, and the pigments for paint were switched from lead compounds to other lead-free compounds. However, a significant number of children suffer even now from chronic lead poisoning; the severe cases include brain damage.

How does a lead compound manifest its toxicity? First, it should be made clear that not metallic lead but a lead(II) or lead (IV) compound is toxic. Pb(IV) is quite reactive (i.e., strongly oxidizing) but is not the major form in the living system. Tetraethyl lead mentioned earlier itself is not toxic, but it loses easily one or two ethyl group(s), resulting in the formation of the very toxic $(C_2H_5)_3Pb^+$. Pb(II) bound with inorganic entities is the most commonly found form in Nature and the living systems. Pb(II) and/or the Pb of $(C_2H_5)_3Pb^+$ are believed to bind to the crucial site of proteins, enzymes, and maybe DNA, and block their functions.

Pb(II) has been demonstrated to inhibit a variety of enzymes. The most interesting aspect of lead toxicity is the fact that Pb(II) has inhibiting effects on the enzymes involved in the blood hemoglobin formation. It may be noted that the oxygen carrying protein hemoglobin consists of a protein called globin and the iron-containing heme group (called Fe(II)-protoporphyrin) (see Chap. 6). Particularly important enzymes are δ-aminolevulinic acid dehydratase (ALA-D) and ferrochelatase (FC). ALA-D, also known as "porphobilinogen synthetase," is a pivotal enzyme in the synthesis of the heme group. FC catalyzes the insertion of iron (Fe(II)) into the heme group. The inhibition of ALA-D results in an increased level of ALA in urine, which is considered to be a sensitive measure of lead poisoning. The level of non-functioning metal-free or Zn(II)-protoporphyrin will increase when FC is inhibited. It has also been reported that the globin protein synthesis is affected by lead. It may be gathered that lead strongly affects the whole hemoglobin formation processes, and a typical symptom of lead poisoning is hence anemia.

Pb(II) binds strongly with phosphate group and forms insoluble lead phosphate $Pb_3(PO_4)_2$. Newly absorbed lead is retained in the body as lead phosphate in liver, kidneys, pancreas, and aorta. This results in a decrease of available phosphate level; phosphate is one of the most important entities in the cell physiology. Pb(II) happens to be similar in size to calcium, Ca(II), and hence often replaces Ca(II) in a bone mineral, which is essentially calcium phosphate $Ca_3(PO_4)_2$. Bones store as much as 90% of the total body lead. It has been estimated that it takes 90 years for about half of the lead to be removed from the bone, assuming no new lead becomes incorporated in the bone meanwhile. That is, it is very slow.

Nuclear inclusion body was first reported in 1936 in hepatic parenchymal and tubular cells of children dying of acute lead encephalopathy. Similar inclusion bodies have also been reported in kidneys of animals (pig, rabbit, etc.) and some plants including moss. A nuclear inclusion body is a dense granular structure localized in the cell nucleus. It has a dense central core that seems to contain a high level of lead, and an outer fibril zone. The protein has a high content of aspartic acid, glutamic acid, glycine, and tryptophan. The formation of inclusion bodies has been suggested to serve as a protective mechanism; that is, sequestering of lead into such a body would make lead unavailable to sensitive locales.

15.3   Cadmium

Cadmium (Cd) is a metal similar to zinc. As a matter of fact, it is located just below zinc in the periodic chart (Fig. 19. 2). The third member of this column in the periodic chart is mercury, which was discussed earlier. In a broad sense, these three elements have similar chemical reactivities. They all make compounds in the oxidation state II, that is, Zn(II), Cd(II), and Hg(II). The only exception is that mercury can also form some compounds in the form of $Hg(I)_2$. It has been discovered recently (2004), though, that a compound containing $Zn(I)_2$ can also be produced.

Zn(II) is one of the most important inorganic elements in the biological systems, as discussed somewhere else (Chap. 6). Cd(II), being similar to Zn(II), can replace Zn(II) of the active site of zinc-dependent enzymes and proteins. Some of the enzymes with Cd(II) bound instead of Zn(II) may retain some enzymatic activities, but the majority of such enzymes and proteins lose their biological activities. This would be very likely due to the fact that Cd(II) is similar to Zn(II) but not sufficiently so. Cd(II) is much larger and less acidic than Zn(II). Hence, replacement of the essential Zn(II) in enzymes and proteins constitutes the basic mechanism of the toxic effects of cadmium.

It might be speculated that Cd(II) may be used as a catalytic element in some very specific enzymes, because Cd(II) is a good Lewis acid after all. Indeed, an enzyme was discovered recently that requires specifically Cd(II) instead of Zn(II).

A widely occurring pollution caused by cadmium was manifested in a malady called "itai-itai" disease. "Itai-itai" means literally "ouch-ouch." It causes a symptom of severe osteoporosis. Even a small movement like sneezing causes fracture of bones and severe pain; thus the name. Many people, particularly middle-aged farming women contracted this malady in the area along Jintuh river in central Japan, apparently since about 1920s. In 1955, a local doctor suggested that the malady might be caused by cadmium that was found in the water and the rice crop. Cadmium eventually was traced to a mining operation in the upstream of the river.

Cadmium seems to be concentrated in the kidney and it disrupts the reabsorption in the tubules of cations such as calcium (II) and anions, particularly phosphate. The mechanism of disruption is not very well understood, but is considered to be an inhibition of some ion-transport pumping enzymes. One of the results is loss of calcium and phosphate in urine. They have to be supplied from bones in order to maintain their concentration in the blood. Hence, the bone is lost; osteoporosis.

15.4   Arsenic

Many wells have been dug in recent years under the auspices of UNICEF for the people of Bangladesh who needed freshwater sources. It turned out that many of such wells are contaminated with arsenic. Arsenic is a natural element, and fairly widely distributed among rocks. Arsenic is often associated with important industrial metal ores containing lead and copper. Arsenic has thus also caused environmental pollution problems, in association with ore mining and smelting.

Unfortunately, arsenic was also used for decorative purposes, as in the so-called Paris Green. It was invented in 1775 and was used as a pigment in paints, wallpaper, and fabrics. Throughout the nineteenth century, there were a number of reports of people becoming ill from living in houses decorated with the poisonous wallpaper. Paris Green was not recognized as a health hazard until the end of the century. Today, some arsenic compounds are used as pesticide and fungicide.

Can you locate element "arsenic" in the periodic chart? It belongs to the column of which the top member is nitrogen, N (see Fig. 19.2). Below nitrogen is phosphorus, and the third member of the column is arsenic. Nitrogen and phosphorus are essential to all organisms as discussed in other pages of this book. Nitrogen makes up amino acids (hence proteins), nucleic acids (hence DNAs and RNAs), and a number of other biologically important compounds. Phosphorus occurs in the biological systems as phosphate (PO_4^{3-}) and its derivatives. Such compounds include nucleotides (both ribo- and deoxyribo-). A nucleotide ATP (adenosine triphosphate) contains condensed phosphate group, triphosphate, and is used as a universal energy source in the biological systems as discussed in Chap. 3. Nucleotides connect themselves through a phosphate group, resulting in the formation of polynucleotides, i.e., DNAs and RNAs (see Chap. 4).

A chemistry textbook will tell you that elements that belong to the same column (of the periodic chart) have properties similar to each other. This is indeed a basis of systematizing the elements present in the universe. Well, then how similar are nitrogen, phosphorus, and then arsenic? They are indeed similar in certain respects, but also different in others. Nitrogen is quite different from either phosphorus or arsenic. However, phosphorus and arsenic are fairly similar. Like phosphorus, arsenic forms readily arsenate (AsO_4^{3-}), which has a structure, size, and the electric charge similar to phosphate. Organisms then would have a difficulty in distinguishing arsenate from phosphate. In fact, the mechanisms to absorb the essential phosphate do absorb arsenate, if it is present in their vicinity.

Arsenic causes a number of health problems including cancer and death by acute poisoning. How these toxic effects are caused and by what chemical forms of arsenic are not very well understood. A brief story of poisonings by arsenic is given in Sect. 17.2.2.

Environmental Issues: Organic Pollutants

16

16.1 The Story of DDT (and Others)

16.1.1 The Effectiveness of DDT as Insecticide

DDT is an acronym for dichlorodiphenyl trichloroethane, but should be named more precisely as 2,2-bis(*p*-chlorophenyl)-1,1,1-trichloroethane (see Fig. 16.1 for its structure). A German chemist, Othmar Zeidler synthesized this compound in 1874 while he was pursuing his PhD. And it had remained as that, a new synthesized compound, for a quite while. Carl Müller of J. R. Geigy (a Swiss Pharmaceutical company, now Chiba-Geigy) discovered in his pursuit of insecticides that the compound synthesized by O. Zeidler was extremely toxic to houseflies. Numerous tests were conducted, and the compound DDT was found to be an excellent insecticide. Besides it is cheap to make. This was the time when the World War II was raging. DDT was then used to control lice on soldiers on the front. In earlier wars, more soldiers died of typhus (born by louse) than by bullets. The WWII was really the first war in history where more soldiers died actually from bullets than the louse-born disease, thanks to DDT. DDT was then considered to be a savior to control many kinds of harmful pests. C. Müller was awarded a Nobel prize in 1948.

A large amount of DDT was sprayed on the field and everywhere. In 1960s more than 3,000 tons of DDT was sprayed in Asia alone to control malaria (borne by mosquito). Its effectiveness is illustrated by the following example. There were 2.5 million cases of malaria in Sri Lanka in 1948 when DDT was not used. The number of cases of malaria was dramatically reduced to only 31 cases in 1962 after DDT had been extensively sprayed on homes since 1948. The number of cases returned to around two million within 5 years after the spraying of DDT on homes had been banned in 1964.

It was estimated that about 1.5 lb (0.7 kg) of DDT had been applied for each acre of the world's arable land by 1964. Let us look at an example of the effects of DDT on a crop. It is a set of data on a cotton field in a Peruvian valley (Cannette Valley). The average yield of cotton there was 406 lb/acre in 1943 (before DDT was

E. Ochiai, *Chemicals for Life and Living*,
DOI 10.1007/978-3-642-20273-5_16, © Springer-Verlag Berlin Heidelberg 2011

introduced in 1949), it went up to 609 in 1954, but then dropped to 296 in 1965; that is, less than the pre-DDT time. The major insect that eats cotton is cotton bollworm. DDT controlled the insect successfully at the beginning. But then something went wrong. One of the reasons was that DDT killed not only cotton bollworm but also other insects, some of which were predators of the cotton worm. Hence the natural control system was jeopardized.

What's more, it started to kill birds and other large animals. In 1962, Rachel Carson, an American naturalist, published a classic titled "Silent Spring." It caused an enormous stir among people, scientists, and ordinary citizens alike, and led to a public inquiry about the issue she raised at the parliament, and led to an eventual ban (1972) of the use of some pesticides including DDT in the United States and others. The title of the book, now a classic, implies that "spring" had become silent as songbirds had been disappearing. She suggested (with an enormous amount of corroborating data and anecdotes) that the causes for disappearance of songbirds were pesticides and fungicides sprayed on the field and the forests. One of the most widely used pesticides at the time was DDT.

16.1.2 The Toxicity of DDT

DDT kills insects such as flies and worms. How does it do that? Then would it not be harmful to also humans and other animals? DDT is a stable, relatively non-polar organic compound. So it has similar characters to fats and oils. As such, it dissolves well in fat and oil. Organisms of any kind are made of cells, and the wrapping (membrane) of the cells are essentially fat; chemically phospholipids (Fig. 1.5). DDT therefore sticks well to the membrane and goes through it readily and eventually is stored in fat tissues.

When applied to small organisms such as flies and worms, DDT goes to fatty tissues, particularly those in neuronal tissues. It kills an insect by disrupting the neuronal system, often through convulsion and suffocation. This is an acute effect.

What effects does it have on humans and other animals? Well, the acute effects on humans (e.g., workers in the DDT-manufacturing plants) are not lethal. It gives headache and fatigue. In an eagerness to promote the DDT's usefulness, some advocates demonstrated its non-toxicity by drinking a cocktail of DDT. The effect was slight. DDT, however, is chemically toxic for human just as for flies and worms, but the differences in body size and physiology make the toxic effects of DDT appear different. We will talk about subtler and, perhaps, more serious effects of DDT later.

16.1.3 Biological Resistance Against DDT

The other reason for the decline mentioned above about cotton yield in Cannette Valley was that DDT became less effective in killing the insects. Larger doses became necessary. The cotton bollworms in Cannette Valley were found in 1965 to be 30,000 times more resistant to DDT than they had been in 1960. In fact, as early as 1946, houseflies resistant to DDT had been noted, and by 1948 the list of resistant insects had grown to 12. This phenomenon, the development of resistance to chemical agents in insects, is one of the basic problems with the chemical agents that are meant to kill insects or other organisms, particularly microorganisms. Another example of resistance was mentioned earlier with regard to antibiotics (Chap. 7).

Antibiotics are natural chemical agents, and hence some microorganisms have evolved to carry resistance factor, which can be spread readily to other microorganisms. However, DDT is not a natural chemical but synthesized by human. How then do insects develop resistance against DDT (or any other non-natural chemical compound)? The mechanisms for this phenomenon are not well understood. However, it must be through the regular evolutionary process. The life cycle is short in smaller organisms, so that they have more chances to evolve than the larger animals like us do.

Some DDT-resistant houseflies were found to contain an enzyme DDT-dehydrochlorinase that removes hydrogen chloride (HCl) from and convert DDT to DDE (dichlorodiphenylethylene). Recall the case of a resistance factor against penicillin (Chap. 7). A resistance factor in the case of penicillin was an enzyme to decompose penicillin. A similar strategy is used here. Other insects seem to convert DDT to 2-OH-2,2-bis (chlorophenyl)-1,1,1-trichloroethane (=dicofol).

16.1.4 Other Chlorinated Organic Compounds and Dioxin

The success of DDT in 1940s and 1950s led to a development of all kinds of chlorine-containing organic compounds for pesticides, herbicides, and disinfectants. Many of them were introduced as commercial products, and used widely. Some of those compounds are shown in Fig. 16.1 along with DDT. These seven compounds (except *o,p*-DDT) are often labeled as "dirty seven" for their toxic effects.

One of the most feared chlorine-containing compounds is "dioxin." Chemically it is 2,3,7,8-tetrachloro dibenzo paradioxin (TCDD). It is reputed to be one of the most toxic substances. It is strongly carcinogenic like many other chlorine-containing organic compounds, but its effects on the endocrine system may be more critical at lower levels, as will be talked about later. This compound, unlike all the other compounds mentioned so far, is not one that is intentionally produced. Instead, it forms as a byproduct in a number of synthetic processes to produce chlorine-containing compounds such as 2,4,5-T (2,4,5-trichloro-phenoxyacetic acid: used as a herbicide) and 2,4-D (2,4-di-chlorophenoxyacetic acid). These herbicides were used during the Vietnam War to clear the obstacles off the way. They happened to be contaminated with dioxin, and it has been believed to have caused numerous health problems among the veterans as well as Vietnamese peasants. High-temperature combustion of chlorine-containing organic compounds such as polyvinyl chloride is said to produce dioxins.

"Dioxin" as a byproduct of many chemical processes is not a single chemically pure compound. The skeleton structure is the same, but the number of chlorine atoms bound and the position of chlorine attachment can vary. The number of chlorine atoms bound can be any between one and eight (Fig. 16.2). The most common and usually the most prominent component is TCDD. It has been demonstrated to be the most toxic among the relatives, and it is known to bind the receptor most strongly among them, as described later.

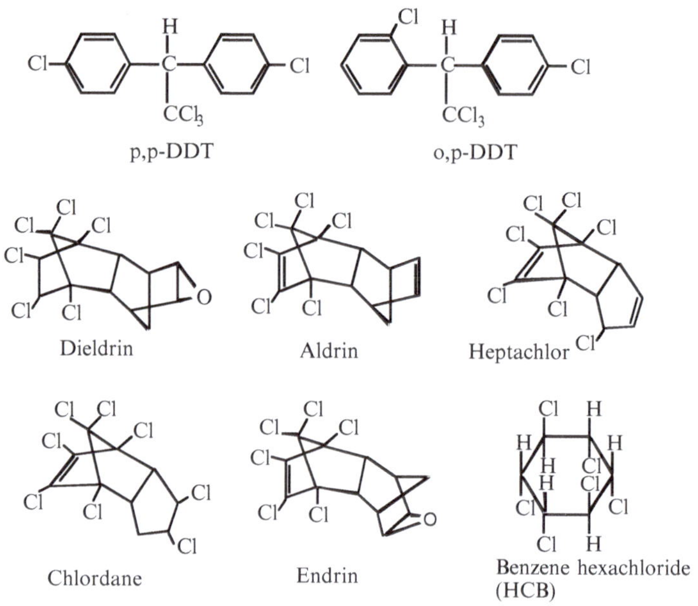

Fig. 16.1 DDT and other similar chlorohydrocarbon pesticides (dirty seven)

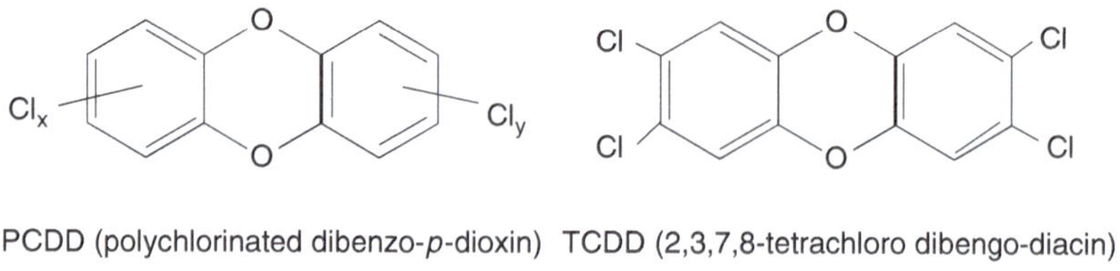

PCDD (polychlorinated dibenzo-*p*-dioxin) TCDD (2,3,7,8-tetrachloro dibengo-diacin)

Fig. 16.2 Dioxins

16.2   Endocrine Disruptors

A book titled "Our Stolen Future" appeared in 1996. The authors Theo Colborn, Dianne Dumanoski, and John Peterson Myers argued that some chemical substances dispersed into the environment have gotten into human bodies and disrupted the endocrine (hormonal) system of the humans (and other animals). They built their thesis based on a number of scattered data.

For example, a Danish reproductive scientist, Niels Skakkebaek started to observe sperm abnormalities as well as a drop in the sperm count (the number of sperms in a unit volume). He also noticed that the number of patients with testicular cancer tripled in Denmark between 1940s and 1980s. He and his coworkers checked literatures available at the time and found to their surprise that the average human male sperm counts dropped by 50% between 1938 and 1990 everywhere including US, European countries, India, Nigeria, Brazil, Libya, Thailand, etc. This rapid change, the authors contended, cannot be due to changes in the genetic composition but must be due to some environmental factors.

Another such symptom is the decreased survival of alligators in Lake Apopka in Florida. An accident at a chemical factory along the lake wiped out about 90% of the alligator population. Yet, long after the lake water seemed to have turned normally clean, the remaining female alligators had a lot of reproductive problems. Only about 18% of their eggs hatched, and many of the hatchlings died early, while the normal hatching percentage is considered to be about 90%.

This second example suggests that even a very small quantity of an environmental contaminant might be effective in disrupting the reproductive system. This (and of course, a lot other reports) led to a thesis that such an effect is very likely due to the effect on endocrine systems, as they are based on low level of specific chemicals, i.e., hormones.

In 1970s, many young women developed a special kind of vaginal cancer in the United States. It turned out that their mothers had taken a medicine called DES, diethyl stilbestrol, during their pregnancy. Many other DES daughters developed malformed organs such as uterus. Apparently, an exposure to DES during the prenatal and neonatal stage of human development, particularly at the earlier stages, has a profound effect on the later development in the affected person, both man and woman. The effects of DES seem to be very widespread, on the reproductive system, immune system, and possibly on the brain functions as well.

DES was not an accidental, environmental pollutant, but it was developed by a drug company, approved by FDA, and touted as a wonder drug, and prescribed widely for preventing miscarriage, morning sickness, etc. in the United States and other countries starting in 1950s. It was supposed to mimic the indigenous hormone, estrogen. Let us take a look at the chemical structures of the two compounds, estradiol and DES, as shown in Fig. 16.3. The overall similarity of their structures is obvious,

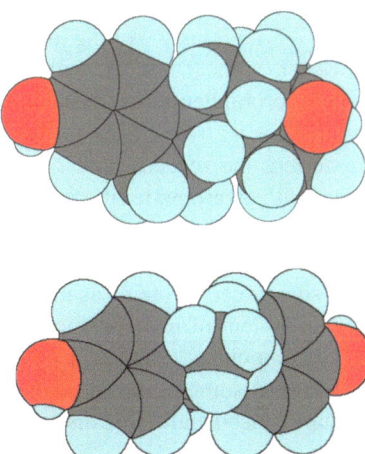

Fig. 16.3 Estradiol (*upper*) and DES

and that makes DES a good substitute for estrogen as far as their interaction with the receptor site is concerned. In other words, the receptor site may not be able to distinguish the two compounds. This is chemistry. However, the manifestations of that interaction with the receptor are physiological issues, and whether it causes good or bad effects depends on many factors including timing of the drug-taking and dosage. The physiological effects, however, are eventually caused by chemical interactions between the chemical compound and the cells/tissues/organs, but this aspect is not very well understood yet, and besides, it is beyond the scope of this book.

Today, an endocrine disruptor is defined as an exogenous substance or mixture that alters function(s) of the endocrine system and consequently causes adverse health effects in an organism or its progeny or (sub)populations. This definition does not say anything about the mechanism of disruption. There are several ways of disruption of the endocrine system by an exogenous substance. Some of them are: (a) a substance interacts with the receptor of an endogenous hormone as in the case of DES mentioned above and disrupts the interaction of proper hormone molecules with the receptor; (b) a substance disrupts storage, secretion, and transport of a hormone; and (c) a substance disrupts the proper processing of production and metabolism of a hormone. The last mechanism can be more general, including such things as inhibition of the enzymes involved in the metabolism of a hormone.

Lists of suspected endocrine disruptors have been published and revised over and over again. It is a rather difficult task to establish the cause–effect relationship with regard to an endocrine disruptor, as the level of presence of such a substance necessary to cause an effect could be fairly low and the relationship could be indirect. We will list several different types of suspected endocrine-disrupting compounds, as follows.

16.2.1 Those Binding to the Steroid Hormone Receptors

These substances bind to the steroid hormone receptors, and disrupt the hormone systems concerned. Synthetic estrogens including DES and those mentioned in Sect. 7.7 obviously belong to this type, because of their structural similarity. However, some other compounds such as bisphenol A, and hydroxylated PCB (polychlorobiphenyl) are also considered to bind to the receptors despite the fact that there is no overall structural similarity (see Fig. 16.4). If they do indeed bind as suggested, the receptor site could conceivably bind them through an OH group along with an aromatic ring in its vicinity. This is a common feature of these different types of compounds. In other words, the receptor site may recognize this feature rather than the overall structural similarity. Many of the steroid hormone receptors are "zinc finger" proteins as mentioned in Sections 6.4 and 7.7.

The system of male hormones, androgens, can also be subject to disruption by some compounds. Such compounds include DDE (a DDT metabolite mentioned earlier) and some esters of phthalic acid. However, whether the esters of phthalic acid are indeed endocrine disruptors is still being investigated.

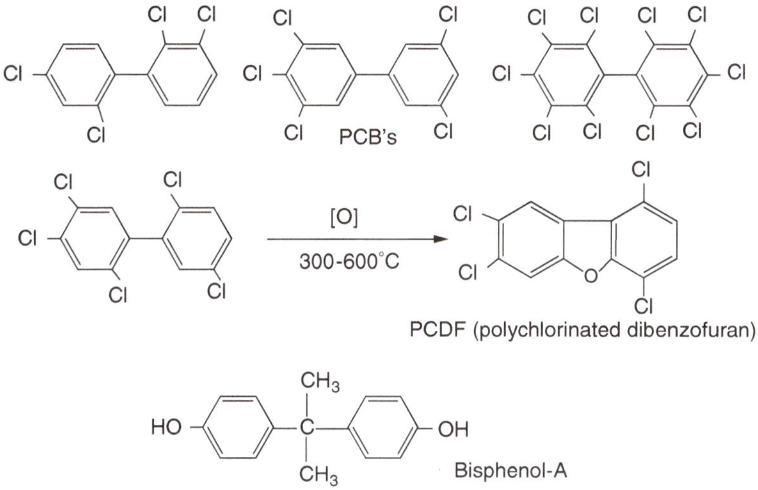

Fig. 16.4 PCB etc.

16.2.2 Those Affecting AH Receptor and Cytochrome-P450 Enzymatic Systems

Dioxin, PCB, DDT, and the like, i.e., chlorinated aromatic hydrocarbons are believed to bind to a special receptor called "AH" receptor (AH stands for aryl hydrocarbon). The AH receptor bound with dioxin (or PCB) enters into the nucleus of the cell, and binds to a specific site of a DNA. One of the effects caused by such a binding seems to be an induction of an enzyme dependent on cytochrome P-450. Cytochrome P-450-dependent monooxygenase was talked about in Sect. 6.2.2, and is involved in the metabolism of a wide variety of compounds including steroids and foreign substances (drugs). Probably the P-450 enzymes induced by dioxin, PCB, and others would then metabolize steroid hormones unnecessarily, and thus disrupt the endocrine system. The details are still not very well understood. However, you see that the processes are all chemical reactions at the deepest level.

Use, Abuse, and Misuse of Chemicals

17

Chemicals surround us. We ourselves are made of chemicals, we need them, and we are absolutely dependent on them. There is no escaping from this fact. Our lives are a festival staged and performed by chemicals.

However, there are also chemicals we can do without. A living organism always tries to enhance its own survival; it is the basic nature of living things. In that effort, an organism may create a substance that is harmful to others; this is a natural defense mechanism. Such a substance is a natural chemical weapon and can be harmful to the human body (i.e., natural toxin), and we can do without it. People have used throughout history a variety of chemicals with an intention to harm others; poisons (of arsenic, cyanide, many poisons (alkaloids and others) obtained from plants and others), explosives and chemical weapons (mostly neurotoxins). We are better off without them. I must hasten to add, though, that some of these chemicals can be used for useful purposes as well.

In addition, we humans have created a civilization in which a vast number of goods that are made of chemicals are produced or synthesized and put into use for enhancing our lives. Altogether about 20 million compounds have been registered by the end of twentieth century. The intended use of most of these substances is legitimate and is beneficial, as we talked about on the pages of this book except for a few chapters. But, after use, they are often discarded, left to deteriorate and decompose and be dispersed into the surroundings. The environment may tolerate the presence of these chemicals to a certain extent, and can even get rid of them by decomposing and converting them to less harmful forms. By the way, the self-cleaning of the environment is done biologically and/or chemically. Most of these processes involved are essentially chemical reactions even if done in the biological systems, i.e., the so-called bioremediation. Some purely physical processes may also be involved in certain cases. The catch is, though, that there is a limit to this self-cleansing capacity of nature (the environments). Beyond this threshold, the environment and the organisms in it may be adversely affected.

E. Ochiai, *Chemicals for Life and Living*,
DOI 10.1007/978-3-642-20273-5_17, © Springer-Verlag Berlin Heidelberg 2011

Some compounds synthesized artificially by mankind are difficult to decompose. They are made to last forever or at least for long. The reasons that they are difficult to decompose are (1) they are usually chemically very inert, and (2) Nature (meaning organisms) does not know how to handle them because they have had no previous experience of handling them. Residual material from such chemicals can be an environmental hazard because organisms that come in contact with them do not have mechanisms to render them harmless. However, these incidents are not intended results, and are simply results of bad management and neglect including disposal of certain chemicals. Some aspects of this issue have been discussed in the last two chapters. It is also to be noted that a lot of efforts are being made to produce the so-called "bio-degradable" products, so that they can be decomposed and be rendered harmless by the microorganisms in the environment.

Chemicals, useful or not, can also be misused because of lack of knowledge or misunderstanding of how to handle them. Let us differentiate "misuse" (incorrect use) from "abuse"; the former involves no intentional improper use whereas the latter does.

Human beings tend to abuse chemicals for certain specific gains, including entertainment. We do so, particularly when there is pressure from others. This is an intentional improper use of a chemical compound other than its proper use. For example, ammonium nitrate is a very good chemical fertilizer, but it happens to be usable as a basic ingredient in an explosive, as exemplified in the bombing of the Oklahoma federal building (see also Chap. 8). (It has to be mentioned that overuse of strictly inorganic fertilizers such as ammonium nitrate can degrade the soil, and nitrate released into water systems pollute the water. Another more commonly used fertilizer, ammonium sulfate is worse than the nitrate in degrading the soil, partly because sulfate combines calcium to form calcium sulfate rendering calcium less available to plants and others). Hormones that enhance muscle and blood formation can be used for the purpose of gaining an advantage in athletic competitions. A number of gold medalists at 2004 Athens summer Olympic Games were stripped of their medals, when they were tested positive for such drugs. We will talk about a few examples of chemicals abused often.

Many of abused chemicals and animal toxins interfere with the functions of brain and its associated organs. Therefore, we will briefly review here the chemistry/ physiology of the brain.

17.1 Chemistry of Brain Function: Basics

The nervous system consists of the central nervous system (brain and spinal cord), the peripheral or autonomous (sympathetic and parasympathetic) nervous systems, and the sensory organs. The brain is where a huge number of signals are brought in from the sense organs and others and processed and bodily functions are effected accordingly. These signals and their interpretation by the brain represent the consciousness; they manifest as "sensation", "emotion", "memory", "thought", etc.

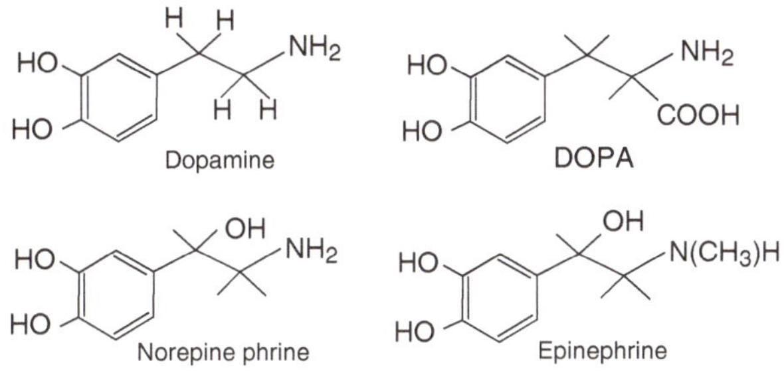

Fig. 17.1 Catecholamines

They elicit some actions such as muscle movement. Hence, disruption of the brain functions can lead to such a result as emotional/mood change or paralysis, etc.

The brain consists of cells called "neurons"; there are several hundreds of billions of such cells in our brain. A neuron makes connections to a number (several thousands) of other neurons. The signal travels within a neuron as an electric signal; the electric impulse is created by ionic movement through ion channels across the cell membrane such as sodium and potassium channel. When the signal comes to a junction called "synapse", it has to be carried by a chemical compound across the narrow gap (which is 5–20 nm wide) between the neurons. Such a chemical compound is called "neurotransmitter". A neurotransmitter molecule binds to the receptor site on the surface of a post-synaptic neuron. The receptor is an ion channel, and the binding of a neurotransmitter opens the channel, hence resulting in the creation of a new electric signal on the postsynaptic cell.

Several different types of neurotransmitter compounds are known. One is acetylcholine, and its mechanism of action seems to be best understood and will be discussed shortly. The second group called "catecholamines" contains rather familiar compounds such as adrenalin (epinephrine), noradrenalin (norepinephrine), and dopamine (Fig. 17.1). The third group consists of several amino acids such as regular ones: glutamic acid and aspartic acid, and unusual ones such as γ-aminobutyric acid (GABA) and N-methyl-D-aspartate (NMDA). Another group contains several small proteins (peptides). Examples are enkephalin, endorphin, gonadotropin, oxytocin, and vassopressin.

The mechanisms of all the neurotransmitters' function have not been well understood. One neurotransmitter, acetylcholine, has been studied intensely and will be discussed briefly here. The structure of acetylcholine is shown in Fig. 17.2, along with those of nicotine, serotonin and muscarine. Acetylcholine is released from the synapse and migrates to the next neuron surface nearby, and binds to a receptor. There are two types of acetylcholine receptor; one is "nicotinic", implying that it also binds nicotine, and the other "muscarinic". Nicotine and muscarine (as seen in Fig. 17.2) have a common feature, i.e., a positively charged nitrogen such as acetylcholine. The receptor itself is an ion channel. When acetylcholine binds with the receptor, the ion channel opens up, and the inflow of Na^+ and outflow of K^+ ensues. The sudden change of the cationic concentration inside and outside of the postsynaptic neuron creates an

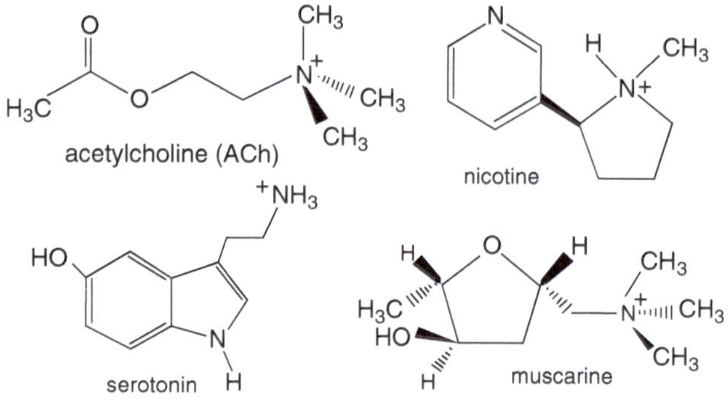

Fig. 17.2 Acetycholine, nicotine, serotonin and muscarine

Fig. 17.3 A schematic representation of (nicotinic) acetylcholine receptor (from Dougherty DA and Lester HA (2001) Nature 411: 252–25)

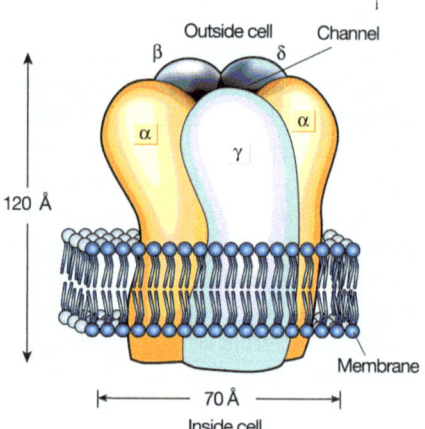

electric potential change; this is the electrical impulse that is then propagated along the neuron. Acetylcholine released to the inter-synaptic space has to be decomposed so that the impulse caused is stopped; otherwise the elicited effect continues and can result in the disruption of the nerve impulse propagation. The decomposition of acetylcholine is carried out by an enzyme called "acetylcholine esterase". We will later look into acetylcholine esterase in connection with chemical weapons.

The structure of a nicotinic acetylcholine receptor has been determined and is shown schematically in Fig. 17.3. It has been demonstrated that acetylcholine binds to a segment of the receptor, in which the binding center is an amino acid residue, tryptophan (149-Trp). Tryptophan has the structure shown in Fig. 17.4, and has a negatively charged portion in the center of the aromatic rings (hexagon and pentagon). The positive charge on the end of acetylcholine binds to this negative charge (Fig. 17.5). Nicotine, having a positively charged nitrogen such as acetylcholine, does bind to the same general area containing the 149-Trp residue, but the major interaction seems to be with different residues in the same area.

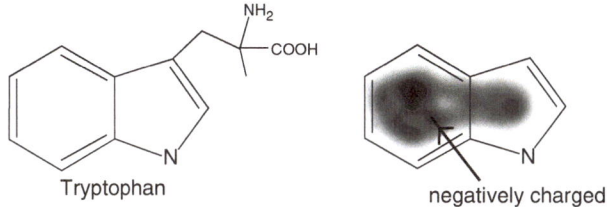

Fig. 17.4   Tryptophan and its electric charge distribution

Fig. 17.5   Binding of acetylcholine to the (149) tryptophan residue (from Zhang W et al. (1989) Proc Nat Acad Sci (USA) 95: 12088–12093)

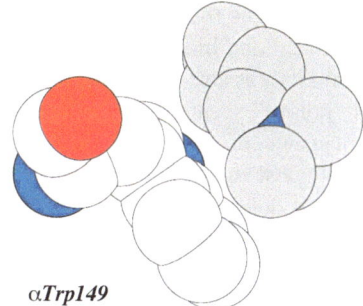

It has been demonstrated also that serotonin (another neurotransmitter, alternatively called "5-HT (5-hydroxytryptamine), see Fig. 17.2) similarly binds to a receptor through the interaction between a tryptophan residue and the positive charge on serotonin.

Catecholamines as depicted in Fig. 17.1 are not electrically charged, but under physiological conditions (pH), a nitrogen atom in each of these compounds gains a proton (H^+) and hence becomes positively charged. This is Lewis acid–base interaction due to the presence of a lone pair on the nitrogen atom and was talked about in Chaps. 1 and 19. It is possible, though it has not been verified, that the positively charged nitrogen entity may be involved in binding to the respective receptor site as in the case of acetylcholine and serotonin. It is further speculated that since alkaloids such as morphine and poison frog's batrachotoxin (see below) also have similar nitrogen atoms they can become positively charged under the physiological condition and, perhaps, bind similarly to the receptors. Such a binding of a foreign compound to the neurotransmitter receptor site obviously disrupts the brain functions.

The workings of brain are very complicated. Hence, there is a complicated, yet to be understood, often convoluted relationship between the basic chemical interactions of a neurotransmitter or a drug with the receptor and the physiological manifestation, e.g., psychic effects. Chemistry can suggest only the basic mechanisms.

17.2   Poisons and Toxins

Harmful substances are everywhere. In fact almost any compound can be harmful. Even water can be used to cause death if administered excessively, as illustrated by a hazing incident (in the Spring of 2003) at an upstate New York college campus where a freshman was forced to drink an excessive amount of water and died. However, the cause of the death is not the toxicity of water but rather because the excessive presence of water dilutes salt (sodium chloride) in the blood, which causes a salt imbalance (hyponitremia).

Poisonous substances can have harmful effects, even when given in a small quantity. As all the living systems live rather precariously, depending on intricate checks and balances (homeostasis and others), they are susceptible to the effects of many chemical and biological (though eventually, chemical) substances. The mechanism of toxicity is also widely varied. Let us look at a few prominent poisons. By the way, a "poison" seems to be defined as a non-biological toxic substance, while a toxic substance of biological origin, particularly a microorganism, is called "toxin". Biological toxins are chemicals that enhance the producers' survivability; they are used as defense and/or mechanism to capture preys.

17.2.1 Common Poisons

Carbon monoxide, CO, is one of the most common poisons. Any carbon-containing substance will produce carbon dioxide CO_2 as one of the ultimate products when burned. In our body as well, carbon dioxide is produced as the final product when we metabolize our food. It is exhaled from the lungs, and is not chemically toxic. A high concentration of carbon dioxide in the atmosphere can cause asphyxia, as carbon dioxide cannot support respiration. But this is not due to the chemical toxicity of carbon dioxide.

When a substance is not completely burned, it tends to form carbon monoxide CO rather than carbon dioxide CO_2. Chemically the difference is subtle; carbon monoxide has only one oxygen atom bound to a carbon atom, while two oxygen atoms are bound with one atom of carbon in carbon dioxide. But this difference in chemical structure causes a big difference in their chemical reactivity. Carbon monoxide binds strongly to the iron atom in hemoglobin (carbon dioxide would not do this). Oxygen (O_2) is to bind to the iron atom in hemoglobin to be carried through the blood. Oxygen and carbon monoxide competes for the iron of hemoglobin. It turned out that carbon monoxide binds about 500 times more strongly to the iron of hemoglobin than oxygen does. Therefore, once carbon monoxide binds to hemoglobin, it cannot carry oxygen. Our body needs a continuous supply of enough oxygen in order to be alive.

Hemoglobin is responsible for the red blood, when it is bound with oxygen or carbon monoxide (or nitrogen monoxide). Some hemoglobin in our blood stream is devoid of oxygen. When hemoglobin is not bound with oxygen or carbon monoxide, that clear red color fades, though it is still reddish. However, carbon monoxide-bound hemoglobin is strongly red. The victim of asphyxiation by carbon monoxide often shows pink skin due to this fact.

Cyanide is another widely known poison. Cyanide (CN^-) usually comes in the form of potassium cyanide (KCN), sodium cyanide (NaCN), or hydrogen cyanide (HCN). The first two are white solid, and hydrogen cyanide is a pungent-smelling gas. Cyanide binds very strongly with a metal ion such as Fe(II), Fe(III), Cu(II), Zn(II), and many others. These ions constitute important portions, that is, the active sites of many enzymes and proteins (Chap. 6). When a cyanide ion binds with a metal ion, the enzyme's function is disrupted. Cyanide that enters into our body almost indiscriminately binds to any metal ions and disrupts their functions, but an especially sensitive place is the last enzyme in the whole series of respiratory chain. The enzyme contains both iron (embedded in the porphyrin as in hemoglobin) and copper. Cyanide binds to both iron and copper in this enzyme, and stops its function, the last step of respiration. What is the result? It stops respiration; hence it stops the production of bodily energy (ATP). As we talked about in Chap. 3, we need to continuously produce ATP. Otherwise, we would die.

17.2.2 Arsenic

Arsenic is reputed to be one of the most widely used poisons to kill people. For example, in the medieval Italy, a potion called "Aqua Tofana" was sold to women who wanted to kill unwanted husbands. It was based on arsenic. Amadeus Mozart died at an early age of 36. The movie, "Amadeus", suggests that somebody (Antonio Salieri in the movie) poisoned him. The poison used has been reputed to be an arsenic compound, though some believe that it was an antimony compound. By the way, it would be interesting to note that arsenic and antimony belong to the same column of the element periodic chart (Fig. 19.2), and that they have similar chemical properties. Murders by arsenic poisoning appeared to be almost fashionable in the earlier Victorian England. Murderers usually could get away, because no good detection method of arsenic was invented until the mid-nineteenth century. This invention by a British chemist James Marsh stopped the situation. Even today, though, arsenic-containing fungicides and pesticides are often used for the purpose of killing somebody. The environmental issues regarding arsenic have been mentioned earlier (Sect. 15.4).

17.2.3 Animal Toxins: Frog Toxins, Snake Venoms, Etc.

Animals have developed means to defend themselves and/or to capture preys easily by poisoning. For example, some 170 frog species are known to be poisonous. Most of them are brilliantly colored: yellow, red, and blue, and live in the tropical rain forests. The deadliest among them is *Phyllobates terriblis*. Its toxin is called "batracho-toxin". One frog of this species has an amount of toxin enough to kill 20,000 mice or eight humans. Another toxin found on poison frogs is pumiliotoxin. These toxins have been used for "poison darts" by the native people in South America.

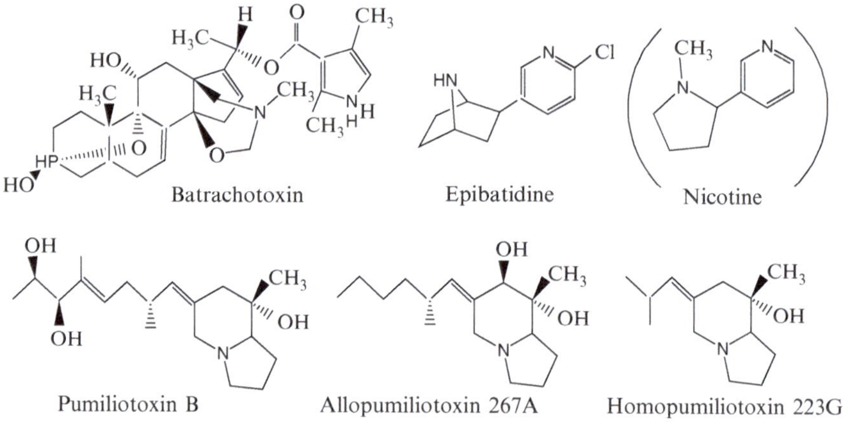

Fig. 17.6 Frog toxins

These are all complicated lipophilic "alkaloids"; "lipophilic" means "oil-like" and "alkaloid" is a collective name of naturally occurring chemical compounds containing nitrogen atom, which exhibits some basicity or "alkalinity". The chemical structures of some of these substances are shown in Fig. 17.6. One of the characteristics of these compounds is the protonation of the alkaline nitrogen; they form a positively charged entity at the nitrogen site. The positively charged nitrogen entity is one of the features of many of neurotransmitters as talked about earlier. Hence these alkaloids could disrupt the trans-synaptic communications (see Sect. 17.1).

Apparently a frog itself does not produce the alkaloid toxin, for a frog kept in a controlled environment does not have the toxic alkaloid. It has been demonstrated that frogs get the toxins from the preys they eat, such as ants, termites, and flies, and they also seem to be able to modify the chemicals to make them more deadly. One of such preys has been identified as *Choresine pulchra*, a small beetle of about a rice grain size. As might be gathered from this description, similar alkaloids including batrachotoxin, have been found in some birds as well.

A poison mushroom, *Amanita muscaria*, contains muscarine. Muscarine (see Fig. 17.2) binds to a receptor of acetylcholine, a neurotransmitter, as was mentioned earlier (Sect. 17.1). However, the primary active component of the mushroom is muscimol that is an agonist at GABA-A receptors.

A puffing fish called "fugu" is a delicacy to many Japanese. Unfortunately, it contains a toxin called "tetrodotoxin". Unless you know exactly where it is located and how to remove it safely, eating the fish can be likened to a Russian roulette. As a matter of fact, a number of people die every year from attempting to help themselves. Tetrodotoxin, shown in Fig. 17.7, is an antagonist for a sodium channel; it blocks the channel, so that the sodium ion cannot pass through the channel. The result is stoppage of the electrical signal conductance. A similar toxin called "saxitoxin" is found in some flagella algae causing red tide. It kills fish and shellfish in red tide. In these compounds, the $NH_2^+=$ group mimics the positive ion, Na^+ or K^+, and binds to the cation channel which is to bind and allow the passage of Na^+ or K^+.

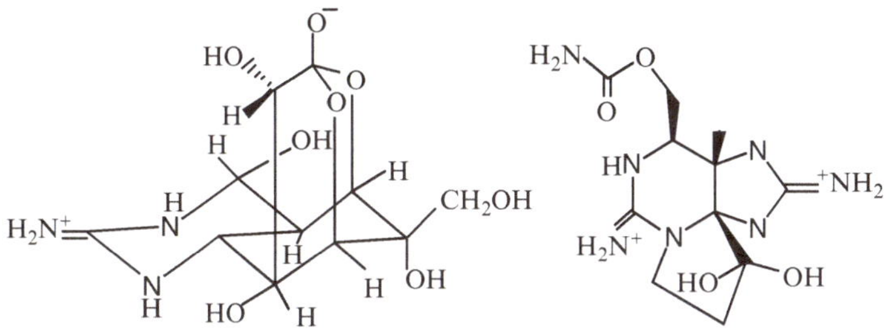

Fig. 17.7 Tetrodotoxin and saxitoxin

Snakes are another group of animals that include a number of poisonous species. The toxins, called snake venoms, are quite different kind of chemicals from those of frogs. They are mostly proteins, i.e., enzymes. Enzymes catalyze biochemical reactions. The enzymes injected by snakebite disrupt the physiology of the animal bitten. They can be grouped into two general kinds. One is "neurotoxins" and the other "hemotoxins". Neurotoxic venom attacks the central nervous system of the victim. For example, the enzyme acetylcholine esterase in the venom decomposes acetylcholine, a neurotransmitter (see Sect. 17.1), and hence disrupts the passage of signal, resulting in relaxed muscles. The victim may have heart failure or difficulty in breathing. Cobra and sea snakes are examples whose venoms contain mostly neurotoxin. Hemotoxic venom attacks the circulatory system and muscle tissue. Rattlesnake and copperhead are examples that use hemotoxic venom. The enzymes known include: amino acid oxidase, hyaluronidase, proteinase, adenosine triphosphatase (ATPas), and phosphodiesterase. Many other enzymes have been isolated from snake venoms but are not very well understood.

17.2.4 Anthrax Toxin

In the aftermath of the 9/11 incident (2001), the anthrax scare killed several people and shut down the major portion of the parliamentary building and several post offices on the Capitol Hill and in other places in the United States. Anthrax is considered to be one of the most popular biological weapons. This toxin is quite different from the toxins talked about above.

Bacillus anthracis, the bacterium responsible for the anthrax toxins produces three proteins, which are collectively known as anthrax toxin. The spores of *B. anthracis* are ingested in the macrophages, one of the immune system cells. The macrophages usually destroy such invading microorganisms, which is their normal function. In this case, however, the reverse happens; i.e., the spores kill the macrophages. The first protein of the three helps the other two proteins to enter a target cell, and the second protein disrupts the cell-signaling process and kills the immune system cell.

The third protein, called the edema factor, is an enzyme catalyzing the conversion of adenyl triphosphate (ATP) to cyclic adenyl monophosphate (cyclic-AMP). The enzyme is called adenyl cyclase. The ordinary adenyl cyclase is an endogenous enzyme involved in the signal transduction. The edema factor's catalytic portion sequesters calmodulin, another protein involved in the signal transduction, and hence makes it unavailable for a proper function. The edema factor is not catalytically active until it binds calmodulin. Once it has bound calmodulin, it starts making cyclic-AMP. In fact it makes cyclic-AMP in excess, and hence causes the death of the cell. The chemical details of these phenomena are yet to be studied.

17.2.5 Chemical Weapons

Chemical weapons are not human invention in the origin. Plants, bacteria, insect, etc. have been producing chemical weapons for billions of years; they are their tools for survival. Some of such examples have been talked about in the previous sections. They are of course natural products; there are not conscious efforts to hurt others. Those who developed such weapons have gained some advantage for their own survival in the biological evolution.

Home sapiens, a conscious being, can judge their own action's consequence and impact on others. Yet, we have developed chemical and other weapons to prevail over others, in addition to the purpose of self-defense.

Ancient Chinese used arsenic together with explosives. It was not until the World War I that the nations earnestly started to develop and use extensively chemical weapons.

The simplest chemical weapon used was chlorine (Cl_2) gas. It irritates eyes and mucous layers of throat, and is highly corrosive. One of the deadliest chemical weapons is phosgene ($COCl_2$). It was used for the first time in 1915, and it accounted for 80% of all chemical fatalities during World War I. It can be produced easily and cheaply. Its vapor is heavy; the density is about 3.4 times that of air. Therefore, it lingers long in the trenches. The main feature of phosgene poisoning is massive pulmonary edema. Both chlorine and phosgene reacts with water (mucous), and forms the strong and corrosive hydrochloric acid: $Cl_2 + H_2O \rightarrow HCl + HClO; COCl_2 + 2H_2O \rightarrow 2HCl + CO_2 + H_2O$. These are sometimes called "choking agents".

Some of the more modern chemical warfare agents are "blistering agents" and "nerve agents". Some representative examples of such agents are shown in Fig. 17.8. The main blistering agent is the so-called mustard gas, sulfur mustard; chemically di(2-chloroethyl) thioether or bis(2-chloroethyl) sulfide. It is cheap to produce and is easily absorbed by skin and other epidermis. It was first produced by the French army in 1917 but was used extensively by the Germans in the World War I. The main chemical effect is the production of corrosive hydrochloride HCl through replacement of Cl by OH of water, like phosgene. That is:

$$ClCH_2CH_2SCH_2CH_2Cl + H_2O \rightarrow ClCH_2CH_2SCH_2CH_2OH + HCl$$

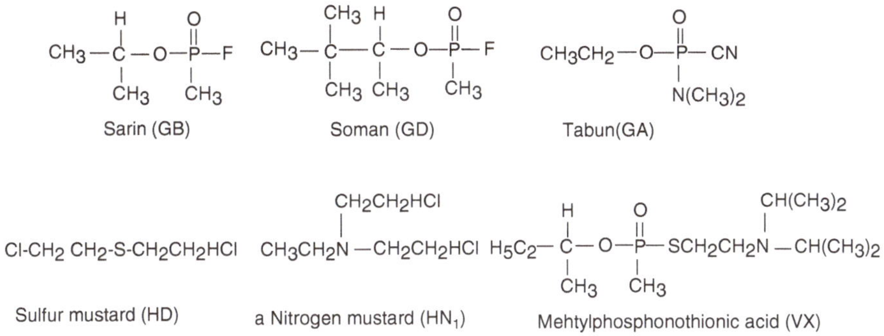

Fig. 17.8 Nerve agents and Mustards

The eye-irritating effect and blistering effect are likely caused by HCl. This reaction is not very fast unlike that of phosgene, and hence there is usually a delay of a few days before its poisonous symptoms show up. The nastiness of mustard gas is due to its small size and relative chemical stability, and hence it can go into almost any place in your body and its effect lasts relatively long. Nitrogen mustard is supposed to work similarly.

Sarin, tabun, soman, and VX agent belong to the nerve agents. Sadam Hussein had allegedly used sarin and other nerve gases against Kurds and other dissidents. A cult group in Japan used sarin to disrupt the routine lives in Tokyo by spreading it at several subway stations and in subway trains during a morning rush hour.

The nerve agents block the function of acetylcholine esterase. Acetylcholine is a neurotransmitter as talked about earlier, and it has to be decomposed once it has done its part. Its decomposition is carried out by acetylcholine esterase. When this enzyme is blocked, acetylcholine would accumulate and the nerve transmission would be disrupted.

Well, let us see briefly how it works. Acetylcholine esterase is one of the so-called serine-dependent esterases and proteases. The catalytic entity is an amino acid serine along with two other amino acids (histidine and glutamic acid in acetylcholine esterase). The serine residue, which has a structure shown in Fig. 17.9, acts as a base catalyst in conjunction with histidine. The oxygen atom of serine binds to the substrate, acetylcholine, and cleaves the C–O (ester bond) of acetylcholine, as shown. This is the basis of catalysis. Now, if you add one of these nerve agents, it binds to the oxygen atom of the same serine residue as shown in Fig. 17.9e, f. This bond between P and O (of phosphate ester) is quite strong, and hence acetylcholine cannot bind to the essential serine, and hence would not be decomposed. That is, the regular nerve conductance will be disrupted.

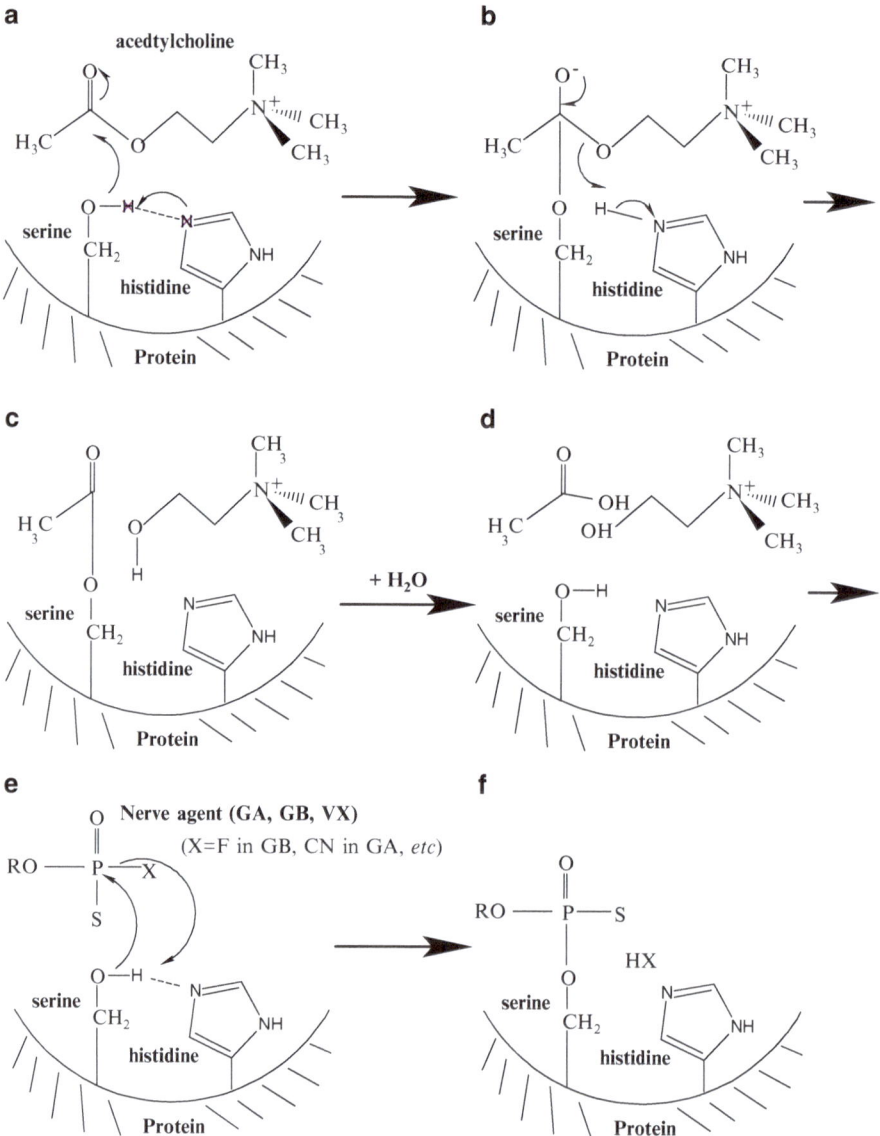

Fig. 17.9 Mechanism of acetylcholine esterase and the binding of nerve agents

17.3 Psychoactive Drugs and Their Abuse

A variety of chemicals influence the brain and its peripheral functions. These are collectively called "psychoactive drugs". Caffeine, nicotine, alcohol, opiates, meth-amphetamine, ecstasy, etc. can all be considered to be "psychoactive drugs". We will discuss here some of the psychoactive drugs that are not talked about in the other places of this book.

17.3.1 Alcohol

Alcohol is perhaps the oldest chemical discovered and produced by humans, which they have used to entertain themselves and have often abused. The pure substance is ethanol C_2H_5OH, and is soluble in both water and organic solvents. [Alcohol is technically a general name for compounds that have an OH group. Methyl alcohol or methanol, CH_3OH and isopropyl alcohol (=propanol-2, used as a rubbing alcohol) C_3H_7OH are two examples of alcohol]. The group OH in it has an affinity to water (through hydrogen bond), whereas the C_2H_5 part is like oil and hence the alcohol can mix with many organic compounds, including fats. When you drink alcohol, it is readily absorbed through the stomach and will be carried throughout the body via the blood circulatory system. This happens because it can readily go through the cell membrane of the lining cells of stomach wall, as the cell membrane is like oil in chemical property, and also because it dissolves readily in blood.

Alcohol, like other foreign substances, is metabolized mainly in the liver. The major metabolic pathway is through an enzyme called alcohol dehydrogenase, which turns ethanol to acetaldehyde: nominally $C_2H_5OH \rightarrow CH_3CHO$ (acetaldehyde) + H. This chemical equation indicates that a hydrogen atom is removed from ethanol in this process; hence the effecting enzyme is called "de (removing)-hydrogen-ase". Removal of hydrogen is a kind of "oxidation".

Acetaldehyde is toxic to the body, but relatively quickly oxidized further to acetic acid: nominally $CH_3CHO + O \rightarrow CH_3COOH$ (acetic acid). In this case, an oxygen atom is added; this is also a kind (or rather a proper kind) of oxidation reaction. This reaction is catalyzed by an enzyme called "aldehyde oxidase". Acetic acid is then incorporated into the regular process of carbohydrate metabolism, and hence alcohol gives a significant amount of calories to the body.

The amount of the enzyme alcohol dehydrogenase is genetically controlled; there is a gender difference also, and so different persons have different amounts of this enzyme. This means that alcohol tolerance varies from one person to another.

By the way, wine, non-distilled, tends to become sour once the bottle is opened. The reason is that some of these enzymes still remain in the wine (because it was not distilled), and hence the ethanol undergoes oxidation processes to turn into acetic acid. It may not be necessary to remind you that acetic acid is the main ingredient of vinegar. Distilled wine or other alcohol beverages are free of such enzymes (because distillation will remove the large molecules such as enzymes and/or destroy (denature) enzymes.

Persons who chronically overuse alcohol would develop another enzyme system in the liver that metabolizes ethanol. The enzyme is dependent on cytochrome P-450, and is called CYP2E1. (Please refer to Chapter 6 for a brief discussion of cytochrome P-450).

Even though ethanol-metabolizing enzymes are available in the liver, there is a limit for the capacity. Only a certain limited amount of ethanol can be dealt with per hour by the liver. Any excess alcohol will pervade the body, particularly the brain. This causes the familiar alcohol effects on the brain. For one, ethanol blocks the excitatory NMDA-type glutamate receptor. This disrupts memory and the learning process. For another, alcohol enhances the inhibitory GABA receptor site; this makes the person feel less anxious and inhibited. The details of how alcohol affects these receptor sites are not yet understood fully.

17.3.2 Tobacco: Nicotine: Death of the Bee

Tobacco is another commonly abused stimulant. It contains nicotine as the major stimulant. As mentioned earlier (Sect. 17.1), our brain has a receptor for nicotine, a part of acetylcholine receptor. Nicotine is an alkaloid and its structure is shown in Fig. 17.2. Tobacco plants produce nicotine as a defense chemical, to kill some insects. Nicotine has indeed been used as an insecticide for long. It is poisonous to human body, too, and a small amount can kill. In smaller quantities (as found in inhaled smoke), it can act as a stimulant interacting with the receptor as mentioned earlier, and it is quite addictive.

Chemical compounds that have some characters similar to nicotine have been synthesized in recent decades and have widely been used as insecticides. They are collectively called "neo-nicotinoid". One of them, called clothianidin, has recently been shown to be a major cause of colony collapse disorder (massive sudden death of bees). It is believed to bind to the brain of bees (and other insects) as nicotine does, and disrupts their brain function. It has been reported that the colony collapse has dramatically decreased once neo-nicotinoid was banned in Germany.

17.3.3 Psychotic Drugs: Ecstasy, Etc.

Catecholamines such as adrenalin (epinephrine), noradrenalin (norepinephrine), and dopamine (see Fig. 17.1) have an amine group and a catechol (benzene ring with two adjacent OH groups). The amine group is very likely protonated and hence is positively charged. This positively charged group is very likely responsible for binding to the receptor. It must be pointed out, however, that the binding mechanisms of catecholamines to various receptors have not yet been understood fully.

Amphetamine, methamphetamine, MDA (3,4-methylenedioxyamphetamine), and MDMA (3,4-methylenedioxy methamphetamine) are examples that have similar structures to catecholamines, as shown in Fig. 17.10. The structural similarity suggests that they may interfere with catecholamines' pathways. Amphetamine is known to bind to and block the re-uptake mechanism (i.e., reabsorb the neurorans-mitter such as dopamine back into the presynaptic cell) and increase the level of the neurotransmitter in the synapse. Methamphetamine seems to have a similar effect but even stronger than that of amphetamine.

Let us take a look at the structure of pseudoephedrine, shown also in Fig. 17.10, and then compare it to methamphetamine. You see only a small difference between them. Pseudoephedrine is the major ingredient of across-counter nasal decongestant drugs. Some people discovered that pseudoephedrine can be converted to metham-phetamine. As you see from the structures in the figure, all you need to do is,

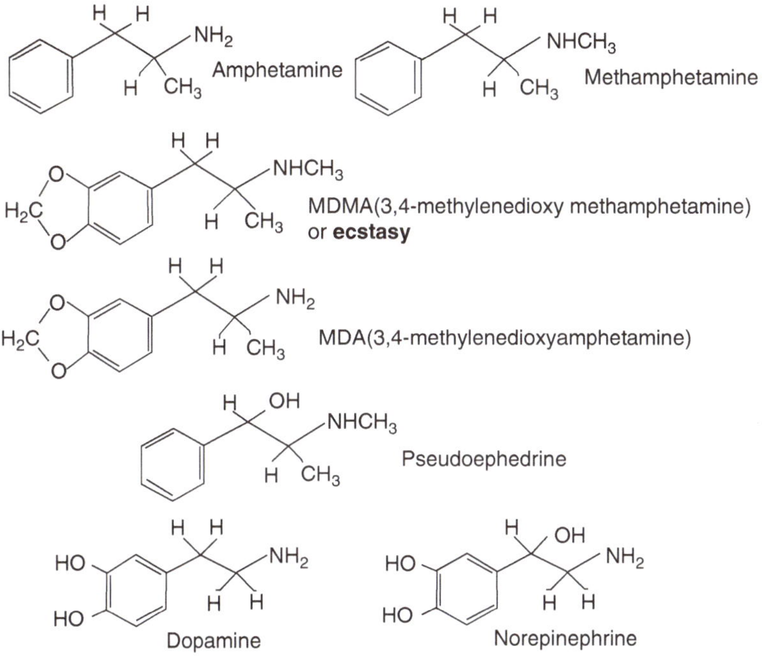

Fig. 17.10 Amphetamine, etc.

on the paper, to convert the "OH" group on pseudo-ephedrine to "H". However, chemically speaking, it is not a simple matter.

MDMA, known as ecstasy among partygoers, has been found to bind to a serotonin re-uptake mechanism and block it. See Fig. 17.2 for the structure of serotonin. Hence it allows serotonin to remain in the synapse longer than required. If you compare carefully MDMA with serotonin, you will not find much structural similarity between them. This implies that the overall structural similarity may not be a sole factor in determining whether a compound is bound with a physiological receptor site.

17.3.4 Opiates, Designer Drugs, and Parkinson Disease

A narcotic, morphine, is one of the ultimate pain killers, and has a legitimate medical use. Morphine is the main ingredient of opium, and a derivative of morphine, heroin, has been widely used as a drug to get "high". One of the basic problems is development of dependence, i.e., addiction. These compounds are called "alkaloids"; they are produced as defense chemicals by plants. Their chemical structures are shown in Fig. 17.11. The presence of the N (nitrogen) atom in them makes them "basic" (or alkaline), and hence the name. We have talked about another group of alkaloids earlier; i.e., frog toxins. These are natural products, though today heroin is produced by artificial means from opium extracts. Another derivative of morphine is known as

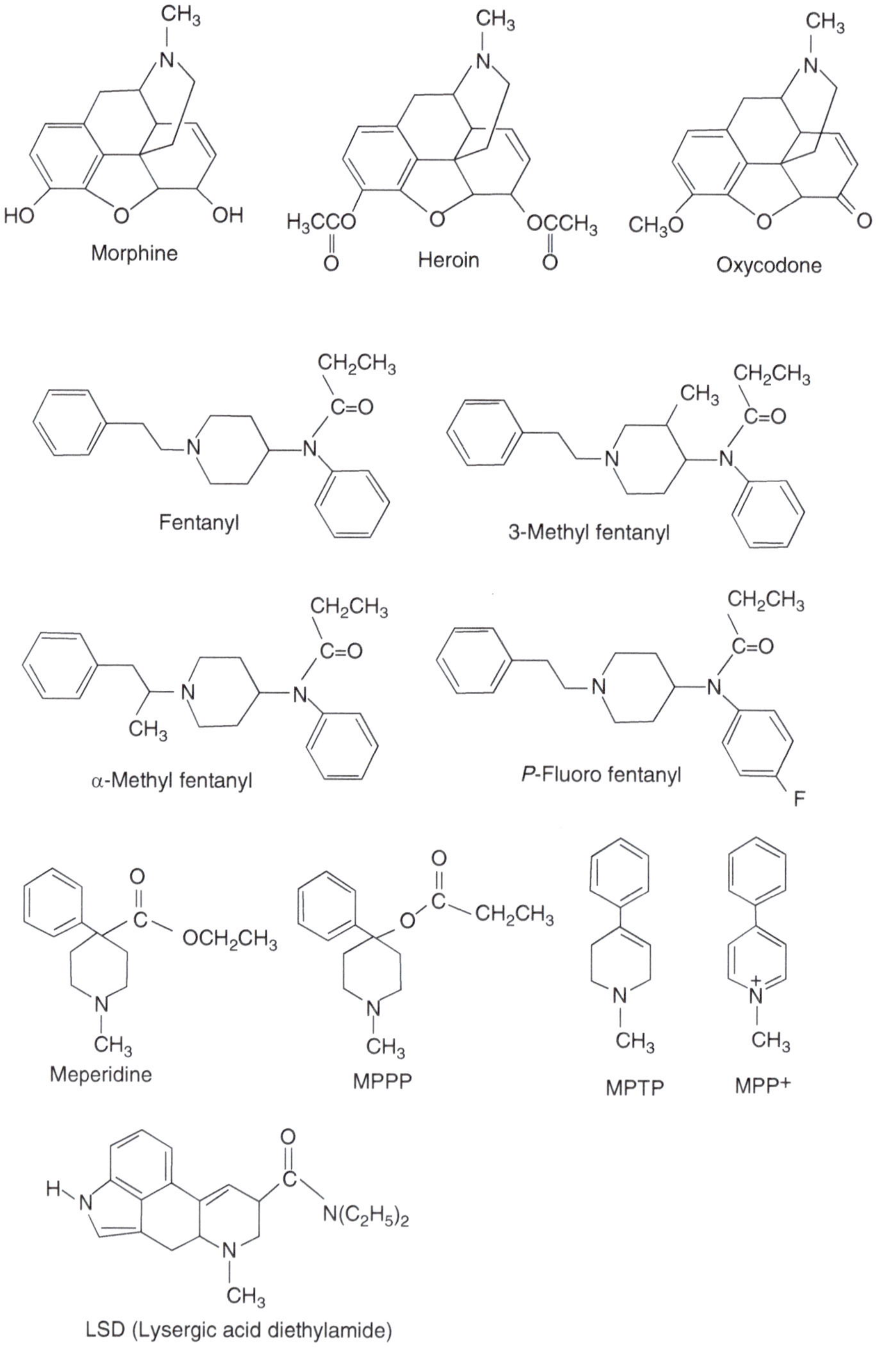

Morphine

Heroin

Oxycodone

Fentanyl

3-Methyl fentanyl

α-Methyl fentanyl

P-Fluoro fentanyl

Meperidine

MPPP

MPTP

MPP+

LSD (Lysergic acid diethylamide)

Fig. 17.11 Opiates, etc.

oxycontin. It is prescribed as a pain killer. An example of its abuse was associated with a certain radio talk show host. Oxycontin is a hydrochloride (HCl addition) derivative of oxycodone, whose structure is shown along with those of morphine and heroin in Fig. 17.11. Its similarity to heroin and morphine is obvious.

Fentanyl, a synthetic chemical compound, was discovered to be about 100 times as potent as morphine or heroin. Its chemical structure shown in Fig. 17.11 indicates that it can also be classified as an alkaloid. This compound has been widely used as an anesthetic in surgical procedures, as the physiological effects are much shorter-acting than morphine. And it gives a "high" like heroin, though short-lived.

To add this group or that to fentanyl by chemical reactions, i.e., making derivatives is what chemists are good at. Unless a chemical compound is specifically classified as "Schedule I controlled substance", it is legal. All the fentanyl analogs, when appeared in 1970s and early 1980s, were legal. Some of the fentanyl analogs that have appeared on streets are shown in Fig. 17.11; these drugs were called "designer drugs". Some of them are easy to make, but some are more difficult. Particularly 3-methyl fentanyl (see the structure in Fig. 17.11) is very difficult to make, but it was produced very cleanly. Apparently a first-class synthetic chemist was behind all the fentanyl analogs. The drug was first used to dope racehorses. Some narcotics are quite species-specific, and fentanyl turned out to act as a stimulant in horses. As 3-methyl fentanyl is called a perfect heroin substitute, these analogs are potent narcotics. They have caused several hundred death of human beings by overdoses.

Another chemical, meperidine, was synthesized and patented in 1939 and was sold under the trade name of "Demerol". An analog of meperidine, MPPP was tried as a substitute for meperidine and was demonstrated to be more effective than meperidine, but was never commercialized. The structures of meperidine and MPPP are shown in Fig. 17.11. MPPP looks easy to synthesize. Some underground chemist(s) tried to synthesize MPPP according to a published procedure. The synthesis requires a specific reaction condition. If it is carried out under slightly different conditions, the synthetic procedure tends to produce a different compound, MPTP (1-methyl-4-phenyl-1,2,5,6-tetrahydropyridine). The product was sold on the street as synthetic heroin, and later shown indeed to be contaminated with MPTP (the structure of which is also shown in Fig. 17.11). People who used this staff were stricken with symptoms similar to those of Parkinson's disease. Fortunately for the medical communities, this provided an opportunity for further understanding of Parkinson's disease, now that a specific compound was serendipitously discovered that caused the disease. Research led to a discovery that MPTP is further converted by an enzyme (called monoamine oxidase, MOA) to MPP$^+$ (1-methyl-4-phenyl-pyridinium ion, see Fig. 17.11 for the structure), and that the latter is the specific toxin for dopaminergic neurons. MPP$^+$ destroys such neurons. More than 80% of the dopaminergic cells must be destroyed for the symptoms of the disease to appear. People under normal circumstances lose several percent (5–8%) of the dopaminergic cells per decades; hence ordinarily people would die before the symptoms show up. MPP$^+$ and perhaps some other compounds similar to it (that occur in the environment) speed up the loss of the dopaminergic neurons, resulting in Parkinson's disease.

17.4   Anabolic Steroids

As athletics, particularly in the Olympic Games and among some professional sports, becomes so competitive, athletes may succumb to a temptation to use some chemicals to enhance their athletic performances. Widely used, such stimulants include anabolic steroids, narcotic analgesics, psychomotor stimulants, and, recently, erythropoietins.

Most of these "drugs" are banned in the Olympic Games, collegiate athletic events, and, nominally, also in professional sports, on both ethical and medical grounds. Medically, any drug can have potential risks. An ethical issue is that the athletes who avoid drugs may be put at a disadvantage in competing with the drug users. The ethical issue may not be so simple as first thought. The athletes in a poor country are definitely disadvantaged against those in the wealthy country where they get an ample opportunity to enhance their ability in terms of both time and resources. In this sense, a competition between them (from different countries) is unfair. However, this is not a right place to get into this kind of arguments, so let us stick to the chemistry issues.

Let us look at anabolic steroids. They are essentially testosterone and its synthetic analogs. Testosterone is a male hormone, and responsible for muscle buildup and development of male sexual characteristics (this hormone is discussed in Sect. 7.7). Synthetic analogs were developed to separate the masculinizing (androgenic) effects and the muscle-building (anabolic) effects of the hormone, so that they can be used therapeutically to correct some hormonal imbalance. Some examples of anabolic steroids are shown in Fig. 17.12. Do not worry too much with chemical structures like these. All you need to see is the similarity among these structures. It turned out that physiological effects of drugs or hormones are often, not always, governed by their overall structures, though minor differences cause some subtle differences in their efficacy and side effects.

You would note that there is a methyl group (CH_3) attached to C-17 position in all (but one) the testosterone analogs shown in Fig. 17.12. The testosterone itself, if taken orally, is rapidly metabolized in the liver, and hence no significant effect will be manifested. It has to be injected for it to be effective. A drug would be more convenient to take, if it is effective in oral administration. Attachment of a methyl group to C-17 position retards the degradation reaction in the liver; hence these compounds can be taken orally and are still effective. Testosterone enanthate, given in Fig. 17.12 is an example of synthetic anabolic steroids that need to be injected. As you could guess, this compound, without a methyl group at C-17, would be rapidly degraded in the liver if taken orally.

Today, there are rapid, fairly accurate, and sensitive analytical techniques available to detect and measure these substances in urine or blood of the athletes suspected of illegal use. How about testosterone itself? You have testosterone naturally in your body, and the testosterone injected is chemically the same compound. Can an athlete cheat and get away with it then? Scientists discovered that testosterone can change itself into epi-testosterone (the positions of the OH group and the H atom (which is omitted in the figure) on C-17 on testosterone are interchanged).

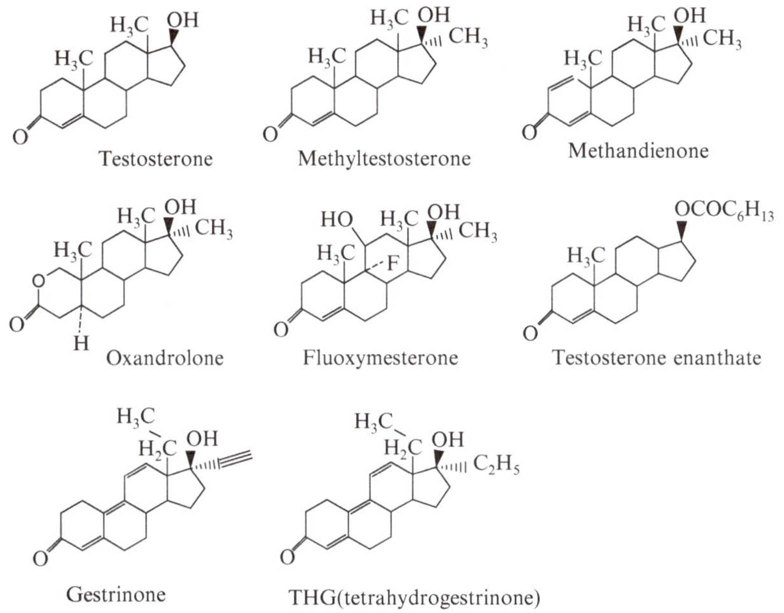

Fig. 17.12 Anabolic steroids

Therefore, under a natural condition there is about an equal amount of testosterone and epi-testosterone in the body fluid, blood, and urine. However, the extraneously added testosterone cannot be converted very quickly to its epimer. Hence if a urine test is conducted after (not too long after) one is doped with testosterone, the test will show an unusually high level of testosterone-to-epimer ratio.

It has been observed that anabolic steroids administered to rats increased the activity of an enzyme, RNA polymerase in the skeletal muscle cells. This will result in an increased synthesis of proteins; hence muscle building effects.

The side effects of anabolic steroids include acne, testicular atrophy, reduction of sperm, and liver tumors. As we talked about elsewhere (Chap. 7), testosterone like other sex hormone is controlled by two hormones, luteinizing hormone (LH) and follicle-stimulating hormone (FH). LH stimulates the production of testosterone in special cells in the testes. When there is too much testosterone in the blood, the hypothalamus tries to reduce it by decreasing the amount of LH and FH it releases. Normal functioning of the testes for sperm production requires the presence of LH and FH. An individual doped with anabolic steroids may have enough testosterone in most of his body, but the testes are virtually shut down to produce sperms.

A bigger problem is liver cancer. Testosterone, unless administered excessively, is not a problem, but the synthetic ones are. Because they persist in liver as mentioned earlier (and this is precisely the very reason that they can be taken orally), the liver is hard-pressed to metabolize it. This is the cause for liver toxicity of these drugs.

Liver is in general the organ where any extraneous substance is dealt with. A dramatic case is that of a young man of 26 years who had taken anabolic steroids for years to build his muscle, but who died of liver cancer. Side effects on female athletes are of course masculinizing effects, including beards, deepening of voice, and menstrual irregularity.

Yet, efforts by some quarters continue to develop anabolic steroids that are more difficult to detect, thus defrauding the authorities and the public alike. In 2003, a coach tipped off the U. S. Anti-Doping Agency at the US Outdoor Track and Field Championship that some athletes both from the US and foreign countries were using a new drug. It turned out to be THG, tetrahydrogestrinone. Gestrinone itself is a synthetic hormone used to treat endometriosis, and has been legally available. THG can be relatively easily produced from gestrinone by a process called hydrogenation. It is the same process by which margarine is produced from plant oil. THG turns out to be unlike any anabolic steroids used legally or illegally by the time of its recognition. Therefore, it was relatively difficult to determine what the chemical structure is. The structures of THG and gestrinone are shown in Fig. 17.12 along with other anabolic steroids. As seen, their overall shapes are the same as other steroids, but the structures, with extra two double bonds, are significantly different from others.

Part VI

Appendix: Essentials of Chemistry

The appendix is provided for those who are not well versed in chemistry and is intended to give a concise introduction to chemistry. Chapter 18 is a survey of the discipline "chemistry," emphasizing how chemistry interacts with adjacent disciplines. Chapter 19 gives a bare minimum of chemical concepts and theories essential for understanding this treatment. Chapter 20 deals with a very common phenomenon, i.e., interaction of chemicals with light, and Chapter 21 tries to show what atoms and molecules actually look like, based on recent progresses in techniques to visualize such small things.

Domain of Chemistry

18

John Horgan, a *Scientific American* staff writer, wrote a book titled "The End of Science" (Broadway Books 1997). The title suggests that sciences are coming to an end in the sense that discoveries of new fundamental principles are not very likely any more, and he talks about the ideas of the prominent practitioners in every major field of science. A major omission in this book is "chemistry." Either he does not know much about chemistry, or he does not think it's worth of his time to talk about it for whatever the reason. No matter what his opinion may be, it may be true that chemistry has very *nearly* come to an end in his sense. Chemistry's basic principles are based on quantum mechanics and statistical thermodynamics, both of which seem to have been well established. I hasten to add, though, that the emphasis is "nearly," and that there could still be a few more basic principles germane to chemistry yet to be discovered. I cannot predict what they may be.

But, a big **BUT**! The fun of chemistry does not lie, in large measure, in making new *fundamental* discoveries. Practitioners of chemistry find fun with making new things (chemicals) or discovering and figuring out things about compounds (chemicals) that nature makes. And there is an infinite variety of compounds possible that can form using the elements that are available in the universe. Imagine how many combinations (i.e., compounds) are possible with combining 100 or so elements in almost any way you want; i.e., infinite. Even with a limited number of elements: carbon, hydrogen, nitrogen and oxygen, the number of possible compounds they can make up is infinite; that is, the number of the possible "so-called" organic compounds is infinite. So far several tens of millions of organic compounds have been registered with "Chemical Abstract Service." That is the beauty of chemistry: variety. As mentioned in the introduction, all the material that exist in the universe and on the Earth including yourself are chemicals, made of at most 100 or so elements. No, we have not quite understood everything (all the chemicals and their behaviors) yet; hence we often find surprises with new compounds synthesized or discovered in nature. Recent such examples include "buckminsterfullerene" and "high-temperature superconductors." But they certainly do not violate the basic principles that we know of, though we still do not fully understand, e.g., the mechanism of the

E. Ochiai, *Chemicals for Life and Living*,
DOI 10.1007/978-3-642-20273-5_18, © Springer-Verlag Berlin Heidelberg 2011

high-temperature superconductors. Particularly puzzling are the interactions of various foreign chemicals with the physiology of organisms. But this is not difficulty at the fundamental level. It arises because of the complexity of the physiology of organisms.

I mentioned in the above paragraph about a group of compounds, called "organic compounds" which contain carbon (and also hydrogen, nitrogen and/or oxygen). The reason that these compounds that contain carbon are called "organic" is historical. Once, until about mid nineteenth century, typical compounds containing carbon such as urea, alcohol, carbohydrates and others had been considered to be made only by organisms, i.e., biological processes; hence the name. However, it was demonstrated by Wöhler in 1828 that urea which is found in our urine could be made artificially by a laboratory procedure heating ammonium cyanate, an inanimate compound. This discovery changed the scene of chemistry and was followed by successful attempts to artificially synthesize compounds after compounds of biological origin that had been considered to be impossible to make in lab. This motivated the chemistry community to change the definition of an organic compound from the one that is associated with organisms to the one that simply contains carbon atom, as the majority of the biological compounds contain at least carbon.

One of the simplest kinds of organic compounds is a group of compounds called "hydrocarbons." They are made of carbon and hydrogen. The simplest among them is methane, which is made of one carbon atom and four hydrogen atoms (technically expressed: that is, chemical formula, as "CH_4"), and is the major component of the natural gas. Gasoline consists of several different hydrocarbons. You must know ethanol, the so-called alcohol, which consists of two carbon, six hydrogen and one oxygen atom(s).

There are 100 or so elements including carbon in the universe. The compounds that are composed of elements other than carbon are collectively called "inorganic compounds" (as opposed to "organic"). Rock salt made of sodium and chlorine and limestone made of calcium, carbon and oxygen are typical inorganic compounds. This will give you an impression that "inorganic compounds" will have nothing to do with organisms, and also that "inorganic compounds" have little to do with "organic compounds." This has turned out to be entirely false. Anyhow, you see that a major division in the chemistry discipline is between "organic chemistry" and "inorganic chemistry."

This is an artificial division, though a fairly good practical one. It is practical because there are indeed some typical characters and behaviors of organic compounds that are distinct from those found with typical inorganic compounds. However, it is artificial and a lot of exception can be expected. After all there is no intrinsic reason why carbon cannot bind with elements other than hydrogen, nitrogen and/or oxygen. It does. It can bind, e.g., with metallic elements such as iron, lead, mercury and nickel (typical inorganic elements). The compounds that contain bindings between carbon and metallic elements have been prepared in an enormous number in recent decades and are collectively called "organometallic compounds."

Another area that borders on biological systems and inorganic compounds is the so-called "bioinorganic chemistry," though British scientists prefer to call it "inorganic biochemistry." This is a very weird term; i.e., it is self-contradictory, as the "inorganic" portion contradicts the "bio" portion. No matter what the term is, "bioinorganic chemistry" is now a very active research area, in which new discoveries are still being made, though they remain within the fundamental principles of chemistry and physics. It is concerned with the interactions in its broadest sense of inorganic elements with the biological systems (Chap. 6).

What these come down to is the fact that there is not two (organic and inorganic) but only one chemistry. Nature, including those compounds synthesized by mankind, does not divide neatly her sovereign into two camps: "organic" and "inorganic." I might say that the distinction is meaningful only for their respective practitioners.

Another basic subdivision of discipline of chemistry is "physical chemistry." It is to study the physical manifestations of chemical compounds; physical properties, interactions of compounds with electromagnetic waves (spectroscopy), and how they react each other, and so on. Therefore, there are "physical organic chemistry," which focuses on the organic compounds, and "physical inorganic chemistry," if you want to subdivide it. Understanding the physical manifestations of chemicals is based on the principles and theories of "physics." Hence physical chemistry is on the borderline between chemistry and physics in a way. If the emphasis of an inquiry is placed more on "physics," it is called "chemical physics."

Biochemistry is to study the chemical aspects of biological systems, and hence is bordering on chemistry and biology. The emphasis here is placed on the chemical and physiological properties of biologically important organic compounds, the biological processing (metabolisms) of them, the workings of proteins/enzymes and the genetic material, etc. If the emphasis is placed on only the chemical aspects of bio-compounds, we call such an inquiry as "bioorganic chemistry." The study of cellular activities in terms of gene and its chemical details (i.e., molecular structures and properties of DNA/RNA) is somehow distinguished as "molecular biology." All these endeavors are sometimes called "chemical biology," looking at biology in terms of chemistry. Now one of the most prominent departments in the US bears this name "chemical biology."

"Geochemistry" studies the chemical aspects of geological material. "Organic geochemistry" is a sub-discipline, and deals with organic geological material such as petroleum. There is also an area called "biogeochemistry"; what do you think it will explore?

Mankind, or I should say, the western scientists have invented the "disciplines": biology, chemistry, geology, and physics, etc. Nature does not concern herself with its subdivisions imposed by us. Nature is a whole; we scientists in a discipline look at only certain aspects of it or only from a certain viewpoint. However, there is an increasing realization of the need to understand the nature as a whole, because everything is related to everything else as seen in environmental issues, for example (see Chap. 22). Thus, we see more and more interdisciplinary endeavors. And all these disciplines deal with materials, and materials are chemicals, and thus chemistry permeates all these other disciplines. Hence, "chemistry" is often called the *"Central Science."*

There is one other important sub-discipline of chemistry; that is, analytical chemistry. You would need to identify and quantify the material you deal with; i.e., chemical analysis. Analytical chemistry borrows ideas from all branches of chemistry, particularly physical chemistry, and devises means to do analysis. It is an applied field of chemistry. In real life, analytical chemistry is, perhaps, the most-often-used chemistry. Depending on the material dealt with, you can have bioanalytical chemistry, environmental analytical chemistry and so forth. Another way of subdividing analytical chemistry is based on the means used: "wet chemical analysis," typically dealing with solutions (hence wet) and using more traditional analytical methods, and "instrumental analysis" which uses a number of different instruments to do analysis: spectrophotometric, electrochemical, etc.

An entirely different kind of chemistry sub-discipline is nuclear chemistry. It deals with chemicals all right, but its concern is quite different from all the sub-disciplines mentioned above. It studies the nuclei of atoms in chemicals. The nuclei obey quite different kinds of rules than the ordinary chemicals do, as the next Chap. (19) explains. It treats the radioactivity, nucleosynthesis (how elements are produced), nuclear fission and fusion, and extends to cosmochemistry.

Chemistry's View of the Material World: Basic Principles

19

This chapter is not meant to be a comprehensive, detailed, and systematic presentation of all chemical concepts/principles and their applications. It tries to present a bare minimum of the essential chemical concepts that may be necessary to appreciate the content of this book.

One of the essentials discussed is the concept of Avogadro number, but its application, i.e., the quantitative calculations involving "moles," is hardly discussed in the main body of the book. The author considers that grasping of such quantitative calculations is unnecessary except for the practitioners of chemistry and the related areas. The inclusion of this important concept is to only emphasize the nature of material world; i.e., it consists of particles, atoms, ions, molecules, etc. And the quantities of any material is fundamentally measured in terms of the number of the particles, though in everyday life the quantities may be measured in terms of mass (weight), volume, length, etc. Occasionally, such calculations do appear in the text for sure, but it is only meant to be a reminder to those who understand or need such a calculation. In general, quantitative calculations that appear in the text are meant to give the readers a sort of feeling toward the quantities talked about. They are not intended to be precise and detailed and often are not shown of the details of assumptions, calculations, and data.

Other major omissions are the concept of atomic orbital hybridization vs shape of molecule and the determination of the molecular structures. These topics would require a lengthy discourse to do full justice to the subject. We entirely omit these aspects and present all the chemical structures as given.

19.1 Atoms/Molecules/Mole/Avogadro Number

According to chemistry, a substance is an aggregate of a very large number of very tiny particles that can be a kind of atom or a kind of molecule. A substance that is made of a single kind of atom is called an "element," while a substance consisting

E. Ochiai, *Chemicals for Life and Living*,
DOI 10.1007/978-3-642-20273-5_19, © Springer-Verlag Berlin Heidelberg 2011

of a large number of a same kind of molecule or the same combination of different atoms is a "compound." We are using terms such as "atom," "element," and "molecule" here without defining them. We are relying on your knowledge of these terms; it does not matter at this stage whether your own definition is precise or not. We will eventually find out exactly what they are.

Now take an ounce (28.35 g) of water. It turns out to consist of approximately 1000000000000000000000000 (there are 24 "0"s) molecules of water. [By the way, 300000 (5 "0"s) is often expressed as 3×10^5; so the large number above with 24 "0"s can be written as 1×10^{24} for short. This kind of expression is called "scientific notation"]. A single molecule of water represents the smallest unit that has the character of water, and hence, weighs only about $3/10^{23}$ g (i.e., 28.35 g (i.e.,1 oz)/10^{24}) [As $1/10^3$, i.e., 1/1000 one-thousandth, is written as 1×10^{-3}, $3/10^{23}$ can be written as 3×10^{-23} and is a very tiny value]. When chemists look at water, they think in terms of literally a sea of these very tiny molecules of water. They are moving around randomly within a confine at ambient temperatures and pressures, and thus the aggregate of water molecules show a character of liquid at ordinary temperatures. Some of the water molecules go out of the liquid into the surrounding space and come back. All of the molecules there always keep going out and coming back, but overall, the majority of water remains in the liquid. However, if molecules that escaped into the space would not come (as in the case where water is in an open container), eventually all water molecules disappear into space.

Let us now go to a little deeper level. That is, the idea of "atom." The atoms constitute molecules and hence are the basis of all materials. For example, a water molecule has been shown to consist of two hydrogen atoms and an oxygen atom. The idea of "atom" is as old as ancient Greece, where several philosophers suggested that the material is made of atoms. Atoms are the smallest units of matter that cannot be divided further. Democritus coined the word "atom" to mean what was just said. He said: *The first principles of the universe are atoms and empty space; everything else is merely thought to exist.* There are different kinds of atom. An aggregate of atoms of the same nature is called an "element"; so atoms of different natures belong to different elements. Some Greek philosophers thought that the whole world was made of four elements: for example, air, water, fire, and soil. We know now that these are not elements in the modern sense.

Today, science recognizes about 100 different elements in the whole universe. "Elements" should be understood as "chemical elements" to distinguish it from "elements" in ordinary usage. Everything that is present in this universe is made of one or more atoms of these 100 or so elements. A water molecule is made of two atoms of element hydrogen (symbol "H") and one atom of element oxygen (symbol "O"). And this fact is expressed as H_2O. Chemists also have learned that in water molecule, two hydrogen atoms are bound with the one oxygen atom as shown in Fig. 19.1. This is the so-called structure of water molecule. In real life we would not be able to deal with a single molecule, for it is too tiny. Instead we will be dealing with, say, a gallon of water. A gallon of water contains a very large number of water molecules as we told you earlier. But the properties of this large body of water or a

Fig. 19.1 Structure of water molecule (often this is further simplified as H–O–H)

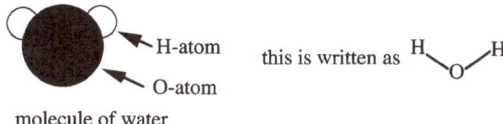

molecule of water

small amount of water, say a drop of water, are determined by the molecule having H_2O composition; therefore, we will use "H_2O" notation even when we talk about this large number of water molecules, i.e., the substance or compound water. This is the chemical formula of water (molecule).

Iron is another element (chemical symbol = Fe). Metal iron, say 100 g, is made of a very large number (about 10^{24}) of atoms of element iron. When iron rusts, it turns into a substance called iron oxide, which may be described as consisting of two atoms of element iron and three atoms of element oxygen (Fe_2O_3). (It might be pointed out here, though it may be confusing at this stage, that there is no such molecular entity having the composition Fe_2O_3. It shows only the composition).

Element carbon (chemical symbol = C) is made of atoms of a single element carbon, but can take different shapes. All carbon atoms in these substances are bound to each other, but their arrangements can be different. As a result, graphite and diamond, though consisting of only carbon atoms, i.e., being elemental carbon, have such disparate properties. Carbon atom(s) make million different compounds (molecules), binding hydrogen atom(s), oxygen(s) and nitrogen(s, chemical symbol = N); they are collectively called "organic compounds." Some of them such as proteins and DNA constitute our body.

All the elements are now arranged neatly in the so-called periodic table, which consists of 18 columns and 7 rows plus 2 extra segments (Fig. 19.2). Each element is given a number (atomic number) and then pigeonholed in each of the boxes. Chemistry deals with all these elements, their compounds, and how they behave. Remember that all the materials that exist are made of these elements.

As we said before, all the atoms and molecules are miniscule and it is very hard to deal with a single atom or a single molecule. Therefore, chemists invented a unit of measurable quantity of elements or compounds (substances).

It is called "mole." One mole of a molecular substance is defined to be a collection of approximately 6×10^{23} molecules. This huge number 6×10^{23} (23 0s after 6) is called "Avogadro Number." Avogadro is an Italian chemist who attempted to determine this number first. The number is frighteningly large, but the idea of "mole" is the same as, say, "dozen," which consists of 12 bodies. So 1 mol of water consists of 6×10^{23} particles of water molecule H_2O. As quarter a dozen is equal to $12/4 = 3$, quarter a mole is $6 \times 10^{23}/4 = 1.5 \times 10^{23}$. By the way, no significance should be attached to the fact that these two quantities 12 and 6 ($\times 10^{23}$) appear somehow related. This is a pure coincidence. "6" in the quoted Avogadro number is an approximation; it has been estimated to be "6.022," to be more precise. This value itself is still an approximation, but practically accurate enough to be useful.

PERIODIC CHART OF ELEMENTS

Legend: Atomic Number → 47; Ag ← symbol for element; 108 ← approximate atomic molar mass

1 H 1.0																	2 He 4.0
3 Li 6.9	4 Be 9.0											5 B 10.8	6 C 12.0	7 N 14.0	8 O 16.0	9 F 19.0	10 Ne 20.2
11 Na 23.0	12 Mg 24.3											13 Al 27.0	14 Si 28.1	15 P 31.0	16 S 32.1	17 Cl 35.5	18 Ar 40.0
19 K 39.1	20 Ca 40.1	21 Sc 45.0	22 Ti 47.9	23 V 50.9	24 Cr 52.0	25 Mn 54.9	26 Fe 55.8	27 Co 58.9	28 Ni 58.7	29 Cu 63.5	30 Zn 65.4	31 Ga 69.7	32 Ge 72.6	33 As 74.9	34 Se 79.0	35 Br 79.9	36 Kr 83.8
37 Rb 85.5	38 Sr 87.6	39 Y 88.9	40 Zr 91.2	41 Nb 92.9	42 Mo 95.9	43 Tc 99	44 Ru 101	45 Rh 103	46 Pd 106	47 Ag 108	48 Cd 112	49 In 115	50 Sn 119	51 Sb 122	52 Te 128	53 I 127	54 Xe 131
55 Cs 133	56 Ba 137	La -Lu	72 Hf 178	73 Ta 181	74 W 184	75 Re 186	76 Os 190	77 Ir 192	78 Pt 195	79 Au 197	80 Hg 201	81 Tl 204	82 Pb 207	83 Bi 209	84 Po 210	85 At 210	86 Rn 222
87 Fr 223	88 Ra 226	Ac -Lr	104	105	106	107	108	109	110	111	112	113	114				

Transition Elements (d)

Metallic Elements

Lanthanides →

57 La 139	58 Ce 140	59 Pr 141	60 Nd 144	61 Pm 145	62 Sm 150	63 Eu 152	64 Gd 157	65 Tb 159	66 Dy 163	67 Ho 165	68 Er 167	69 Tm 169	70 Yb 173	71 Lu 175

Actinides →

89 Ac 227	90 Th 232	91 Pa 231	92 U 238	93 Np 237	94 Pu 244	95 Am 243	96 Cm 247	97 Bk 247	98 Cf 251	99 Es 254	100 Fm 257	101 Md 260	102 No 259	103 Lr 262

Fig. 19.2 Periodic Chart. (Ac, actinium; Ag, silver; Al, aluminum; Am, americium; Ar, argon; As, arsenic; At, astatine; Au, gold; B, boron; Ba, barium; Be, beryllium; Bi, bismuth; Bk, berkelium; Br, bromine; C, carbon; Ca, calcium; Cd, cadmium; Ce, cerium; Cf, californium; Cl, chlorine; Cm, curium; Co, cobalt; Cr, chromium; Cs, cesium; Cu, copper; Dy, dysprosium; Er, erbium; Es, einsteinium; Eu, europium; F, fluorine; Fe, iron; Fm, fermium; Fr, francium; Ga, gallium; Gd, gadolinium; Ge, germanium; H, hydrogen; He, helium; Hf, hafnium; Hg, mercury; Ho, holmium; I, iodine; In, indium; Ir, iridium; K, potassium; Kr, krypton; La, lanthanum; Li, lithium; Lr, lawrencium; Lu, lutetium; Md, mendelevium; Mg, magnesium; Mn, manganese; Mo, molybdenum; N, nitrogen; Na, sodium; Nb, niobium; Nd, neodymium; Ne, neon; Ni, nickel; No, nobelium; Np, neptunium; O, oxygen; Os, osmium; P, phosphorus; Pa, protactinium; Pb, lead; Pd, palladium; Pm, promethium; Po, polonium; Pr, praseodymium; Pt, platinum; Pu, plutonium; Ra, radium; Rb, rubidium; Re, rhenium; Rh, rhodium; Rn, radon; Ru, ruthenium; S, sulfur; Sb, antimony; Sc, scandium; Se, selenium; Si, silicon; Sm, samarium; Sn, tin; Sr, strontium; Ta, tantalum; Tb, terbium; Tc, technetium; Te, tellurium; Th, thorium; Ti, titanium; Tl, thallium; Tm, thulium; U, uranium; V, vanadium; W, tungsten; Xe, xenon; Y, yttrium; Yb, ytterbium; Zn, zinc; Zr, zirconium)

The Avogadro number has been decided in such a way so that 1 mol of a substance, say, water, weighs an ordinary everyday number, such as 18 g in this case. It turned out that 1 mol of many elements comes close to an integer number if we use the Avogadro number of about 6×10^{23}; for example, 1 mol of hydrogen (symbol H) is about 1 g, carbon (C) 12, oxygen (O) 16, iron (Fe) 56, and so on. Note that when you combine two hydrogen atoms and one oxygen, you will come up with $2 \times 1(H) + 16(O) = 18$ g for 1 mol of water as indicated above. This mass is called the "molar mass" of a substance. For example, the molar mass of the rust Fe_2O_3 would be approximately $2 \times 56 + 3 \times 16 = 160$ g. There is a deeper meaning to this fact; i.e., that the mass of 1 mol of an element comes close to an integer number. In order to understand this, we need to look into the structure of an atom.

19.2 Structures of Atoms

A lot of research on the structures of atoms has come up with a picture shown in Fig. 19.3. An atom consists of a nucleus at the center and electrons surrounding it. The nucleus carries a positive electric charge and is very minute, while electrons are negatively charged. And the magnitudes of the positive charge on the nucleus and of the negative charge on electrons are the same (in a neutral atom). Since an electron is regarded as a particle (one of the fundamental particles), we can assign an electric charge to a single electron and use that charge as a unit of electric charge; let's say it is "e." Suppose that a neutral atom has "n" electrons, then the nucleus must have "n" of something that carries a positive electric charge of magnitude "e." This something is called "proton." A proton, though carrying the same magnitude of electric charge positive (instead of negative), turned out to be about 1,800 times as heavy as an electron.

The simplest atom consists of a single proton in the nucleus and an electron surrounding it. This is hydrogen, symbolized as "$_1H^1$." It turned out that another atom exists that has one proton and one electron, but, in addition, has one more particle called "neutron" in the nucleus. "Neutron" is approximately the same in mass as proton but carries no electric charge. This atom is about twice as heavy as the regular hydrogen atom and is called "heavy hydrogen" $_1H^2$ (or often called "deuterium D" ($_1D^2$)). In this notation, the superscript is the total number of protons and neutrons in the nucleus and the subscript is the number of protons (or the positive electric charge in units of "e"). The former is called "atomic mass number" and the latter "atomic number," or more generally electric charge number. The reason that this is

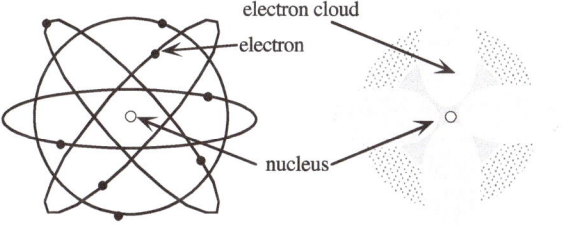

Fig. 19.3 The atomic orbitals that the electrons will occupy. These are actually more like fuzzy clouds of different densities and shapes

atomic orbitals (schematic)   Atomic orbitals (more realistic)

still regarded as a kind of hydrogen is that deuterium (D) behaves almost like hydrogen, as the chemical behaviors of an atom are determined mostly by the number of electrons and the positive charge of the nucleus. Scientists invented a word "isotope" to designate two or more atoms that have the same number of protons but different number of neutrons in the nucleus. The isotopes with an atomic number (i.e., the same number of protons) are chemically similar and hence designated as belonging to the same element. Thus, both "$_1H^1$" and "$_1H^2$" (or $_1D^2$) are isotopes and belong to the element "hydrogen." A third hydrogen isotope $_1H^3$ [how many protons and neutrons here?] is also known. This is often called "tritium, T." Not all the isotopes of an element exist in an equal quantity in nature (on the Earth). In hydrogen case, the predominant isotope is $_1H^1$, whose natural abundance on the Earth is 99.985%, while the heavy hydrogen is present only 0.015% in nature (on the Earth). Tritium can be produced only artificially.

Now let us move on. The second simplest element is helium, He or rather $_2He^4$; that is, its nucleus contains two protons and two neutrons. The third element is $_3Li^7$ (lithium, how many neutrons are here?). There is also an isotope $_3Li^6$. The natural abundance of the former is 92.58% and the latter is 7.42%. Element iron, Fe, consists of isotopes of $_{26}Fe^{54}$ (natural abundance = 5.82%), $_{26}Fe^{56}$ (91.66%), $_{26}Fe^{57}$ (2.19%), and $_{26}Fe^{58}$ (0.33%) (how many protons and neutrons here?). Since a proton and a neutron weigh about the same, the mass of an isotope, say $_{26}Fe^{56}$ would weigh 56 g (per mole), if we assign 1 g for 1 mol of hydrogen (i.e., $_1H^1$). It usually comes close to that, i.e., an integer number, but it is not quite an integer number. This is because some mass (weight) is lost when protons and neutrons gather together to form a nucleus. This weight lost is turned into an energy, through the famous Einstein's equation $E = mc^2$, where E is the energy, m is the mass, and c is the speed of light.

Nevertheless, this fact that a nucleus is made of an integer number of particles (protons and neutrons) is the basis for the choice of the Avogadro number. Avogadro number has been chosen in such a way that the molar quantity of an element (an isotope to be more precise) comes out as a number very close to an integer [because the mass loss in the form of mc^2 is relatively small]. Historically, the Avogadro number has been reworked several times; currently the Avogadro number (i.e., 6.022×10^{23}) is chosen so that 1 mol of the isotope $_6C^{12}$ weighs exactly 12.000 g. The mass of 1 mol of an atom is called "(molar) atomic mass" (given in units of g/mol). It should be and is close to an integer number (i.e., the mass number, the superscript figure of isotope), if it is defined for a single isotope. However, the "atomic mass" is usually defined for an element. Hence the term "(molar) atomic mass" is a misnomer and should be called "(molar) elemental mass." The atomic mass for an element can differ significantly from an integer number, because the atomic mass is the average of 1 mol of all the naturally occurring isotopes of an element. [Try to calculate the (molar) atomic mass of iron based on the isotopic distribution data mentioned above for a fun, assuming that each isotope has an integer number for its atomic mass, though this assumption is not quite right. Compare your result with the approximated value of molar atomic (elemental) mass given in the periodic chart, Fig. 19.2. If you find a discrepancy, think why].

19.3   Ordinary Chemistry and Nuclear Chemistry

An atom consists of a nucleus that is made of positively charged protons and neutral neutrons, and electrons surrounding it, as we outlined above. Two entirely different types of chemistry stem from this structure. One is concerned with the nucleus and the other with how electrons behave. The former is "nuclear chemistry," with "radiochemistry" as its important sub-discipline. The latter is the ordinary "chemistry." The basic reason for this division is that the nuclear forces binding protons and neutrons in the nucleus are enormously stronger than the electrostatic force binding the electrons to the nucleus. When one applies a force to a substance and induces a change, a certain amount of energy may be expended or gained. Hence, an energy change always accompanies a change in substance. "Energy" is often used as a measure of a change in science, particularly in chemistry (see Sect. 19.8). In terms of energy, then, a nuclear reaction (change in general) is greater by several orders of magnitude (typically a million times) than a typical chemical reaction, as the nuclear reaction involves changes in protons/neutrons in the nucleus while chemical reactions involve changes in electrons. Therefore, ordinary chemical reactions would not be able to cause a change in nucleus (i.e., nuclear reaction). As a result, it is quite safe to deal with nuclear chemistry as separate from "ordinary" chemistry. As a corollary, all isotopes that belong to an element, though they have different atomic masses, can be assumed to behave (approximately) the same chemically. However, isotopes behave very differently in terms of nuclear reactions. It is now obvious that principles governing nuclear reactions are quite different from those operating in the ordinary chemical reac-tions.

19.4   Nuclear Chemistry

19.4.1 Radioactivity: Spontaneous Nuclear Reactions

A nucleus that consists of neutrons and protons can be stable or unstable. Understanding of the factors that affect the nuclear stability is beyond the scope of this discourse, and it is not necessary to really understand for our purpose here. However, we have the facts; i.e., which nuclei (isotopes) are stable and which are unstable. The unstable isotopes would not remain as such and spontaneously change into stable isotopes. In this process, the unstable nucleus emits particles and/or energy in the form of radiation (gamma (γ)-ray). Particles that are emitted include electron (called β-particle), and helium nucleus (α-particle). Hence, these unstable isotopes are called "radioactive," emitting α, β, and/or γ radiation. There are other kinds of radiation, as well, but these three are the major ones.

There are exactly 264 stable isotopes known on the Earth. They lie on or slightly above a line that corresponds to the composition of equal numbers of neutrons and protons, as shown in Fig. 19.4. The area that is occupied by stable isotopes is coined as "stability peninsula" (the darkened area; the solid black curve represents a sort of central area where most of stable isotopes are distributed). In the figure, each of the

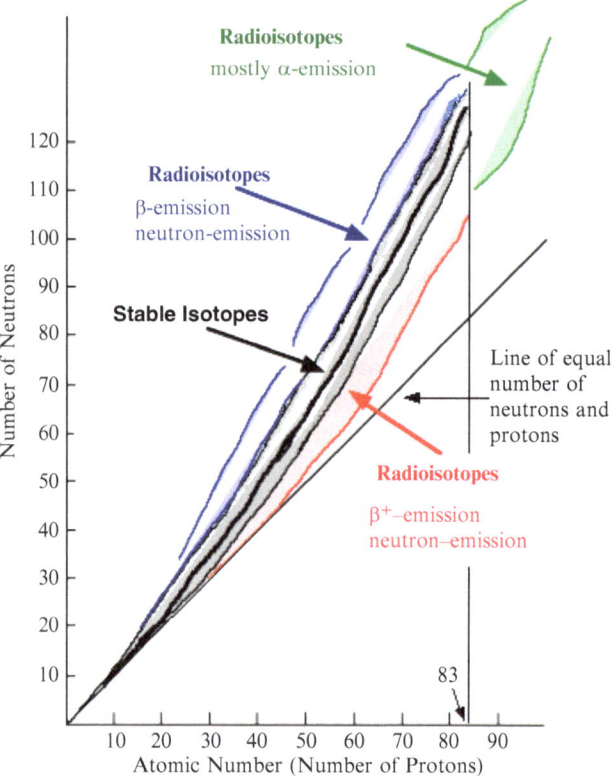

Fig. 19.4 Stable isotopes and unstable (radioactive) isotopes

colored areas consist actually separate dots, each of which represents an isotope (stable or unstable), and it represents an approximate area where these isotopes are distributed.

In understanding the radioactivity, we need to recognize some changes that can occur with neutron and proton. That is, they can interchange in the following manner:

$$_0n^1 (\text{neutron}) \rightarrow {}_1p^1 (\text{proton}) + {}_{-1}e^0 (\text{electron}) \tag{19.1}$$

$$_1p^1 + {}_{-1}e^0 \rightarrow {}_0n^1 \tag{19.2}$$

$$_1p^1 \rightarrow {}_0n^1 + {}_{+1}e^0 \left(\text{positron}\right) \tag{19.3}$$

[In these equations we omitted another particle "neutrino"].

Unstable isotopes that lie above the stability peninsula have too many neutrons. Hence they become stable by reducing the relative number of neutrons. This can be accomplished by emitting an electron (i.e., converting a neutron to a proton through (19.1)).

Therefore, most of these neutron-rich unstable isotopes are β-emitters. What would be an alternative way to reduce the neutron/proton ratio?

On the other hand, the unstable isotopes that lie below the stability peninsula would want to reduce the excess protons, by e.g., either (19.2) or (19.3). Process (19.2) is called electron capture and would not emit a particle, but usually emit γ-radiation. Process (19.3) would emit a positron, which has the same character as an electron except for the positive electric charge (instead of negative); this is called β+-radiation.

As you see in Fig. 19.4, there are no stable isotopes for elements whose atomic numbers are 84 or higher. These heavy elements usually decay by emitting α-particles. An example is $_{92}U^{235}$, which is used for the nuclear power generation.

The speed of the radioactive decay (disintegration) is governed by only the nature of the nucleus, and often is characterized by the time that it takes for one half of a sample of a radioactive isotope to disintegrate. This "time" is called "half-life." Some isotopes have very short half-life: e.g., 2.4 s in the case of $_6C^{15}$. They disintegrate very quickly. Others have very long half-life: e.g. 4.5×10^9 years in the case of $_{92}U^{238}$. This means that it takes 4.5 billion years (approximately the life span of the Earth) for 1 g of radioactive $_{92}U^{238}$ to become half a gram of radioactive $_{92}U^{238}$; in this process, the end product is lead $_{82}Pb^{206}$, through a multistep process.

19.4.2 Induced Nuclear Reactions: Modern Alchemy

Nuclei, stable or unstable, behave on their own. It is difficult to make them change their courses of destiny. This is because the energies involved in the change are quite large. And this is the reason that alchemists had never succeeded in converting lead into gold. However, if large enough energy is provided, one can cause changes in a nucleus. Such nuclear reactions are taking place in nature (in the universe) and also can now be induced by artificial means.

In the upper portion of the atmosphere on the Earth, nitrogen isotope $_7N^{14}$ is constantly bombarded by (energetic) neutrons of the cosmic ray and is undergoing the following nuclear reaction:

$$_7N^{14} + {}_0n^1 \rightarrow {}_6C^{14} + {}_1p^1$$

[Note that the sum of the superscripts on the left-hand side $(14+1=15)$ is equal to that on the right-hand side and also that the sum of the subscripts on the left-hand side $(7+0=7)$ is equal to that on the right-hand side $(6+1=7)$. This conservation of mass number and electric charge always holds true for nuclear reactions. Check this with (19.1)–(19.3) above as well]. By the way, the resultant nucleus $_6C^{14}$ is radioactive, though $_7N^{14}$ is not, and the radioactivity of $_6C^{14}$ is used to determine the age of the archaeological artifacts.

British physicist Rutherford succeeded in 1919 for the first time in human history in changing an element into another. He bombarded $_7N^{14}$ with α-particles and converted it into oxygen. That is:

$$_7N^{14} + \alpha \, ({}_2He^4) \rightarrow {}_8O^{17} + {}_1p^1 \, [\text{Check the balance of super/subscripts}].$$

This was the beginning of the modern "alchemy," though it should rather be called "nucleosynthesis." One element that is 43 in the atomic number had never been found on the Earth. Technetium, $_{43}Tc^{97}$ was made only artificially for the first time in 1937 by the following nuclear reaction:

$$_{42}Mo^{97} + {}_{1}H^{2} \rightarrow {}_{43}Tc^{97} + 2{}_{0}n^{1}$$

The term "technetium" reflects this fact. All isotopes such as $_{43}Tc^{99}$, $_{43}Tc^{98}$, and $_{43}Tc^{97}$ are unstable (radioactive) and so have to be created artificially, if one wants to study them or use them.

In 1938, German scientists L. Meitner, O. Hahn and F. Strassman discovered that uranium, when bombarded with neutrons, splits into smaller nuclei, such as barium and krypton. For example:

$$_{92}U^{235} + {}_{0}n^{1} (\rightarrow {}_{92}U^{236}) \rightarrow {}_{56}Ba^{141} + {}_{36}Kr^{92} + 3{}_{0}n^{1}$$

A heavy nucleus uranium here splits into smaller nuclei, and hence a nuclear reaction of this type is called "nuclear fission" reaction. The reaction here yields a lot of energy (heat), too. This energy comes from the mass of the material; in this case the total mass on the right-hand side of the equation is smaller than that on the left-hand side. And this mass lost is converted into energy (through $E = mc^2$). This particular reaction produces three neutrons using only one neutron, as seen in the equation. The neutrons produced then react again with the remaining U^{235}, and as a result, more neutrons will be produced. This kind of reaction in which active reactants (neutrons in this case) is reproduced (in equal quantity to or more than the quantity consumed) is called "chain reaction." In many chain reactions the reaction would soon become explosively fast because more and more active reactants are produced as the reaction proceeds. It can explode if not controlled. This is the basic principle of the atomic bomb, and of the current nuclear power reactor under a controlled condition. The other types of nuclear reaction are discussed in Chap. 13.

19.5   Behaviors of Electrons–Atoms (Elements)-Ordinary Chemistry

An atom consists of a nucleus (composed of protons and neutrons) and electrons surrounding it. The chemistry of elements and their compounds (molecules made of atoms) is dependent on the behaviors of the electrons in atoms. The nucleus is very, very small (though it is quite heavy), something like 10^{-15} m in diameter. On the other hand the size of an atom is of the order of 10^{-10} m, i.e., about 100,000 times larger than the nucleus. This size (atomic radius) is largely determined by the extent by which electrons are distributed around the nucleus.

How are electrons distributed around the nucleus? Electrons are held around the nucleus by the electrostatic attractive force; that between the positive electric charge of the nucleus and the negative charge of electrons. Early in the twentieth century,

Danish physicist, Niels Bohr proposed a model in which an electron goes round and round in a circle (orbital) about the nucleus, just like the Earth going around the Sun (see Fig. 19.3). In this model he made an assumption called "quantization" for the angular momentum (momentum of the orbital motion). "Quantization" is a difficult issue. Please be patient, we will explain the concept shortly. His model very successfully explained the then-known atomic spectrum of hydrogen atom. Hydrogen atom at high temperatures gives off a number of lights of certain definite frequencies; this is the "atomic spectrum" of hydrogen. These lights are given off when an electron in the hydrogen atom changes its states, suggesting that the electron is in different discrete states that are characterized by having different discrete energy values. This situation is said to be that the energy of the electron is "quantized." Bohr imposed a quantization condition on his model, though not directly on the energy but its related angular momentum, as mentioned above.

The idea "quantum" evolved slowly about the turn of nineteenth century to twentieth century, because no traditional (classical) idea was found to be able to explain a number of phenomena, including the light emitted when a black body (say, charcoal) was heated. This phenomenon (black body radiation) could be explained only if one assumes that the light is a particle (quanta) with a discrete energy (so argued by German physicist, Max Planck). Einstein also contributed to the idea that light is a "quanta." However, light is a wave (an electromagnetic wave), as everybody knows. This led to another intriguing idea that small particles such as electrons have dual characters: of particle and wave. French physicist de Broglie proposed that a particle which is travelling with a momentum "p" ($p = mv$, where m is the mass of the particle and v the speed) would behave like a wave which has a wavelength $\lambda = h/p$, where "h" is a constant called "Planck constant." This relationship has been found to hold true in many situations. For example, an electron beam (stream of particles) was found to be bent, just like a light (wave) is bent by a prism, and hence an electron microscope has been devised. The relationship between the speed of electron beam and its wavelength has also been verified.

So what would happen if we treat an electron around the nucleus as a wave? This idea was formulated in the celebrated "Schrödinger equation," and the theory of atom or rather the behavior of electrons in an atom based on this equation was originally called "wave mechanics," though later came to be known as "quantum mechanics." To find out the behaviors of electrons, you have to solve the "Schrödinger equation." This is the hardest part of chemistry, and, to tell you the truth, not many "real" chemists can do this. So do not worry. Today, various computer software are available for solving the "Schrödinger equation," not necessarily rigorously but in ever-increasingly accurate approximation.

However, we need to know a little bit of the results of such calculations in order to proceed. One of the strangest things of the results is that the energy of an electron going around a positive nucleus is indeed "quantized." That is, the energy of an electron can take only certain discrete values, and that the state of the electron is characterized by a certain set of integer numbers, called "quantum numbers" (n, l, m_l), e.g., ($n = 3$, $l = 2$, $m_l = +1$). We call, in analogy to the Bohr model, a state characterized by a set of these three numbers as an "orbital," though the picture of an electron going around in an orbit is not quite valid. The electron is simply distributed about the nucleus, as described by the so-called wave function that is a solution of the equation when we ignore the time-dependence part of the "Schrödinger equation." That is, this picture, the distribution of an electron, is a sort of time-averaged one, and in reality the electrons are moving rapidly indeed, though not necessarily in the orbital motion, as described by the Bohr model.

Scientists devised a name for the orbital (atomic orbitals) for each set of (n, l, m_l), as nx_y orbital. "n" is simply the same as the first (called principal) quantum number, x is "s" for $l=0$, "p" for $l=1$, "d" for $l=2$, "f" for $l=3$, "g" for $l=4$, etc., and $y=m_l$. So there will be 1s, 2s, 3s, 4s (and so on) orbitals; 2p, 3p, 4p, etc., orbitals (p starts with $n=2$); 3d, 4d, 5d, etc. (you guessed right!, d starts with $n=3$); and 4f, 5f, etc. There is only one kind of s-orbitals, but there are three different p-orbitals with different m_l values: $+1, 0, -1$ in the case of p-orbitals. You can write them out as p_{+1}, p_0 and p_{-1}. As you guess, there will be five d-orbitals: $d_{+2}, d_{+1}, d_0, d_{-1}, d_{-2}$ [how many for f-orbitals?]. The atomic orbitals do not look like orbits as in Bohr model or the planetary motion. Instead, atomic orbitals are fuzzy blobs of electron clouds; some of them are depicted in Fig. 19.3 (right-hand side).

One more thing needs to be said before we talk about the atoms. That is, an electron is not only electrically charged, but also turns out to be a tiny magnet. This magnet behaves strangely, as any small particle in the quantum world does. It can take only two directions, that is, one pole being directed either up or down. In quantum theoretical terms, an electron is said to have a spin whose quantum number is $s=1/2$, and can take either $m_s=+1/2$ (up) or $-1/2$ (down).

Now let us summarize what we have said so far. An electron around a nucleus behaves in the manner characterized by the atomic orbitals specified by the quantum number set and whose energies take discrete values. The energy depends on both "n," the principal quantum number, and the nuclear positive charge value Z in the manner that the energy is proportional to $(Z/n)^2$, if you want to know. An electron's behavior can be specified by a set of four quantum numbers, n, l, m_l, and m_s.

Pauli, an Italian physicist, asserted (and this is now known as "Pauli's exclusion principle") that no two electrons (in atoms or molecules) can take exactly the same quantum number set. Or we may say that if two electrons occupy the same space (orbital) they must have different spins. This suggests that an orbital (n, l, m_l) can accommodate up to two electrons (but no more); i.e., an electron with $m_s=+1/2$ and another with $m_s=-1/2$. This is the basis for building up atom(s). What we do essentially is to put electrons into the orbitals available around the nucleus. Since we are interested in the most stable such atom, i.e., of the lowest energy (technically called "ground state"), we place electrons, starting with an orbital of the lowest energy, up to two electrons in it, and then going up to orbitals of the next lowest energies. In the atoms of the first 20 elements, the order of energy of different atomic orbitals is known to be: $1s<2s<2p's<3s<3p's<4s$.

Let's start with the simplest: hydrogen atom, which has one electron. This electron would occupy the 1s orbital; we say that the electron configuration of hydrogen atom is $1s^1$, meaning one electron in 1s orbital. This electron can occupy other orbitals for sure, i.e., 2s, 3p, or whatever. But all these configurations would be higher in energy than the ground state hydrogen $1s^1$. Next, helium atom has two electrons. Those two electrons can occupy 1s orbital, i.e., $1s^2$, provided that they have different m_s values: one is $+1/2$ and the other is $-1/2$. Next comes lithium with three electrons; two electrons in 1s and one electron in the next lowest orbital, 2s; hence $1s^2 2s^1$. Now you can do it. Try carbon, nitrogen, oxygen, fluorine, and neon [they are $1s^2 2s^2 2p^2$, $1s^2 2s^2 2p^3$, $1s^2 2s^2 2p^4$, $1s^2 2s^2 2p^5$, $1s^2 2s^2 2p^6$]. Note that 2p orbitals, consisting of three orbitals, can accommodate as many as six electrons. This completes the $n=2$ shell consisting of 2s and 2p orbitals. Now we have to start putting electrons into $n=3$ shell: 3s, 3p. So, sodium Na would have $1s^2 2s^2 2p^6 3s^1$; magnesium Mg $1s^2 2s^2 2p^6 3s^2$; aluminum Al $1s^2 2s^2 2p^6 3s^2 3p^1$; sulfur S $1s^2 2s^2 2p^6 3s^2 3p^4$; chlorine Cl $1s^2 2s^2 2p^6 3s^2 3p^5$; argon Ar $1s^2 2s^2 2p^6 3s^2 3p^6$ [Complete the list by consulting the periodic chart, Fig. 19.2]. We have come up to the 18th element.

Let's review what we have done so far and compare it with the periodic chart. First, the row corresponds to "n" values. H and He in the first row use 1s orbital. The elements starting with Li through Ne use 2s and 2p orbitals, and the third row 3s and 3p. Secondly, you will note that the elements in the same column have the analogous electron configuration in the outermost shell. That is, H $1s^1$, Li $2s^1$, Na $3s^1$ in the first column (refer to Fig. 19.2); C $2s^2 2p^2$, Si $3s^2 3p^2$ in the fourth column (or fourteenth column); F $2s^2 2p^5$, Cl $3s^2 3p^5$ in the seventh (or seventeenth column), etc. The outermost shell is called "valence shell." Electron(s) in the outermost shell (which can be termed as "valence electron") has the highest energy among all the electrons in an atom. Hence the valence electrons are the most reactive; as a matter of fact, they are the ones involved in chemical bonding and chemical reactions. Then you can deduce that the elements in the same column would behave similarly, as they have a similar electron configuration. This is the great organizing principle in the study of all elements; and this is essentially "chemistry."

When we move further, the situation becomes a little more complicated, though the basic principles remain the same. Let's try. The 19th element is potassium, K, in which the last electron would occupy 4s orbital rather than 3d; so $1s^2 2s^2 2p^6 3s^2 3p^6 4s^1$. Ca is likewise $1s^2 2s^2 2p^6 3s^2 3p^6 4s^2$. Next come 3d orbitals and then 4p orbitals. So the 21st element Sc (scandium) would have $1s^2 2s^2 2p^6 3s^2 3p^6 4s^2 3d^1$, but it should be written as $1s^2 2s^2 2p^6 3s^2 3p^6 3d^1 4s^2$.

The reason for this involves too much detail and cannot be explained fully here. It would be sufficient to say that in elements of atomic number higher than 21 the 3d orbitals are a little lower in energy than the 4s orbital. The ten elements that utilize 3d orbitals come before the next six elements that utilize 4p orbitals after completing 3d subshell ($3d^{10}$). Note that d orbitals can accommodate as many as ten electrons, as there are five d-orbitals. These ten elements are called "transition elements" and have very interesting properties. The rest of the periodic chart can be understood by these principles. However, two extra series of elements, called "Lanthanides" and "Actinides," each of which consists of 14 elements intervene and they use f-orbitals (4f and 5f, respectively). A recently developed periodic chart accommodates these anomalous series more smoothly as shown in Fig. 19.5.

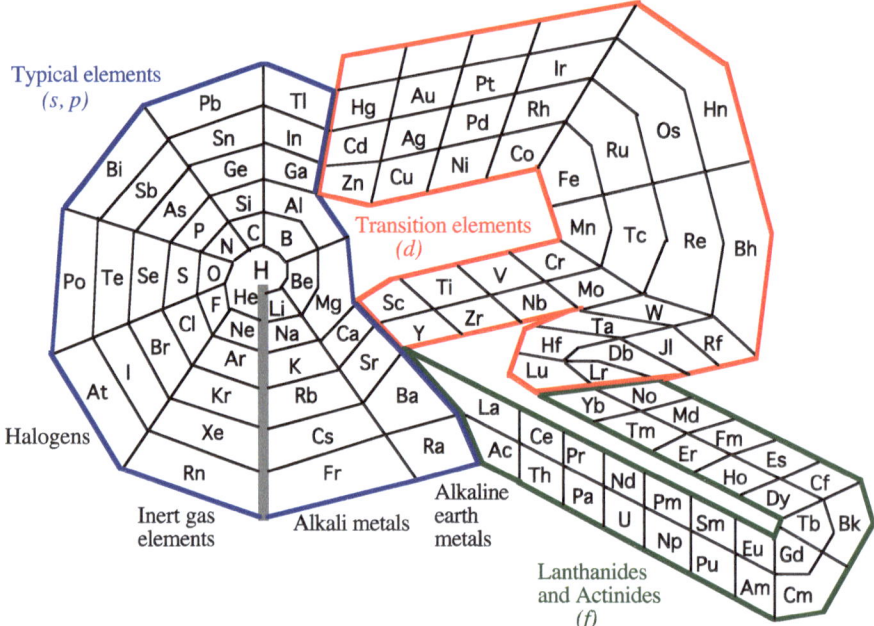

Fig. 19.5 An alternative Periodic Chart. Start at the center (H) and then move counterclockwise to the immediate right. A period starts at the right of the purple line (e.g., K) and ends at just the left of the purple line (e.g., Kr) (you have to go through the red portion after Ca, and the green portion after Ba)

19.6   Ions

You can remove an electron from or add one to a neutral atom. As an electron is bound by the positive electric charge of the nucleus, it requires a certain amount of energy to remove an electron. The energy required for removing an electron from an atom is called "ionization energy." The removal of an electron from a sodium atom results in the formation of a positively charged species, which is described as Na^+, because now there are one-less negative charges (electrons) than the positive charges (protons), so the net is plus 1 charge. This kind of entity is called "ion"; more specifically "cation" (positively charged ion) in this case. Removal of electrons becomes more and more difficult as you remove more electrons, because the attractive force from the nucleus increases. As a matter of fact, in the case of sodium cation Na^+ it is so difficult to remove another electron that the next possible ion Na^{2+} has never been found under ordinary circumstances.

Let us look at this situation a little more closely. The first electron to be removed from Na is the one in orbital 3s, which lies farthest (on the average) from the nucleus, hence the easiest to remove. The next electron in line is in an orbital 2p, which is much more tightly bound to the nucleus. Take the case of Mg. It has two electrons in 3s orbital. The second electron can be removed with quite a bit more energy than that required for the removal of the first electron, but not prohibitively so, because it is still in 3s orbital. Hence it is relatively easy to form Mg^{2+}. The third one has to come from the much low-lying 2p orbital; hence Mg^{3+} cannot form under ordinary circumstances.

How about adding electron(s) to an atom? Certain elements such as nitrogen, oxygen, fluorine, and chlorine accept an electron willingly. That is, it does not require energy to add an electron to these (neutral) atoms; actually it gives off a small amount of energy. This means that it becomes more stable or lower in energy releasing that extra energy. The result is the formation of a negatively charged entity such as F^- or Cl^-. These are called "negative ions" or "anions." However, the majority of elements would not do so. It requires energy to force an electron onto neutral atoms. By the way, the energy change upon addition of an electron to a neutral atom is called "electron-addition energy." [Historically, the concept was expressed as "electron affinity" but it is confusing. Hence we abandon the concept here]. This energy will be negative in the case of fluorine, chlorine, or a few other elements, and positive for many other elements.

The tendency of an atom to attract electron(s) toward itself (in a molecule) is called "*electronegativity*." Fluorine, chlorine, and oxygen that are located at the upper right-hand corner of the periodic table (Fig. 19.2) are strongly electronegative, because they have relatively more (effective) positive charges at the nucleus compared to those located toward the left in a row. The order of electronegativity is $F>O>Cl>N \sim Br>S$. A significance of electronegativity will be discussed in Sect. 19.7.4. On the other hand, the elements at the lower left corner are least electronegative.

You might wonder where that energy required for the removal of electron(s) (and/or addition of electron(s)) would come from. Well you can bombard atom(s) with high-energy electrons, for example. This will either knock an electron out of an atom, resulting in a cation, or add to form an anion. But ordinary compounds such as NaCl which consists of Na^+ and Cl^- form without the application of such a violent means. By the way, a compound made of cation(s) and anion(s) is called an ionic compound (see below). NaCl and MgO are two examples.

Then how? As we learn soon, the ionic species of opposite electric charges bind strongly through the so-called electrostatic interaction, and the interaction releases a fairly large amount of energy. Typically, this energy compensates the energies required to make ions. For example, when you put sodium (Na) metal in chlorine (Cl_2) gas, a strong reaction (implying that a lot of heat (energy) is released) takes place, resulting in the formation of NaCl solid. In this reaction, Cl_2 will remove an electron each from two Na atoms. Magnesium burns in air (i.e., reacts with oxygen in the air) to form magnesium oxide MgO. Magnesium was used for photographic flash; magnesium foil burns quickly in the air and gives off that flash. We will discuss the issue of energy and chemical reactions shortly.

19.7   Molecules/Compounds: Chemical Bonding and Structures

19.7.1 Molecular Compounds and Ionic Compounds

This is the crux of chemistry. Chemistry is about molecules and compounds. "Molecule" is the smallest unit of a compound, but the smallest unit of a compound may not necessarily be a "molecule." Molecule is an aggregate of atoms that are covalently bound together. Water is an example and is made of two hydrogen atoms and an oxygen atom bound in the manner of H–O–H, where the line between atoms represents a covalent bond. We will explain the "covalent bond" shortly. Methane, the major component of the natural gas is another example, which can be written as CH_4; the carbon atom is bound with four hydrogen atoms in the manner below. This simply shows that a C-atom binds 4H-atoms. All the so-called organic compounds are usually made of molecules.

$$
\begin{array}{c}
\text{H} \\
| \\
\text{H} - \text{C} - \text{H} \\
| \\
\text{H}
\end{array}
$$

However, there are still a large number of compounds in which atoms are not bound by covalent bonds but rather by ionic interactions or electrostatic interactions. These are ionic compounds; typical examples include table salt $NaCl$ (made of Na^+ and Cl^-) and calcium carbonate (limestone, $CaCO_3$). In these compounds no molecular species is found. The formula $CaCO_3$ simply implies that the compound consists of Ca^{2+} ions and CO_3^{2-} ions in one-to-one ratio; they are stacked together by the attractive forces between the positive and the negative electric charges and are arranged in an orderly manner (crystal). $CaCO_3$ would not behave as a unit like a molecule does. Hence the chemical properties of $CaCO_3$ are those associated with Ca^{2+} and CO_3^{2-}. (I hasten to add, though, that the atoms in an ion, such as CO_3^{2-}, are bound by covalent bonds). So there are two types of compounds: molecular compounds and ionic compounds.

19.7.2 Ionic Bonding and the Structures of Ionic Compounds

The force between two electric charges is called "electrostatic force." The associated energy is "electrostatic energy," and the energy can be expressed by the equation:

E (electrostatic energy) $= q_1 q_2 / 4\pi\varepsilon_0 r$ (the corresponding force $= q_1 q_2 / 4\pi\varepsilon_0 r^2$)

That is, it is directly proportional to the products of electric charges q_1 and q_2, and inversely proportional to the distance r, as you can intuitively guess. The other

Fig. 19.6 NaCl rock salt
structure: (**a**) *red balls*
represent sodium ion (Na⁺)
and *blue balls* chloride ion
(Cl⁻); (**b**) shows the crystal
lattice. (from P. Atkins and
L. Jones, "Chemical
Principles", 3rd ed
(Freeman W.H., 2005))

a   **b**

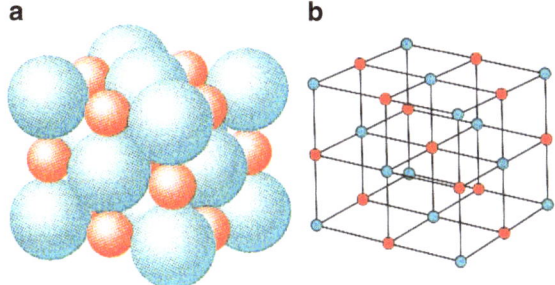

items are some constants. [ε_0 is called the permittivity of vacuum and π is the ratio of the circumference to the diameter of a circle]. Note that the energy is negative when the electric charges of two entities are opposite and that the negative energy means an attractive interaction, whereas that the interaction energy between the electric charges of the same sign is positive and the interaction is repulsive, as you expect.

Take sodium chloride (NaCl) solid as an example. The structure of NaCl crystal is shown in Fig. 19.6. The electric charges of Na and Cl are +1 and −1, respectively; these values are expressed in units of "e," the magnitude of electric charge of a proton or an electron. The distance between them has been estimated as 283 pm (pm = picometer = 10^{-12} m). The ions are represented by balls of different sizes and colors. They are arranged in a regular manner. In this case, the cation (Na(+I)) and the anion (Cl(−I)) occupy alternate positions (called lattice point). Sodium chloride is said to take a lattice structure of rock salt type. The rock salt type crystal structure is one of the most commonly found ionic crystal structures. Other types of crystal structure are also known.

In the table salt, i.e., solid NaCl crystal of 1 mol (58.5 g), a very, very large number of Na⁺ and Cl⁻ interact electrostatically, and the energy of such an assembly is expressed by a slightly more complicated equation than the one shown above. We can actually perform such a calculation relatively easily, and the result is about −760 kJ/mol. This means that 1 mol of crystalline NaCl is stable by 760 kJ than 1 mol each of separate Na⁺ and Cl⁻ ions. You can also say that it requires 760 kJ of energy to separate completely 1 mol of NaCl solid into separate Na⁺ ions and Cl⁻ ions. The energy of this kind is often called the "lattice energy" and can be regarded as the ionic bonding energy. Other typical lattice energies are: −701 kJ/mol for KCl, −912 for AgCl, −2,393 for MgBr₂ (Mg²⁺ and two of Br⁻ io) and −3,398 for CaO (Ca²⁺ and O²⁻). The details of these calculations and the magnitudes of the ionic bonding energy are not particularly important, but it should be noted that the energy is larger between more highly charged species and that the magnitude of ionic bonding energy is fairly large. You do not know yet how large the typical bonding energies in chemical compounds are, and you may not be able to make a judgment about the second point. You will see them shortly.

19.7.3 Covalent Bonding and Structures of Covalently Bound Compounds

The basis of covalent bonding is also "electrostatic," i.e., interaction between electric charges. But in this case the interactions that are important are those between the negative charges of electrons and the positive charges of the nuclei.

Let us consider the simplest molecule that forms using a covalent bonding. That is, hydrogen molecule, H_2, H–H. A hydrogen atom consists of one positively charged (+1) nucleus and one negatively charged (−1) electron. The electron is attracted by the nucleus by the electrostatic force. When two hydrogen atoms (atom 1 and atom 2) come closer, the electron (1) in atom 1 also comes under the influence of the positive charge of the atom 2 nucleus. The other electron (2) in atom 2 on the other hand is now attracted by the nucleus of the atom 1 as well (see Fig. 19.7). The net effect is that atom 1 attracts and binds with atom 2.

You might picture this as the two electrons distributing and acting as glue between the two nuclei (Fig. 19.7). Of course when two particles of the same sign come closer, they repel one another. In this case there are repulsive interactions between the two electrons and also between the two nuclei. The latter is relatively weak because of their distance, and the first is also relatively weak, because the distance between the electrons is relatively large on the average though they are constantly moving. In other words, the attractive force overcomes these relatively weak repulsive forces, and results in the binding of the two atoms. One other thing that needs to be said is

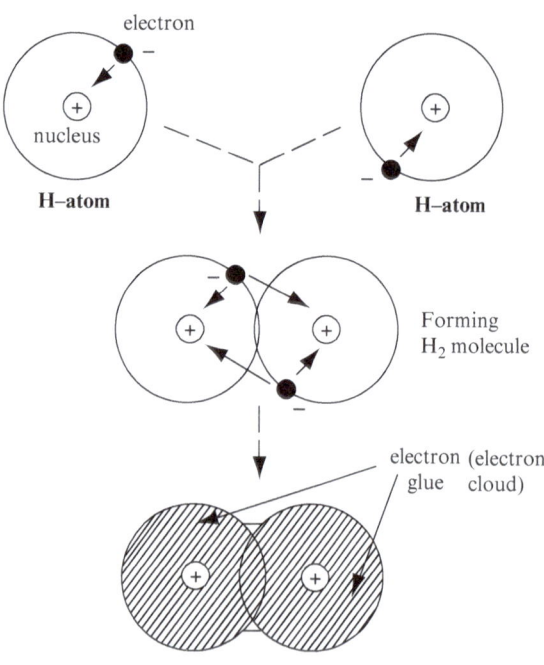

Fig. 19.7 How a covalent bond is formed between two hydrogen atoms

that the two electrons now occupy the same space around the two nuclei and hence they have to have the opposite spins for this to happen; that is, one electron has +1/2 spin and the other −1/2, just like the case where an atomic orbital is filled up (see above).

This is the basic principle of covalent bonding. We may say that two atoms form a covalent bond which consists of a pair of electrons shared by the two atoms; hence this type of bond is called "covalent," "sharing electrons." We often use a single line connecting two atoms to indicate a bond. The bond of this type, i.e., in which the contributing electrons occupy the space between the two nuclei is called "sigma, σ" type. The shared electrons in H_2 molecule are equally distributed between the two atoms, because the attractive forces of the two atoms are equal in this case. The molecule H_2 is about 432 kJ/mol more stable than the two separate hydrogen atoms as a result of this electron sharing. Since the molecule is formed by a single H–H bond, 432 kJ/mol is required to break the H–H bond. Such an energy is defined as the bond energy of the H–H bond.

Now let us see some other examples. CH_4, methane, is made of four C–H bonds, as the carbon atom has four electrons available for bonding and each hydrogen atom has one. There are three C–H bonds on each carbon atom and a C–C bond in the case of ethane C_2H_6 (see Fig. 19.8 for the structures of these molecules). All these bonds are of σ-type. In Fig. 19.8a, a methane molecule is shown to consist of a carbon atom bound to four hydrogen atoms through σ-bonds, and the four hydrogen atoms appear to be arranged in a square (with the carbon atom at the center). In reality, this is not the case. The hydrogen atoms arrange themselves around the carbon atom in a so-called tetrahedral manner. In other words, if one connects the adjacent hydrogen atoms by a line, one gets a tetrahedron (four-faced). Therefore, it would be expressed better by a three-dimensional structure such as the one shown in Fig. 19.8b. This kind of structure is called "ball and stick" model, where balls represent the atoms and sticks the bonds. This model is still not quite representative of the actual shape.

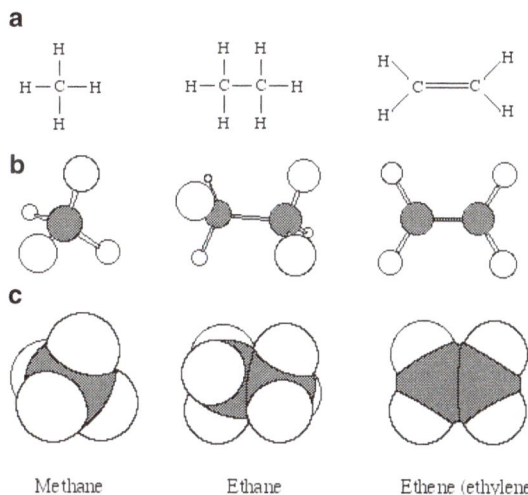

Fig. 19.8 Structures of simple hydrocarbons

The electrons around the nucleus actually fill up the space between the atoms. So, another way of representing a molecular structure is to fill the space that represents roughly the shape of an electron cloud in the molecule. This model is called "space-filled." They are exemplified by those in Fig. 19.8c. We will often use the simplest structural expression shown in Fig. 19.8a, because it shows clearly the connections (bonds) between atoms. The "ball and stick" representation is good at showing the three-dimensional relative arrangement of atoms, but often it can obscure some of the atoms, unless we rotate the molecule. The "space-filled" model is good at showing the overall three-dimensional structure.

How about C_2H_4, ethylene ("ethane" in technical terms)? It has two less hydrogen atoms than ethane. If you remove two hydrogen atoms from ethane, you would end up with C_2H_4 but you would also see that one electron on each carbon will be left. The carbon–carbon bond present in the original ethane remains as such, but the odd electron on each carbon that is created in the process is distributed in the direction perpendicular to the C–C bond. The direction of the C–C bond is defined as z-axis. The odd electron, technically speaking, is in a p_x atomic orbital. The electrons in p_x orbital, one in each carbon atom, also contribute to binding the two carbon atoms. The distribution of electrons in this type of bonding, which is designated as π-bond is shown in Fig. 19.9. Hence the connection C-to-C in C_2H_4 molecule is made of two different types of bond, σ and π and is contributed by two pairs of electron. The bond between the carbons in this molecule is said to be a "double" bond and is indicated by a double line (see Fig. 19.8). You may guess that the binding through the π-bond is weaker than that of σ-type bond, as the electrons in π-bond are less effectively overlapped.

Let us consider another simple-looking molecule, N_2, dinitrogen, the major component of the air. Each nitrogen atom has five valence electrons available. One electron on each nitrogen atom binds in the manner of the σ-bond. The second electron in p_x (refer above) forms a π-bond. The third electron is in another p-orbital, say p_y, which is perpendicular to the σ-bond as well as p_x. This electron in an N atom forms another π-bond together with that on another N atom. So now, there are one σ-bond, and two π-bonds that are perpendicular to one another (Fig. 19.10). Two more electrons still remain in each nitrogen atom. These are not involved in the bonding and instead remain in an orbital in each N-atom. The two electrons have to have opposite spins as they occupy the same space, and they are called "lone pair" (not involved in bonding). The entire situation of electrons in an N_2 molecule is shown in Fig. 19.10. So we may say that the bond in N_2 is a "triple" bond.

How about the next molecule, O_2? We now have two more electrons. These will remain as independent electrons with the same spin, but have an effect to cancel the

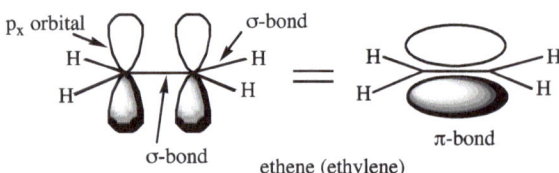

Fig. 19.9 Sigma (σ)-bond and pi (π)-bonds

Fig. 19.10 Bondings
in (**a**) dinitrogen and
(**b**) dioxygen; each of π_x, π_y,
and sp-orbitals (one on each
nitrogen atom) contains a pair
of (coupled) electrons; in the
case of dioxygen molecule,
each of the two additional
electrons occupies the π_x^*
and π_y^* orbitals, respectively.
In both of these, the sigma
bond between the two atoms
are omitted

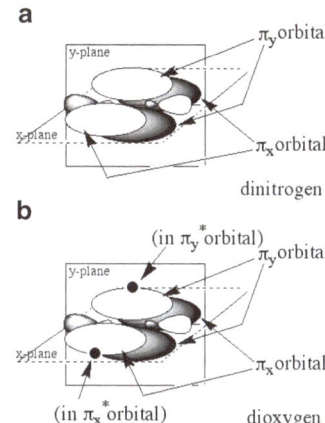

Fig. 19.11 Paramagnetism
of dioxygen molecule (from
P. Atkins and L. Jones,
"Chemical Principles", 3rd
edn (2005, W.H. Freeman))

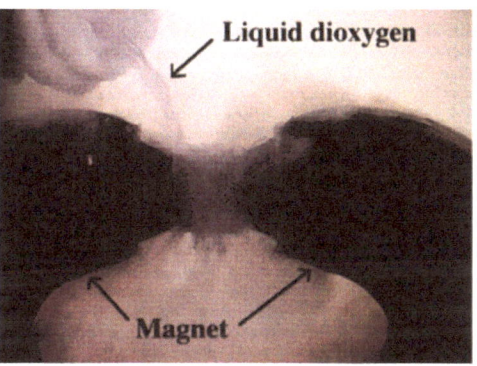

boding effect of a π-orbital. Then the bonding in O_2 is equivalent to a "double" bond. The resulting electron configuration of O_2 is shown in Fig. 19.10b. Note that the O_2 molecule has two unpaired electrons; that is, those two electrons have the same spin (either both are $+1/2$ or both are $-1/2$). This has an interesting consequence. As a spin behaves like a magnet, O_2 behaves as if it is a magnet. This is called "paramagnetism." All the other molecules mentioned above in this section are, instead, diamagnetic. The electrons in these compounds are all paired up as in σ- and π-bonds or lone pairs. One of a pair of electrons has a spin of $+1/2$ and the other $-1/2$, so that their magnetic characters cancel one another and, as a result, these compounds behave as non-magnets. The fact that O_2 is paramagnetic can readily be demonstrated by pouring liquid O_2 through a magnet. The liquid stream will be attracted toward and sticks to the poles of the magnet (Fig. 19.11). Nothing like this happens with liquid nitrogen N_2; it will stream through the poles of magnet without sticking.

The overall shape of these molecules, i.e., N_2 and O_2, would be similar to that of H_2 shown in Fig. 19.7; that is, a dumb bell, two balls bound together. Indeed such a shape has been observed for O_2 recently by a technique called scanning tunneling microscope. It is shown in Fig. 21.7.

Fig. 19.12 Structure of benzene: (**a**) emphasizes the p-clouds (above and below The hexagonal plane); (**b**) represents each atom by a space-filling sphere, hydrogen atoms bound with each of 6 carbon atoms are omitted in (**a**), and the white sphere represents hydrogen atom in (**b**)

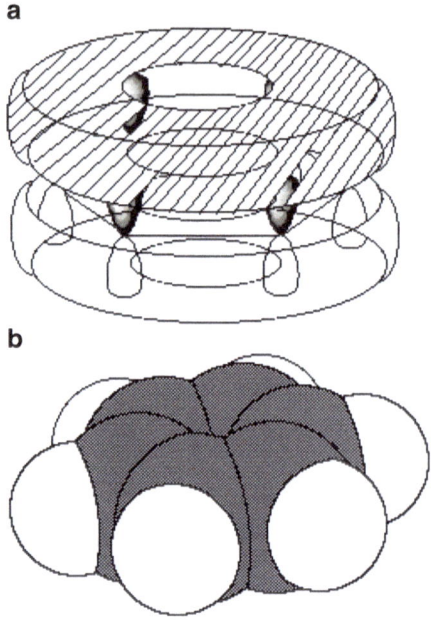

One more interesting mode of bonding needs to be mentioned before we leave this topic. It is illustrated by benzene, C_6H_6. The six carbon atoms form a skeleton of hexagon (see Fig. 11.3). This is formed by σ-bonds between two adjacent carbons. Each carbon has used three electrons for these bondings, and has still one more electron left. These electrons are in orbitals (p) that are sticking up and below of the hexagonal plane. These electrons can be combined to form three π-bonds between adjacent carbons (see Fig. 11.3). But as Fig. 11.3 shows, there are two ways of doing this, and neither of them should be preferred over the other. In fact the two are equivalent, and the reality may be said to be a sort of mixture of both forms. This situation is called (chemical) "resonance." Unfortunately, "resonance" in everyday use or physics has a different connotation. It may be better described as a six-electron cloud being distributed equally along the hexagon perimeter above and below the plane, like a sandwich of doughnuts (see Fig. 19.12). The atomic structure of benzene has recently been imaged also by a technique called scanning tunneling microscope. The image that is shown in Fig. 21.8 indeed shows that this description of the benzene's structure is valid.

19.7.4 Polarity of Covalent Bond

The electrons in H_2, N_2, or O_2 molecule will be distributed equally between the two atoms, because the electron-attracting strength of the two atoms is the same. Thus the bonding in these molecules can be said to be "purely covalent," and such a bond is called "non-polar." That is because each atom in say N_2 would share ten valence electrons equally and hence would have five electrons about it and would appear to be electrically neutral, no net electric charge on it.

The electron-attracting strength of an atom (or rather element) is called "electronegativity." Let's consider methane, CH_4. There are four C–H bonds (see Fig. 19.8). The electronegativities of elements carbon and hydrogen happen to be about the same (though C is slightly more electronegative than H). Therefore, the electrons are almost equally distributed between C and H, and hence this bond is considered to be non-polar.

What happens if we replace one of the hydrogen atoms of methane with chlorine, for example, i.e., CH_3Cl (monochloromethane)? As chlorine is more electronegative (attracts electrons more strongly) than carbon, the two electrons located in between carbon and chlorine are not equally distanced from the two atoms. They are located closer to the chlorine atom. So it looks that the chlorine is negatively charged and the carbon is positively charged. It is expressed as: $C^{\delta+}$–$Cl^{\delta-}$. The value of δ here is called partial charge and less than 1. And the bond C–Cl is said to be "polar." If the value of δ is 1, the two electrons are now completely located in the chlorine atoms and are not shared by carbon and chlorine atoms. The bonding then would be essentially an ionic bond. The polarity of a bond is perhaps the most important character, which will dictate the behavior and reactivity of a compound (molecule). Polar bonds will form between atoms of different electronegativity, but important ones among carbon compounds (i.e., organic compounds) are C–X, where X is F (fluorine), Cl (chlorine), Br (bromine), and I (iodine). By the way these four elements are collectively called "halogen." The other important ones are C–O, as in CH_3–OH (methanol), and C–N and C–S. And, of course, O–H bond itself in methanol is also polar. [Which carries a small positive charge, O or H here?] On the contrary, when a carbon atom binds with a metallic element such as Mg, the carbon atom will carry a negative charge and the metal (Mg in this case) will be positive, because metallic elements are much less electronegative than carbon.

Let's now take a look at carbon dioxide: O=C=O. The carbon–oxygen bond here is polar (negative on oxygen and positive on carbon). The polar character, polarity, is defined in terms of dipole moment which is directed from the positive toward the negative and defined as "the partial charge x the distance between the charges." In this case, one bond (say, the right-hand C=O) has a dipole moment directed to the right and the other bond (the left-hand O=C) has the dipole moment of the same magnitude but directed to the left. Therefore, the dipole moments cancel one another, and the molecule CO_2 does not have a dipole moment, and hence it is non-polar as a molecule. The polarity of a molecule is thus determined by the polarities of individual bonds and the entire structure of the molecule. The former, i.e., bond polarity is important in determining the chemical reactivity of a molecule, and the polarity of the entire molecule is important in determining its physical behavior. Polar molecules, because of having separate electric charges in themselves, attract each other more strongly than non-polar comparable molecules. And this will be reflected in their physical state, as will be discussed in the next section.

19.8   Energy: How and Why Changes Occur

19.8.1 Energies

Chemicals, whether atoms, molecular compounds, or ionic compounds, are made of particles: nuclei and electrons, atoms, ions, and molecules. The particles affect each other; i.e., they interact. Energy is associated with any interaction. For example, a positive ion and a negative ion attract one another, and the magnitude of interaction can be expressed in terms of energy that will be required to separate these two ionic species. We have already talked about covalent bonding between atoms; this is another example of interaction. The energy that will be required to separate two atoms bound by a covalent bond is called "bond energy." Different bonds would have different bond energies.

A bond or rather atoms bound by a bond are not static. For example, H-atoms in an H–H molecule do not sit still. They are moving all the time. Rather, they are vibrating about a certain distance. This distance is a time-averaged value and called "bond length (or bond distance)." Vibration requires a certain amount of energy. If you heat up (i.e., add energy to) a molecule, the vibration will become more vigorous. A molecule is also tumbling (rotation), and this motion is also associated with energy. The motions mentioned so far are those within a molecule; we can say, these are intra-molecular motions.

We do not, however, deal with a single molecule ordinarily. We will be dealing with an ensemble of a very large number of molecules (and ions). So we have to consider the interactions between molecules as well. Again the force is "electrostatic." Then if the interacting entities are ions, the interaction energy will be large and depends on their electric charges and the distance between them. Polar molecules will interact through the partial charges as well in them. The partial charges in a molecule are represented by a value called "dipole moment," which depends on both the magnitude of the partial charges and the distance between the charges in a molecule. The interaction between polar molecules depends on the dipole moment and the distance between molecules.

The interaction energy will be small between neutral non-polar entities. There is a fundamental electrostatic interaction, however, even between neutral non-polar entities. This is called "London dispersion" force, and it depends on the size (hence the molar mass) of the molecule. The elements in the last column of the periodic chart are called inert gas elements; they are neutral and non-polar (spherical). Therefore, the smallest, helium He, has the lowest boiling point, because the interaction energy between helium atoms is minimal. By the way, these inert gas elements are literally inert; that is, they can neither form a molecule with itself (like He–He) nor with other elements (though Xe-containing compounds have been made). The relationship between the inter-particle interaction and the boiling point (and others) will be discussed in the next section. As you go down the column, Ne–Ar–Kr–Xe, their boiling points go up; the boiling point of Xe is $-107°C$ as compared with that of He, $-269°C$. This trend reflects the increase of London dispersion force as you go down the column, i.e., the elements become larger and heavier, and hence the inter-particle energy (London force) becomes larger, as you go down.

Ordinarily, the interaction energy is largest between ions, then between polar molecules, and weakest between non-polar molecules. However, the non-polar interaction (London dispersion) can become fairly large when it comes to larger molecules, even if they are non-polar.

There is a special type of interaction between an electronegative atom (such as F and O) in a molecule and a hydrogen atom bound to an electronegative atom in another molecule. Such an interaction is called "hydrogen bond." For example, a hydrogen atom of a water molecule interacts with an oxygen atom of another water molecule through a hydrogen bond. The nature of the hydrogen bond has not yet been understood fully but it is certainly a mixture of ionic interaction and covalent interaction. Hydrogen bonds are crucial in the biological systems, i.e., our lives, as we discuss in several other chapters.

So an ensemble of molecules or compounds (let's call such an ensemble a "system") would have a certain amount of energy under a certain condition. This energy is called "internal energy," which depends among other things on the pressure under which the system exists. It will be more convenient to use energy value under a constant pressure, because we normally operate things under the ambient constant pressure. Such an energy is technically called "enthalpy." We mean "enthalpy" when we say "energy" in this book, unless otherwise stated. If a system changes (in a widest sense), the accompanying energy would also change, because the interactions among all particles inside molecules and/or between molecules in the system would change.

19.8.2   Phase Changes and Energies

Suppose we have water at 25°C. It is a liquid, in which water molecules are rapidly moving around but are always in close contact with each other. If we want to separate a molecule from another, we need to put some energy, perhaps in the form of heat. Heat would encourage more vigorous movements of molecules and may overcome the attractive interactions between them. If this happens, water molecules get out of the confine of liquid, and move far apart from each other (except for occasional collisions). Now they are in the form of "gas" (vapor). This is a change of state, from liquid state to gas state. Since you added heat (a form of energy) in the process, you can imagine that the energy content (enthalpy) of a gas is larger than that of the corresponding liquid. This is an example of the so-called phase change. It is accompanied by a change in energy content. As a liquid changes to a gas, the system would gain energy; we say that the energy change (of the system) in this case is positive. This means that the system's energy will increase (i.e., positive) as it turns from liquid to gas. The opposite process, i.e., vapor turning into liquid (condensation), would then release energy which goes to the surrounding; hence the system loses energy and hence the energy change in the system accompanying the condensation process is negative. The change from liquid water to solid ice is another example of phase change, and accompanied by a negative energy change.

How high temperature you have to bring a liquid to in order to vaporize would depend on how strongly molecules are interacting in the liquid. Technically speaking, the temperature at which the vapor pressure (gas pressure above liquid) becomes 1 atmospheric pressure is defined as boiling point. So the boiling point depends on how strongly molecules interact. In the case of water, the boiling point is 100°C, reflecting a relatively strong interaction between water molecules. Water (H_2O) is polar, resulting in a strong dipole–dipole intermolecular interaction, and in addition there is a strong hydrogen-bond interaction. On the contrary, comparable molecules such as CH_4 (methane) and NH_3 (ammonia) have much lower boiling points, −164 and −33°C, respectively. Ammonia is polar but less so than water and the hydrogen-bond interaction is not as strong. Methane is a non-polar molecule and the only possible interaction between methane molecules is the weak London dispersion force.

Melting point, the temperature at which a compound melts, is also dependent on the strength of intermolecular (inter-particle in general) interactions. Naphthalene, the smelly stuff used as an insect repellant (mothballs) in the chest of drawers, is a non-polar neutral molecule and hence the intermolecular interactions are relatively weak. Therefore, its melting point is rather low, 80.5°C, and thus it relatively easily turns into gas (hence the smell). In contrast, table salt, sodium chloride, melts at a high temperature, 801°C, because the components, sodium ions (Na^+) and chloride ions (Cl^-), attract each other very strongly. In general, ionic compounds have high melting and boiling points, while molecular compounds melt and boil at relatively low temperatures. The strength of interaction between molecules is also a determining factor for the other physical properties such as surface tension and viscosity. The main principles and concerns of chemistry up to this point (i.e., except for the issue of chemical reaction which is discussed below) can be summarized as in Fig. 19.13.

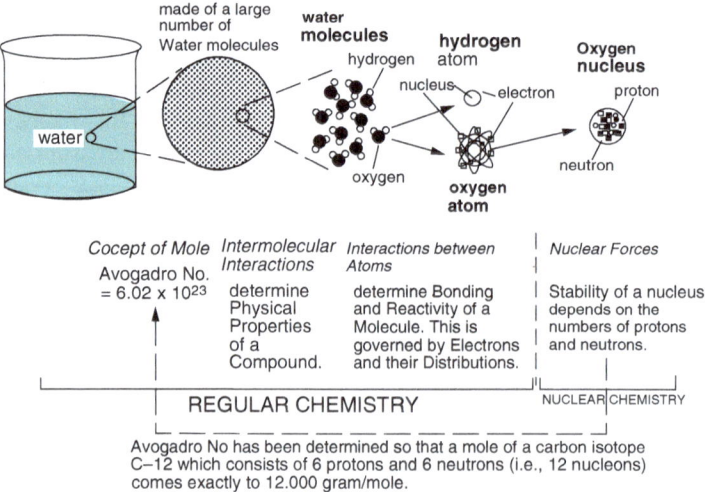

Fig. 19.13 Chemistry in a Nutshell

19.8.3   Writing Chemical Reaction Equations

A chemical reaction involves any number of chemicals, say, chemical species A, B, C, D. Let us suppose that A and B react chemically and, as a result, C and D would form. In other words, a chemical reaction involves change in chemical species. There is one more thing that needs to be added in order to be able to write this chemical reaction equation. That is the so-called stoichiometric relationship. It is about the relationship among the quantities of these chemical species when they react. For example, two molecules of A react with three molecules of B and form two molecules of C and one molecule of D. This chemical reaction can then be represented by the following type of equation, though the equal sign "=" is now usually replaced by an arrow sign:

$$2A + 3B = 2C + 1D \text{ ("1" is often omitted) or } 2A + 3B \rightarrow 2C + D$$

In this particular case, the reaction is supposed to proceed from the left-hand side to the right-hand side, and the species on the left-hand side (A and B in this case) are called "reactants" and those on the right-hand side are "products." The stoichiometric relationship implies only the overall (relative) quantitative relationship among the chemical species involved and would not imply how the reaction takes place. The latter issue is called "reaction mechanism." That is, the chemical reaction equation such as shown above does not necessarily imply the reaction mechanism. The reaction mechanism is a very difficult issue, and its delineation involves a lot of detailed studies on the reaction, and currently is one of the most interestingly studied in the chemical/biological science.

The stoichiometric balance is based on the principle that in a chemical reaction no change (both quantity and identity) should take place regarding all the individual atoms involved. In other words, the total number of atoms on the left-hand side should be equal to that on the right-hand side. And this applies to all the elements involved.

Let us try a few examples.

1. Sodium metal (represented by the symbol Na) burns (reacts with) in chlorine gas (Cl_2) and forms sodium chloride (NaCl) and nothing else:

$$Na + Cl_2 \rightarrow NaCl$$

Chemical species and their arrangements are correct, but the stoichiometric relationship is not correctly represented. In other words, the total number of atoms in chemical compounds should be equal on both sides. You have the same number of Na on both sides, but there are two Cls on the left-hand side and only one Cl on the right. To adjust the number you can do the following:

$$Na + \left(1/2\right)Cl_2 \rightarrow NaCl$$

Now you have the same number ("1") of Cl atoms on both sides. Therefore, this is a correct chemical reaction equation. However, the fractional number (in front of a chemical) should be avoided. So you can multiply the whole quotation by2, and will get:

$$2Na + Cl_2 \rightarrow 2NaCl$$

This process of adjusting the stoichiometric relationship is called "balancing the equation," and the resultant equation is said to be "balanced."

2. Try this. Magnesium metal (Mg) would be burnt in air; that is, it reacts with oxygen (O_2) in the air and forms only magnesium oxide (MgO). Write a balanced equation.

3. When you dissolve copper sulfate ($CuSO_4$) in water, it forms a light blue solution, and the solution contains two ions, Cu^{2+} and SO_4^{2-} (sulfate). This reaction is simply expressed as

$$CuSO_4 \rightarrow Cu^{2+}(aq) + SO_4^{2-}(aq)$$

"(aq)" implies the ionic species are interacting with water (aquo) molecules, and "$Cu^{2+}(aq)$" is responsible for the light blue color. Now if you add an aqueous ammonia solution (i.e., consisting of NH_3 and H_2O) to this light blue solution, a light blue precipitate, copper hydroxide ($Cu(OH)_2$), and ammonium ion (NH_4^+) form. This reaction can be expressed as

$$Cu^{2+}(aq) + NH_3 + H_2O \rightarrow Cu(OH)_2 + NH_4^+$$

That is, only $Cu^{2+}(aq)$ is involved in this reaction, and sulfate (SO_4^{2-}) is simply present in the solution. Is this equation balanced? If not, try to balance it. (You can ignore (aq) in balancing an equation).

4. Here is one of the most complicated reactions. MnO_4^- (permanganate) reacts with $C_2O_4^{2-}$ (oxalate) in the presence of acid (H^+) and the products are Mn^{2+}, CO_2 (carbon dioxide), and H_2O (water). That is:

$$aMnO_4^- + bC_2O_4^{2-} + cH^+ \rightarrow dMn^{2+} + eCO_2 + f\,H_2O.$$

The stoichiometric coefficients $a, \ldots, f$ are to be determined so that the equation is balanced. In this case the chemical species are electrically charged (i.e., ions), and you need to balance the electric charge as well. If you are comfortable with algebra, you can do something like the following. Because the numbers of Mn (manganese) on the left- and the right-hand side should be the same, "a" has to be equal to "d"; $a=d$. Likewise, $2b=e$ with regard to C; $4a+4b=2e+f$ for O; and $c=2f$ for H^+.

   The electric charge on the left-hand side is $-a-2b+c$ and that on the right-hand side is $2d$. These equations constitute a set of simultaneous equations. All you need to do is solve them. You will realize that you have only five equations (relationships) for six unknown values, so that you cannot solve them. It is OK, because all we need is relative values. One thing you can do in this situation is assume, for example, $a=1$.

Then $d=1$. Since $e=2b$, the balance equation for O will become $4+4b=4b+f$ so that $f=4$. This means $c=8$. From the electrical balance equation you get $c-2b=3$. Therefore, $b=5/2$. Now you have obtained: $a=1$, $b=5/2$, $c=8$, $d=1$, $e=5$, $f=4$. Because b is a fractional number, you multiply everything by 2, and finally obtain

$$2MnO_4^- + 5C_2O_4^{2-} + 8H^+ \rightarrow 2Mn^{2+} + 10CO_2 + 8H_2O$$

Balanced chemical reaction equations are necessary when quantities of reactants and products are to be calculated and/or the energy relationship as below is to be evaluated. Otherwise chemical reactions, even if not balanced, are usually sufficient to provide the necessary information about the chemical reactions.

19.8.4 Reactions and Energies: Free Energies

Now let's take a simple reaction: i.e., $2H-H\ (H_2) + O = O\ (O_2) \rightarrow 2H-O-H$. [This is usually written as $2H_2 + O_2 \rightarrow 2H_2O$]. This reaction involves splitting two H–H bonds and one O=O bond, and then recombining them to form two molecules of H–O–H which has two O–H bonds. Bond energies of H–H, O=O, and H–O are 432 kJ/mol, 494, and 459, respectively. Therefore, the energy change accompanying the reaction will be: $2 \times 432 + 494 - 4 \times 459 = -478$ kJ. This means that 2 mol of water is more stable by 478 kJ than the starting material, 2 mol of H_2 and 1 mol of O_2. [By the way, the end product is assumed to be gaseous water (water vapor) in this calculation]. Another reaction $C + CO_2 \rightarrow 2CO$ is accompanied by an energy change of +172 kJ.

Why are we concerned about the energy change accompanying a phase change, a chemical reaction, or a change of other kinds? Because the energy change is the driving force for a chemical change. *Ordinarily*, a change would occur in the direction in which its energy content decreases or becomes lower, i.e., the system becomes lower in energy or more stable. In other words, a system will change if the change brings the system to a lower, more stable state, as water flows from a higher place to a lower place.

However, it has been demonstrated that a system can turn into a higher-energy state if a certain condition is met. There is a deeper criterion for the direction of chemical change. The most basic principle in determining the direction of change is called "2nd law of thermodynamics." In order to understand it, we need to introduce one more concept, entropy. "Entropy" is a measure of randomness and related to the change in enthalpy divided by the prevailing temperature (as expressed in absolute temperature scale). The 2nd law of thermodynamics can be expressed in many ways. One version is: "the entropy of the universe cannot decrease." That is, the randomness of the universe as a whole should either remain the same or increase, and never decrease. Then, the entropy of the universe eventually will reach a maximum; at that point no further change can occur. This is called the entropy death of the universe.

How does this apply to chemical changes? When a substantial amount of energy is released from a system, it goes to the surroundings (universe) and increases the entropy of the surroundings, and hence the overall entropy, i.e., the sum of the entropy of the system and that of the surroundings usually increases. This is usually so even if the entropy of the system decreases, because the entropy increase in the surrounding is large enough to compensate the decrease in the system entropy. However, when the system's energy change is positive (i.e., energy comes into the system, rather than is released), the entropy of the surroundings has to decrease. So unless the entropy of the system itself increases to such an extent to overcome the decrease in the surroundings' entropy, such a change would not occur, because the overall entropy decreases in this case. This implies, however, that when the entropy increase in the system is large enough, a change could occur even if the change is accompanied by an increase in the enthalpy of the system; i.e., the system can go uphill, provided that the overall entropy change is positive. This is another expression of the second law of thermodynamics.

To combine the effects of energy (enthalpy) and entropy, a Yale physical chemist, Henry Gibbs, introduced the concept of "free energy" (now called "Gibbs free energy"). It is a sort of available energy of a system to its surroundings; that is why it is called "free." The second law of thermodynamics can now be stated in terms of free energy as: "a change of a system can occur if its associated free energy change is negative." The free energy value is usually heavily dependent on temperature.

Let's take an example. Iron ore usually comes in the form of iron oxide Fe_2O_3. To produce iron metal from it you have to remove oxygen from it (reduction reaction in technical terms). This is commonly accomplished by using coke (coal, i.e., carbon). The reaction can be expressed as $Fe_2O_3 + 3C \rightarrow 2Fe + 3CO$. The free energy change of this process at room temperature is positive (+328 kJ at 25°C); i.e., it would not occur at room temperature. But if you raise the temperature above 672°C, the free energy change of the same process becomes negative (e.g., −45 kJ at 700°C); hence the reaction becomes possible. That is why the iron-producing furnace has to be operated at such a high temperature.

Another interesting application of the free energy change of chemical reaction is "battery." The free energy change (negative for the system) that is available from a chemical reaction can be turned into an electric energy, which lights a light bulb or operates a radio. For example, the car battery is taking energy from the chemical reaction: $PbO_2 + Pb + 2H_2SO_4 \rightarrow 2PbSO_4 + 2H_2O$, where Pb stands for "lead," PbO_2 lead dioxide, $PbSO_4$ lead sulfate, and H_2SO_4 sulfuric acid. The last mentioned is that corrosive acid, which is consumed as you use (discharge) the battery. The free energy change for the above reaction is −415 kJ, and you are converting this chemical free energy to an electric energy. The amount of this free energy change should give rise to about 2 V. So by combining six of such components in series you will get about 12 V, and that is the car battery.

We discussed this topic, because this is the crux of chemistry; that is, why a phase change or chemical reaction can take place or in what direction it will go. In order to apply this concept fully, we have to go into the energy values and their calculations. However, in the main portion of this book, we would use only the concept but not go into actual quantitative calculations. That is, we will remain to be only qualitative in our discussion, except for a few simple, fundamental cases.

19.9   Speed of Chemical Reactions

The free energy change as expounded above will tell us whether a reaction will go or even how far the reaction will go, but would not tell us how fast the reaction will take place. These are two separate issues. The former factor that we discussed above is the "thermodynamic" factor, and the latter is the "kinetic" issue.

There is a barrier to go over for a reaction to proceed. We can picture the progress of a reaction as a hiker moving from a base camp over a mountain and then to another base camp on the other side of the mountain (Fig. 19.14). The difference in the height between the starting base camp (the initial state in technical terms) and the final camp (final state) is the free energy difference. The height of the mountain to climb over is called "activation energy." The higher the activation energy, the slower the reaction will be. In general the activation energy is not related to the free energy difference. So, even if the free energy change predicts that a reaction should take place, it may not proceed at an appreciable speed; that is, no reaction appears to occur. Such is the case for example a mixture of hydrogen (H_2) and oxygen (O_2). They should react strongly to form water (the free energy change tells us so). But nothing happens; in other words, the reaction barrier, activation energy, is so high that the reaction speed is virtually zero.

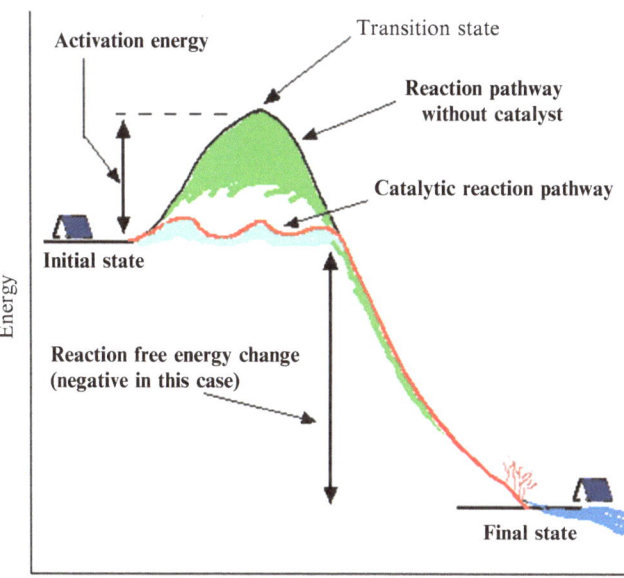

Fig. 19.14 How a chemical reaction proceeds, and the effect of a catalyst that allows a reaction goes over a lower barrier route

There are two ways to increase a reaction rate. The first is to raise "temperature." It must be an everyday observation that chemical reactions, e.g., in cooking (cooking is all chemical reactions), become faster as you raise temperature. By raising temperature you are raising the level of the initial state in a sense, so that the height to overcome would become lower.

Another is to somehow lower the height of the barrier, activation energy. The chemical substance that can do this is called "catalyst." In reality, it is not simply reducing the energy barrier. In most of the cases, a catalyst allows the reaction to go through a pathway that has lower energy barriers (see Fig. 19.14).

Now let's take the mixture of H_2 and O_2 mentioned above. When you throw a flame (of a match) into the mixture, the reaction that takes place is so fast that it explodes. The flame contains a sort of catalyst, a radical initiator in this case. A tiny bit of this free radical brings about a so-called chain reaction, which becomes very fast soon. On the other hand if you flow a mixture of H_2 and O_2 over a fine powder of platinum metal in a glass tube, the glass tube glows indicating that a rapid reaction is taking place and releasing a lot of heat (remember that this reaction should release a lot of energy). The platinum powder is a catalyst in this case. These two reactions are the same in producing water as the only product, but obviously proceed quite differently. One is explosively fast, while the other is fast but smooth. The way in which a reaction proceeds (this is called "reaction mechanism") in general depends on the catalyst used.

Thousands of chemical reactions are taking place constantly in our body. Without catalysts, these chemical reactions would not be fast enough for our body to function properly. Virtually all the chemical reactions in living organisms are dependent on catalysts. Biological catalysts have a special name: enzyme(s). They are crucial for biological functions. Enzymes are one of the most studied groups of substances today.

Chemicals and Light

20

We have five senses: touching, seeing, hearing, smelling, and tasting. All these senses are chemical in their operations. Chemicals will bind some receptors on the surface of tongue to cause "bitter" or "sweet" sensation. Smelling is caused by similar mechanisms. Hearing is more mechanical in its cause, but relayed to the brain through chemical processes in the nerve systems. Seeing is perhaps the most important sense. It is caused by light hitting the eyes. That light comes from or through a filter of substance. Light does something to the substance (a chemical) and it is changed as a result. This is a very general phenomenon and has a wide-ranging implication to the everyday life. We will see how light and chemicals interact with each other.

You are now looking at this particular page; the paper looks white and the letters black. From the window you see tree leaves; they are green. These are visible. Well, there are a lot of molecules floating around you: air. You do not see it. Some chemicals are visible, i.e., colored, while others are invisible. Why is that so?

20.1 Characters of Light

Light is a kind of wave called "electromagnetic wave." It consists of an oscillating electric field and the associated oscillating magnetic field. You can imagine it as two electrode plates and its voltage varies with time from a positive value to a negative value and then back and the electrodes are continuously moving along. An electromagnetic wave is then characterized by how often the field oscillates; this is called "frequency" (often symbolized by v). The radio signal is also conveyed by an electromagnetic wave from the radio station to your radio set. A particular radio station uses an electromagnetic wave of a particular frequency, say, 91.5 MHz (MHz = mega hertz, 10^6/s). We cannot *see* this electromagnetic wave. X-ray you would be subjected to in order to have a disease or a broken bone diagnosed is also "electromagnetic wave"; the frequency of X-ray is very high, something like 10^{17}–10^{20} Hz. Light wave (electromagnetic wave of any frequency) travels at a very high speed. The speed (c) is constant in vacuum, irrespective of the frequency, and is about 3×10^8 m/s.

E. Ochiai, *Chemicals for Life and Living*,
DOI 10.1007/978-3-642-20273-5_20, © Springer-Verlag Berlin Heidelberg 2011

Since light is a wave, it also can be characterized by "wavelength" (distance between the peak to peak of wave or valley to valley). It is often symbolized by a Greek character λ. And these three parameters are related by an equation $c = \nu\lambda$ (c = speed). The light that our eyes can respond to, that is, visible to human eyes, has wavelengths between about 800 and 320 nm ("nm" is nanometer, i.e., 10^{-9} m, one-billionth of meter). Eight-hundred-nanometer light appears "red" to human eyes, and 350–320 nm light "violet." We humans cannot see lights of other ranges. Lights that are longer than 800 nm in wavelength are called "infrared" (infra = below), and those shorter than 320 nm are called "ultraviolet" (beyond violet).

20.2   Interactions of Light with a Chemical Compound

What would happen when light (of, say, the visible to ultraviolet range) hits a compound? As the light is a vibrating electric field, the electrons in the compound would be moved around by the changing electric field (of light). The compound has discreet energies, as we discussed earlier, and ordinarily it is in the most stable (ground) state, i.e., of the lowest energy. There is a gap in energy between the ground state and a higher energy state. When the energy of light (which is defined by $h\nu$, h = Planck's constant) matches the energy gap between states, the vibrating light wave causes the electron in the ground state to jump ("excitation" in technical terms) to that particular higher energy state (excited state). Simultaneously, the light energy would be absorbed by the compound. In other words, it should be said that the energy of the light is used to excite the molecule from the ground state to the excited state. And the light that has come though this compound now lacks that component which has been absorbed. Suppose that the light shone on the compound is sunlight which consists of lights of all wavelengths (this causes sunlight to look "white") and the compound has absorbed red light. Then the light that comes from the compound lacks the red component, and that is what we see. As a result, the compound looks "green" (Fig. 20.1). This is the reason that a certain compound is colored.

Many compounds are colored, as they absorb light in the visible range. However, a lot more compounds absorb light not in the visible but in the ultraviolet region.

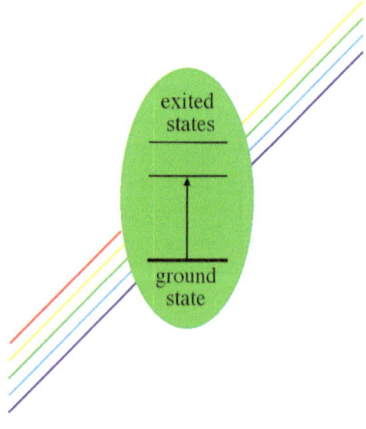

Fig. 20.1 Light absorption by a molecule

Such a compound, though it absorbs light, appears to human eyes as "colorless," as the absorption would not disturb the visible range.

20.3 Basis of Spectroscopy

We can use the phenomenon of interaction between light and a compound to explore the details of its energy states and structure. Scientists have devised means to measure the wavelength and the intensity at which light is absorbed by a compound. The device is known as spectrophotometer, and the technique is "spectroscopy." Absorption of light in the visible and ultraviolet region has been shown to be caused by the movement of electron between different energy states.

The nuclear motions in a molecule take the form of vibration as mentioned earlier. The vibrational motion's energy also takes only discrete values (quantized). Light can also cause the transitions between vibrational energy levels in a molecule. The light that does it occurs in the so-called infrared range. So there is "infrared" spectroscopy that gives us information about the geometrical structures of molecules through the vibrational motions.

Electromagnetic waves of the FM radio frequency range cause flipping of the electron spin and hence is utilized to study how electrons behave. This is the basis of the so-called electron-spin resonance spectroscopy. The most widely used spectroscopy, though, has to do with the magnetic nature of the nucleus of a certain atom, particularly hydrogen or carbon. This is called "nuclear magnetic resonance (NMR)" spectroscopy and is very useful for determining the structures of many compounds, particularly those of organic compounds. The technique has now been extended to other elements, including cadmium, mercury, and others. A medical diagnostic tool "MRI" (magnetic resonance imaging) is essentially the same as "NMR" in principle. The electromagnetic wave used here is in the so-called microwave range, the same as that used in the microwave oven.

The microwave oven, however, does not play with the nuclei in the compound, but it interacts and encourages the rotational movement (tumbling) of some molecules (particularly water) in the material. As the tumbling motion of water in a meat, for example, becomes more vigorous, the meat temperature goes up. So the tumbling motions of molecules can then be studied with microwave spectroscopy. Molecules floating in the interstellar space are emitting light due to the tumbling motions. A microwave telescope receives these signals from molecules and can convert them to spectra, which reflects the rotational motions of molecules (Chap. 13). Hence, we can identify those molecules.

20.4 Emission of Light

What would happen to a molecule which has gone onto a higher energy level as a result of absorption of light? A higher energy means a less stable state. So the molecule will seek ways to go back to the lower (and eventually the ground) states. It can do this in two ways. It transfers its extra energy as heat to its surroundings.

This is what usually happens. An alternative is to emit a light. This happens often but not always. For example, a compound called "fluorescein" gives a blue-green solution in ethanol (due to absorption of light). But under sunlight, it glows emitting yellow-green light. This phenomenon is called "fluorescence." Some other compounds give emission long after the light source is turned off. This phenomenon is called phosphorescence.

Excitation to higher energies can be brought about not only by light but also by many other means. Neon sign uses emission of light from neon (Ne) in the glass tube; it is excited by electric discharge. Fluorescent lamp is another such device. TV screens and monitor screens of computers (of older type) are coated with light emitting metal compounds (such as europium and copper). These metallic ions are excited by the electron beams and emit light. Laser, the red light to read the price bar code on the cash register counter in a grocery store for example, is also dependent on the phenomenon of emission of light. That red laser light comes from a ruby (chromium (III) in alumina) crystal. The excitation is done electrically in this case. Firefly uses chemical reaction energy to create that ephemeral light in summer evening. The same principle is used in light sticks (see Chap. 9).

Sun is burning hydrogen atoms (nuclear fusion), and is at very high temperatures. Hydrogen atoms are there in excited states. When hydrogen atoms come down to lower energy states, they emit light. And that is essentially what we get here on Earth as Sunlight. Sunlight consists of far ultraviolet, ultraviolet to visible light.

Are Atoms and Molecules for Real? Can We See Them?

21

Well, we have been talking about "chemicals" (the material world in general) and their behaviors, assuming that they can be understood in terms of atoms and molecules. Hopefully, you will agree with chemists by the time you finish reading this book, that yes, the ideas of atoms and molecules make a very good sense and that they seem to give a very good basis for understanding the material world in terms of chemistry. However, can you believe them, i.e., atoms or molecules? How can we believe something we cannot see?

We do see water; an aggregate of tiny water molecules. We are able to see the molecules, all right, but not an individual molecule. It is too minute to see with our naked eyes. So, are the atoms and the molecules for real? Chemists assume and also believe that they are. What do atoms or molecules look like? Chemists picture them based on the atoms and their combination, as we did in the last two chapters and do throughout this book. Can we see them? "Seeing is believing," isn't it? But the atoms and molecules are too small to be seen with our naked eyes. So what should we do? Magnify them!

We have been trying to magnify things for a long time, by telescopes and microscopes. We can believe what we see under a microscope, because our experience with such a simple devise as a magnifying glass tells us so. However, there is a limit to magnification by a light microscope. Or we should say that its resolution is limited; resolution is the smallest separation that can be distinguished. The resolution is limited by the wavelength of the light used and is of the same order of magnitude as the wavelength. In the case of light (visible light, more precisely), the wavelength is in the range of 320–800 nm. "nm" is a unit of length and is one-billionth of 1 m, 10^{-9} m. (Let us review the units of length. The traditional units for atomic level distance is Å, which is 0.1 nm, but we will not use Å, as the general trend is to use the SI units, meter. 1/100 of meter is centimeter (cm), 1/1,000 of meter is millimeter (mm), 1/1,000 of millimeter or one-millionth of 1 m is 1 μm, 1/1,000 of micrometer (or one-billionth of 1 m) is 1 nm, and 1/1,000 of a nanometer is 1 pm). A separation smaller than, e.g., 500 nm cannot be distinguished by light microscopy. A typical bacterial cell is of the size of the order of 1 μm (1,000 nm or 10^{-6} m), and hence the

E. Ochiai, *Chemicals for Life and Living*,
DOI 10.1007/978-3-642-20273-5_21, © Springer-Verlag Berlin Heidelberg 2011

bacterial cell and relatively larger cell bodies may be seen by light microscopy. But more details of the cell structure need a higher magnification or higher resolution to be seen. We need a light of shorter wavelength. A light of shorter wavelengths can be produced by a stream of electron. This is the "electron microscope."

21.1   Electron Microscope

One of the weirdest things and mysteries in the material world is that a particle behaves also like a wave, and vice versa. For example, a visible light (wave) can also behave as if it is a particle. This particle is called "photon." On the contrary, a particle electron can act also like a wave, an electromagnetic wave. This dual character of material is universal. An ordinary body, say a golf ball weighing 20 g, can behave also like a wave when it is flying at a high speed. Only because of its weight, the wavelength of such a flying ball is so small, and hence we do not recognize it as behaving like a wave. Therefore, the dual character manifests only with very small (light weight) particles such as electrons.

An electron beam can be accelerated by applying a high voltage, and then it behaves like an electromagnetic wave (light) of a short wavelength. This light (of electron) can be used as the basis of a microscope; this is the "electron microscope." The wavelength of a typical electron beam is of the order of 5 pm (pm=picometer= 10^{-12} m). The resolution in this case is limited not only by the wavelength but also by other factors. As a result, the practical limit of resolution with electron microscope is of the order of 100 pm (0.1 nm) in an (ordinary) electron microscope. The sizes of the cell organelles are of the order of 10–100 nm, and hence they are seen clearly in ordinary electron micrograms. Even the inner structure of a virus or a cell organelle such as mitochondrion or chloroplast can be seen.

How about molecules? Well, molecules vary widely in their sizes. The smallest hydrogen molecule, H_2, is of the order of 100 pm (or 0.1 nm) (in length). The individual hydrogen molecule has not been seen by electron microscopy; it is too close to the limit. But the larger molecules may fall in the range greater than the resolution of electron microscope. A relatively simple molecule called uranyl acetate was photographed by electron microscopy (see Fig. 21.1), because uranium is fairly large as an atom. The "uranyl" is made of an atom of uranium and two atoms of oxygen, and "acetate" is the anion of acetic acid which is the ingredient of vinegar. In this case, the uranium atoms are separated by about 0.37 nm from each other and hence they are clearly shown separate, but the other atoms involved, carbon, oxygen, and hydrogen are not clearly seen.

Among the largest molecules are some proteins (enzymes) and DNAs. Proteins, though fairly large in molecular weight, often coil up and take globular shapes. For example, the diameter of serum albumin (the most abundant protein in your blood serum) is about 2 nm and that of myoglobin (the protein in the muscle to pick up oxygen) is about 10 nm. Some of larger proteins have been photographed by electron microscope. They usually look like blobs and reveal no detail. So, yes, the protein molecules can be seen. But whether they are constructed with amino

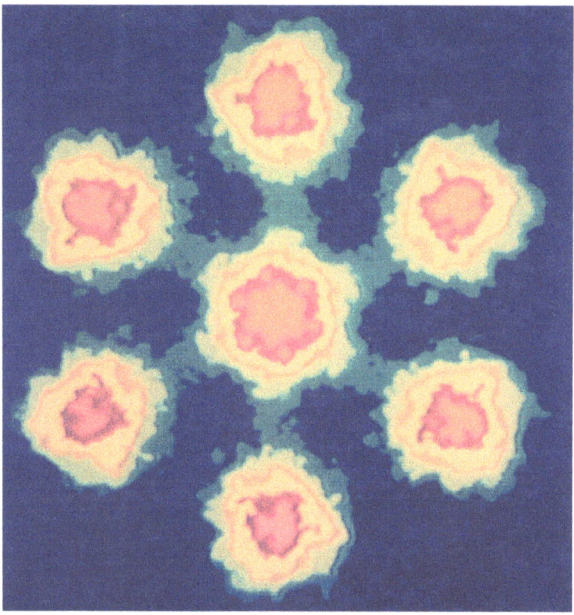

Fig. 21.1 An electronmicrogram of uranyl acetate (from R. H. Petrucci and R. K. Wismer, "General Chemistry with Qualitative Analysis" (1983, Macmillan))

acids to take a definite shape has not been ascertained from the ordinary electron microscopy.

DNAs are very large molecules. One of the smallest DNAs, that of bacteriophage xΦ174, consists of a single strand of 5,375 nucleotides (see Chap. 4 for a discussion on DNA). When stretched, it should be about 1.8 μm (1,800 nm) long and about 2.5 nm wide. Let us try a little math here. The interval between two nucleotides in a helix has been determined to be about 0.34 nm (3.4 Å); hence, $0.34 \times 5,375 = 1827.5$ nm. So it should show up as a string in electron micrograms. Well, an example is shown in Fig. 21.2. It shows two strings (actually circular), i.e., two separate DNA molecules. The length of the circular string is estimated from the magnification factor to be about 1.8 μm. This is exactly what chemistry has been predicting it should look like. It is supposed to be a helical strand, but those minute details cannot be seen here. But, *this is a DNA*!. This is the stuff that governs the inheritance and all biological actions. Isn't it incredible?

More recently techniques have been developed to enhance the resolution of electron microscopy; the technique is called "High resolution transmittance electron microscopy" (HRTEM). Figure 21.3 shows KI (potassium iodide) incorporated in a single-walled nanotube (see Chap. 11 for nanotubes). Individual potassium (K^+) and iodine (I^-) atoms are clearly seen embedded in a tube of about 1.6 nm (1,600 pm) diameter. You can see clearly individual atoms (actually ions) in the crystalline solid of $SrTiO_3$ as shown in Fig. 21.4. What you are seeing in these pictures is the image of the electron cloud around a nucleus, and that is essentially an atom.

Fig. 21.2 An electronmicrogram of DNA molecules

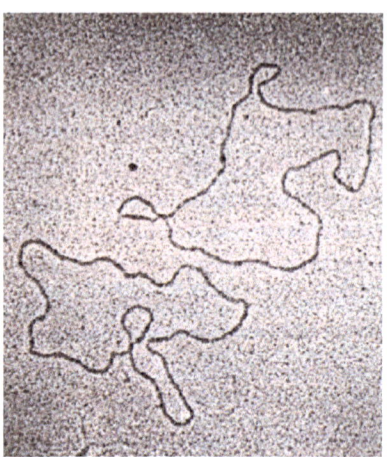

Fig. 21.3 HRTEM image of potassium iodide (K) embedded in a carbon nanotube. As the inset indicates, the atoms of K^+ and I^- are clearly seen as separate balls (from R. R. Myer et al., Science, **289** (2000), 1324–1327)

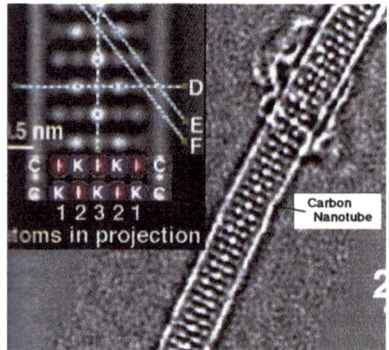

Fig. 21.4 HRTEM image of strontium titanate ($SrTiO_3$); atoms of Sr, Ti, and O are seen as separate balls (from Z. Zhang et al., Science, **302** (2003), 846–849)

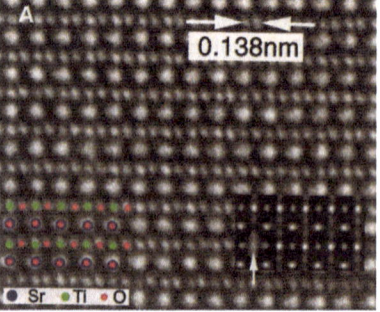

21.2 Scanning Probe Microscopy

Figures 21.2–21.4 illustrates the best you can do with the electron microscope. In 1980s, two scientists, Heinrich Rohrer and Gerd Binnig at IBM's Zürich Research Center in Switzerland developed a revolutionary method of microscopy. They were awarded a Nobel prize in 1986 for this invention. The principle is simple, though technologies required to realize them are quite advanced.

Figure 21.5 illustrates the principle of scanning tunneling microscope [STM, a kind of scanning probe microscopy (SPM)]. The probe tip is sharpened so that it is made of a single metallic atom (to make such a probe is not an easy task). A solid sample (a metal or a semiconductor) is placed flat and the probe tip is brought very close to the surface of the sample. If the probe comes within a few atomic distance from the atom just below (of the sample), a strange thing happens. Electrons jump over that distance from the atom of the probe to that of the sample. This phenomenon is known as "tunneling" (of electron). This corresponds to a flow of electric current between the probe and the sample, though they are not in contact. It is known that the electric current flow due to tunneling is dependent on the distance and the nature of atoms of probe and sample.

The probe is scanned across the surface of the sample, and the height of the probe is adjusted so that the tunneling current flow is kept constant as it is being scanned, and the height can then be recorded as the probe is scanned all over the surface of sample. This movement of the probe must necessarily be minute and precise and is accomplished by a devise based on piezoelectricity. The trace of height of the probe may represent the shape and arrangement of atoms and vacancies in between. The scanning result can then be treated by a computer and be visualized.

Such an example as shown in Fig. 21.6 then can be regarded as that of the atomic arrangement on the surface. This is gallium arsenide (GaAs); the blue represents arsenic atoms and the red gallium. Each hump represents a single atom. This is the way chemists had pictured the structure of this compound (and others) long before a devise like STM gave them the picture of atoms in a substance. In other words, the chemists' depiction of atoms and molecules was not very bad.

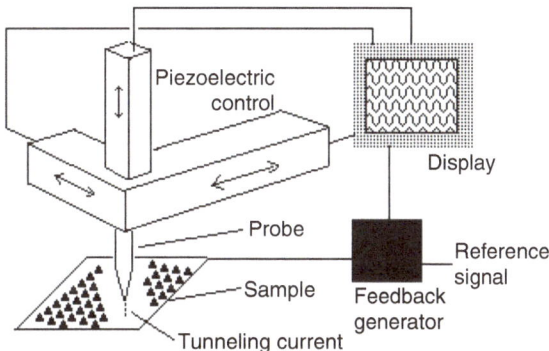

Fig. 21.5 A schematic presentation of STM

Fig. 21.6 STM image of
gallium arsenide (GaAs) (from
D. P. Kern et al., Science, **241**
(1988), 936–944)

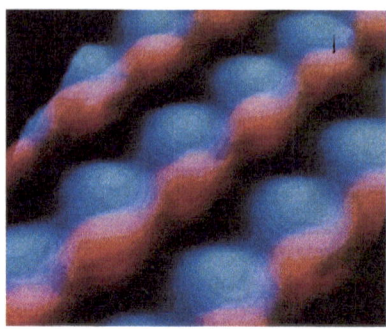

Fig. 21.7 STM image of
dioxygen molecule (from B. C.
Stipe et al., Science, **279**
(1998), 1907–1909)

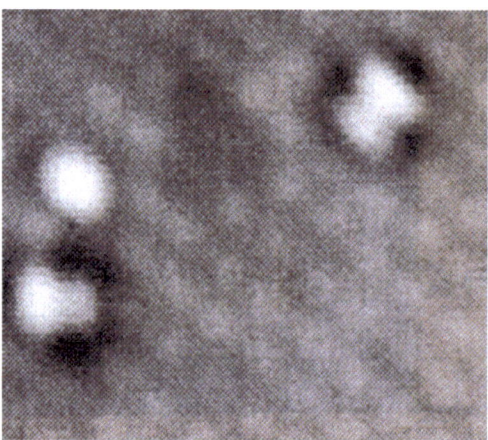

By the way, gallium arsenide is expected to function as a semiconductor like
silicon, which is the basis of the today's high-technology industry. The diameters
of gallium and arsenide can then be estimated from this result. They are about 0.2
and 0.4 nm (2 and 4 Å), respectively, and are in agreement with the data obtained
by other methods. These results suggest that the resolution of STM is of the order
of 0.01 nm (=0.1 Å = 10 pm).

Astonishingly, a small molecule like dioxygen O_2, which is the form of oxygen
found in the air, has also been imaged by STM. Figure 21.7 shows two O_2 molecules
bound (adsorbed) on the surface of silicon. The extra blob just above the oxygen
molecule at the left-hand corner is a defect in the underlying silicon crystal. As you
see, the dioxygen molecule looks like a dumbbell; i.e., each dioxygen molecule is
made of two spheres bound together. These molecules do not look like being bound
flat to the surface, but being bound slanted. The sphere that represents an atom is
actually the shape of the electron cloud around the nucleus. Again this is the same
as the picture chemists have been using to represent the oxygen molecule O–O.
Figure 21.8 gives an STM image of benzene molecule adsorbed on a rhodium metal
surface. The positions of six carbon atoms and six hydrogen atoms are shown by

Fig. 21.8 STM image of benzene molecule (from P. Saulet and C. Joachim, Chem. Phys. Lett., **185** (1991), 23)

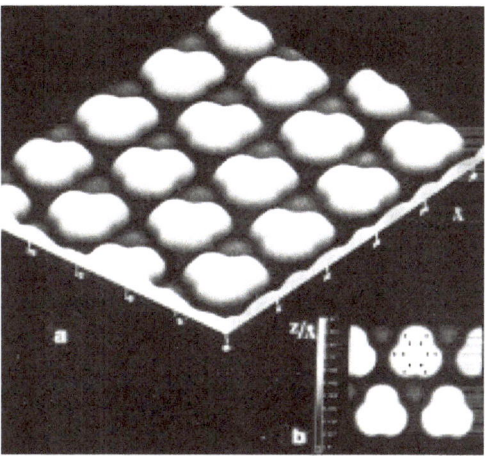

Fig. 21.9 AFM image of DNA (from J. Mou et al., FEBS Lett., **371** (1995), 279–282)

small dots in the inset portion. Because it is adsorbed on a metal, it is slightly distorted from the free benzene molecule. However, the overall picture corresponds well to the structure that chemists have been imagining based on the theories.

STM is but an example of scanning probe microscopes. Another important scanning probe microscope is called atomic force microscope (AFM). When a probe is brought close to a sample, a force will be exerted between the probe and the sample. If this force (atomic force) is of a general character, i.e., London dispersion force, then the force will be dependent on the distance and the nature of the probe and the atomic nature of the sample. Hence, a geometrical structure of a sample will be imaged at atomic level by scanning such a probe that responds to atomic force.

The double helix structure of a DNA has been imaged by AFM. That is shown in Fig. 21.9. A DNA molecule looks like a right-handedly wound ropes. The ropes are the (alpha) helically coiled double strand. These images (Figs. 21.6–21.9) give us very good pictures of molecules with atomic level resolution in some cases, and really convince us that, yes, atoms and molecules do exist, and moreover, that they do look like how chemists have been saying they do. What more do you need to be convinced?

21.3   X-Ray Diffraction: Atomic Structure of Large Molecular Compounds and Ionic Compounds

The scanning probe microscopes are still unable to resolve atomic structures of such complicated large molecules as proteins and DNAs. The atomic structures of these large molecules (as well as smaller molecules and ionic compounds) are now almost routinely determined by X-ray crystallography. This technique is much older than SPM. However, the way this technique produces the atomic structure of a molecule is not as straightforward as electron microscopy or scanning probe microscopy. Yet, the structures obtained by this techniques, particularly those of smaller molecules and ionic compounds, have well been verified by many other methods and experience. Therefore, the atomic structure of a large molecule obtained by this method is also now well believed to represent the correct structure of the compound. It is true, though, that the structure is valid only for the solid crystal state to which this method is applied.

We will explain very briefly how it is done with a few examples. X-ray is a light of very short wavelength. For example, when an electron beam is directed to a copper metal, the copper metal produces an X-ray of 154 pm (1.54 Å). This is the source of light (X-ray, that is). The light (any light) shone on a compound will be scattered by the electron clouds (of atoms) of the compound. When the atoms are arranged regularly as in a crystal, this scattered light will show a pattern. This is called diffraction pattern, and is related to the crystal structure. The diffraction pattern can be photographed or now digitally recorded. The pattern consists of spots (where the diffracted X-ray hits) of differing intensities. The intensity of a spot is related to the electron cloud density that scatters X-ray. (The more dense electron cloud scatters X-ray more strongly; thus the spot will be more intense). Therefore, we can reconstruct how the electron clouds are distributed throughout space (in a crystal) by analyzing the diffraction pattern including intensities. The result can be expressed as an electron density map (in three dimensions), and that can then be reinterpreted as an aggregate of atoms. Softwares have now been developed to convert the diffraction pattern to the atomic arrangements in a molecule. The resolution obtained is now routinely about 1.5–2 Å (150–200 pm = 0.15–0.2 nm). Hence, we can distinguish atoms in molecules, as the inter-atomic distances in molecules are around 1–2 Å (0.1–0.2 nm).

The actual process is now fairly automated, but one of the basic problems is to obtain a single crystal of an appropriate size. Many biologically interesting samples such as proteins and DNAs are difficult to crystallize, and researchers would spend days, months, and even years to get single crystals suitable for X-ray crystallographic studies. Many other problems prevent X-ray studies of crystal structures from becoming a truly routine technique. But it is not appropriate to pursue this issue further here.

Below we display a few protein structures (Figs. 21.10–21.12) determined by X-ray crystallography. Figure 21.10 shows three different modes of expressing

a   **b**   **c**

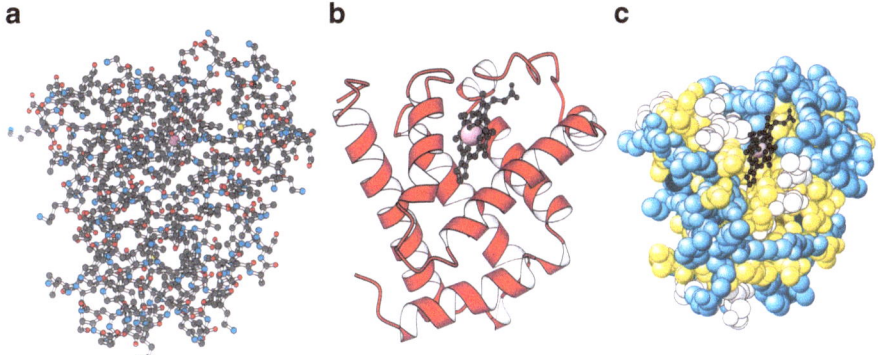

Fig. 21.10 Structures of myoglobin as determined by X-ray crystallography; myoglobin is the oxygen-binding protein in muscle. (**a**) shows all the atoms except for hydrogen as small balls; (**b**) ribbon presentation of secondary structure – the spiral is helix; (**c**) space-filling representation of constituting amino acids (from J. M. Berg, J. L. Tymoczko and L. Stryer, "Biochemistry 5th ed" (W. H. Freeman and Co, 2002))

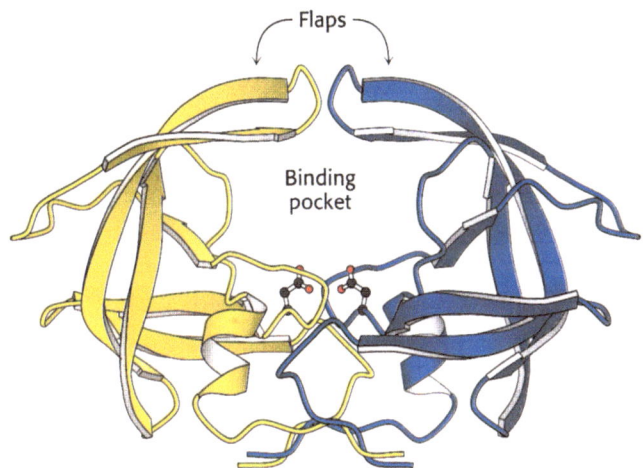

Fig. 21.11 HIV-protease that consists mostly of β-strand (from J. M. Berg, J. L. Tymoczko and L. Stryer, "Biochemistry 5th ed" (W. H. Freeman and Co, 2002))

a protein structure. Mode (a) is called "ball-and-stick" representation, and uses colored dots to indicate the position of an atom where different colors represent different elements, and the bonds are shown as sticks. In mode (b) a helical strip tape is used to schematically represent a helical structure and a string is a random coil portion. Another major secondary structure is β-strand, and it is represented by flattened tape as seen in Fig. 21.11. In mode (c), amino acids of different characters are represented by colored balls; in this case yellow: hydrophobic amino acid, blue: hydrophilic (electrically charged) amino acid and white: others. An account of the protein structure, including the secondary structures α-helix and β-strand, is found in Fig. 5.4.

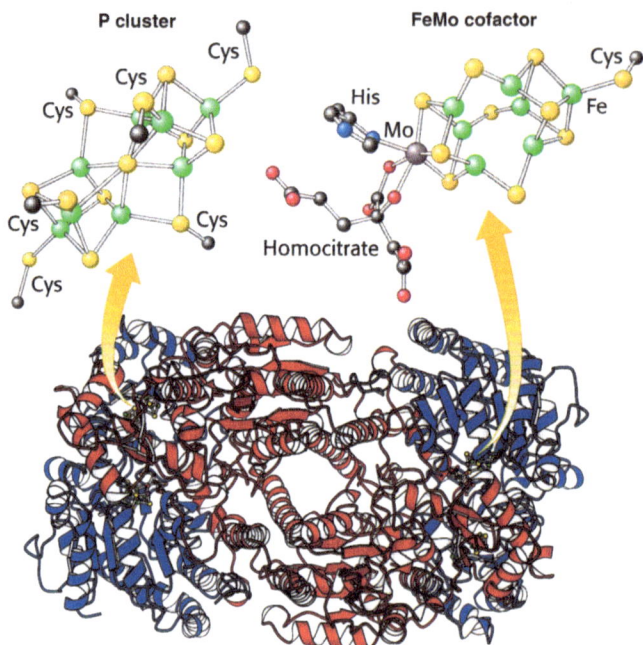

Fig. 21.12 Nitrogenase; its active sites P-cluster and FeMo cofactor are shown schematically (from J. M. Berg, J. L. Tymoczko and L. Stryer, "Biochemistry 5th ed" (W. H. Freeman and Co, 2002))

HIV-protease (Fig. 21.11) is the target of one kind of AIDS drugs, as discussed in Chap. 17. Figure 21.12 is nitrogenase, one of the most complex enzymes, and is talked about in Chap. 6. Determination of protein (enzyme) structures as illustrated here and others help enormously in understanding their functions and the mechanisms in which they play crucial roles.

Part VII

Postscript

Holistic Chemical View of the World

22

The readers have hopefully been convinced by now that our material world is indeed made of chemicals, and that our entire universe, including us human beings, would not exist without chemicals. Chemistry as a scientific discipline has been fairly well advanced to give us a nice consistent picture of the material world. But this is a chemical view of the world; i.e., the world is viewed only through "chemistry" eyeglasses and in terms of chemical concepts and theories advanced so far.

Chemistry as a discipline is an analytical (reductionist) endeavor; i.e., it tries to understand things by dissecting them into compounds (molecules and ionic ensembles) and then taking account of interactions among atoms, ions, and molecules. The reality of the world consists of a whole set of these compounds interacting with each other in diverse ways. Compounds and molecules involved may or may not have been known to chemistry (i.e., *Homo sapiens*), and chemistry may or may not have fully recognized or understood the diverse ways they interact. In other words, we should recognize that chemical understanding of the world as practiced by today's chemists is far from perfect, both in the analytical sense (i.e., we do not yet know all the compounds existent or emerging in the whole universe and all their interactions) and also the"synthetic" sense. Well, language here is inadequate because chemical synthesis as a technique is widely practiced by chemists, but "synthesis" as a philosophical concept is alien to the discipline of chemistry. In other words, chemistry may not concern itself with synthetic understanding of things taking into account all the material presents (known and unknown) and all the interactions at chemical, physical, biological, and (maybe geological) levels.

This kind of limitation exists implicitly in all the so-called disciplinary sciences or modern (western) science, but particularly acute in chemical science (and also in many of the so-called "social science" disciplines including economics and political sciences). The entirety of a phenomenon (material world, society, world economy, etc.) is there as a whole, but each disciplinary science dissects it and recognizes only an aspect (or portion) of the whole. Understanding of the world then requires more than disciplinary understandings, and technology based on disciplinary thinking has its limitation.

E. Ochiai, *Chemicals for Life and Living*,
DOI 10.1007/978-3-642-20273-5_22, © Springer-Verlag Berlin Heidelberg 2011

Well, let us take some examples and see what this implies. The issues chosen are organic foods and GM plants/foods, and synthetic chemicals in general and their effects on everything they interact with: human bodies, ecosystems, and environments. These deal with chemicals and at least certain aspects of the issues can be regarded to be within the realm of chemistry. Some specific examples have been discussed in earlier chapters.

22.1   Organic Foods

"Organic" farming and their products, "organic" foods, are in vogue now. Having read this book, you will realize that in fact all foods based on plants and animals are indeed made of organic compounds, irrespective of the farming or raising methods. From the chemistry point of view alone, therefore, there is no such thing as "non-organic" food; all foods are organic. The "organic" designation in today's market is a misnomer, and is based on non-use of any artificial chemicals, both inorganic and organic in the production.

Farming earlier was done without adding many artificial substances; it was indeed "organic" faming. Fertilizers were natural products or their derivatives, and no weed killer was used; weeds were removed by hand. Some natural products were used for insect repellants. It was so only because no synthetic chemicals were available. Since the early twentieth century, synthetic fertilizers, pesticides, and so on have become widely available and used extensively in "conventional" farming. People have now come to a vague realization that the product made without artificial treatment (i.e., "organic food") seems to be better (in quality) than the same produced using artificial chemicals. Indeed some research data seem to demonstrate that the so-called "really" organically produced food, tomato and fruit in particular, do contain more nutritious compounds than their conventional counterparts (Benbrook et al., 2008; see note below). Let us look at the potential problems of the "conventional" farming and its products as against organic farming in several areas.

Plants require a number of nutrients for healthy growth as discussed in the main body of this book. However, it had been recognized since the final days of nineteenth century that the major essential elements for plant growth are nitrogen, phosphorus, and potassium (N–P–K). This knowledge has led to the development of artificial fertilizers that supply these essential elements. Particularly important are nitrogen fertilizers such as ammonium sulfate and ammonium nitrate; these are synthesized in modern factories and are now used far and wide in large quantities. For example, the so-called "Green Revolution" of rice (efficient production of rice) relied heavily on synthetic nitrogen fertilizers and others. Indeed, use of cheap synthetic fertilizers in general has increased the efficiency of grain and other production (quantity/unit area); a success story of chemistry. However, an excess of nitrogen fertilizer tends to encourage vegetative growth only, producing mainly carbohydrates and carotenes, and less of other nutrients such as ascorbic acid. One serious drawback of ammonium sulfate (most widely used fertilizer) is to reduce the fertility of soil as discussed earlier; other problems will be discussed later. As mentioned in Chap. 6, a number of nutrients other than N–P–K are indeed required for healthy growth, and these micronutrients are not taken account of when applying synthetic fertilizers.

Plant growth will usually be hampered by a number of factors other than shortage of nutrients (fertilizer); among them are insects, disease-causing microorganisms, and weeds. The chemical industry has developed chemicals to reduce the effects of such factors: pesticides, fungicides, and weed killers (herbicides), etc. These are widely used in today's farming, again to increase yield. Seeds and fruits are often eaten by animals and birds as well. Chemicals are often sprayed to deter such eating . The adverse effects of some of these chemicals (to the environment and human health) are discussed in Chap. 16.

Interestingly, nature herself, i.e., plants themselves, often produce some chemicals to deter insects, birds, and animals. Use of artificial chemicals often could discourage plants to produce their own defense chemicals, as the artificial products applied may provide such a defense. This is a virtually general reaction of living systems; i.e., they try to economize their own production of chemicals as much as possible, if feasible. For example, a substitute chemical such as an artificial hormone ingested would tend to suppress the indigenous production of the hormone as discussed earlier in the human body (Chap. 17). Likewise, insect repellants applied to a plant may reduce the indigenous production of repellant such as polyphenolic substances, which, if ingested by humans, can function as an antioxidant, a useful nutrient.

A chemical kills an insect attacking a plant; it is made for that purpose. The chemical can, however, interact not only with the target insect, but also some other insects or chemical substances that it gets in contact with. Such interactions may be innocuous, or the chemical sprayed on may be easily washed away or may not be. Here lies the basic problem; the whole picture of even a single chemical's interactions with all the components of the system (in this case, insect(s) and plant, plus soil, the other surrounding plants and humans eventually eating it) may not be sought or even be considered often in the process of developing such a chemical.

Use of a chemical is directed at a specific problem: supply of a nutrient, killing bugs, or killing weeds. This is a typical chemical approach; finding out the cause of a problem and then correcting that problem with a specific chemical agent. To obtain an overall good result may require a number of chemicals targeted at different problems. The purpose pursued here in this efficiency-conscious society is often the overall yield of product; i.e., increased quantity. The overall better yield of a plant can often be measured as an increase in the bulk components (such as starch or cellulose), which in turn dilute relatively minor components. Some of these minor components may be involved in enhancing the nutritional value and the taste for human consumption; examples are the content of vitamins, particularly vitamin C or sugar component. This is particularly true in seed bodies, i.e., fruits and vegetables such as cucumber. Because such an individual body starts with a fixed number of cells, a larger fruit body contains on average larger cells and/or larger air space between cells. And such intercellular spaces would contain few nutrients. This means that the larger fruit (higher quantity in general) tends to contain less density of nutrients.

Now let us get back to the so-called organic farming/food. The rationales for such farming are not well spelled out; it is based on the vague notion that "natural" is better than "artificial" in any way. The farming practices must have been developed naturally to the best degree attainable before the introduction of artificial chemicals: fertilizers, insecticides, etc. Synthetic chemicals developed as above should have exhibited their specific effects and most of them did so, but many of them have turned out to cause a number of undesirable consequences. Hence it might be wise to return to the practice without artificial chemicals.

Well, does this notion make sense? Does the so-called organic farming produce better food? First of all, the organic farming practice is based on the accumulated knowledge and wisdom of farmers over several millennia. The particular species and the farming practice have been selected to best suit the local condition. Fertilizers were produced from plants' and animals' wastes, which might contain all the necessary ingredients for healthy growth of a plant, not only the major nutrients such as nitrogen, potassium, and phosphorus, but also some of important minor ones such as calcium, iron, magnesium, and zinc. A chemically pure, factory-synthesized fertilizer would not contain these important micronutrients. Besides, the organic portion of such traditional fertilizers would enrich the soil. In other words, the traditional fertilizers not only provide necessary nutrients for plant growth but also make the soil more fertile for future farming. This has something to do with the interactions of the organic substances and the structure of soil. That is, whereas a synthetic fertilizer each provides a specific nutrient, the traditional fertilizer provides most of the necessary nutrients. In the philosophical term, the synthetic fertilizer represents the analytical way of nutrient-provision, while the traditional one represents a holistic nutrient provider. Hence the traditional fertilizer would encourage overall healthy growth, while an individual synthetic fertilizer, e.g., ammonium sulfate, encourages only vegetative growth.

This is the ideal of organic farming, but the reality may be much less ideal; the traditional, natural fertilizers applied may not be sufficient in both quantity and quality. Of course, it depends on the practice of farming.

There is a lot of anecdotal evidence for the superiority of organic foods: "organically grown tomato tastes much better than those grown in the conventional manner"; etc. Is there a sound scientific basis for such a comparison? It turned out that such a comparison is not easy to conduct satisfactorily in the scientific sense. In other words, it is rather difficult to compare different farming methods because the basic condition (other than applied material) has to be set to the same level in scientific sense; soil fertility, water, sunlight, temperature, plant arrangement, etc.

A survey by Benbrook and coworkers (see note below) of the reports on comparison of "organic foods" vs. "conventional products" carefully evaluated the scientific validity of each one. They found 236 scientifically valid pairs with regard to 11 nutrients (potassium, phosphorus, the total proteins, vitamins, and antioxidants), among literatures published in 1980–2007. The organic foods were nutritionally better in 61% (145 pairs) of the cases of all the matched pairs. On the one hand, the conventionally raised food contained more nutrients than its organic counterparts, among 87 matched pairs (37%). In three quarters of these 87 cases where the conventional ones were more nutritious, the nutrients found more in the conventional products were potassium, phosphorus, and total protein levels. These nutrients are usually adequately provided by other foods and are not critical. On the other hand, the major nutrients more abundant in the majority of the organic food of the 145 pairs were polyphenols and antioxidants including vitamin C and E. In terms of quantity of nutrients, one quarters of the organic foods contained 31% or more of the nutrients than the conventional counterparts, and almost one-half contained 21% or more. These researchers concluded: "Yes, organic plant-based foods are, on average, more nutritious."

(Note: Benbrook, Charles; Zhao, Xin; Yáñez, Jaime; Davies, Neal; Andrews, Preston, (2008), "State of Science Review: Nutritional Superiority of Organic Food," http://www.organic-center.org).

This study indeed indicates the general superiority of organically produced foods in terms of important nutrients, but of course the organic foods may not necessarily be superior in all kinds of nutrients. Besides, careful and conscientious application of organic farming methods and material is necessary to assure a better quality of the products, as in any case. Issues to be carefully weighed for all farming to rely on this method (organic) include: (1) whether the method is sufficient for preventing or tolerating attack by insects and disease-causing microorganisms, and (2) whether organic farming alone can provide enough nutrition for all the human beings on this planet today and in the future. The latter depends heavily not only on the technical details but also on the social conditions including distribution methods of food and others.

22.2   GM Plants/Foods

GM stands for "Genetically Modified." Unraveling the secret of the functions of gene has prompted some quarters to attempt to manipulate the genes in plants (and some animals) for our purposes. Several important factors are mentioned in the previous section, which facilitate plants' healthy growth; nutrients (fertilizers), and defense chemicals (insecticide, fungicide, and herbicide, etc.). The intention of those who would use the knowledge of genetics has been to produce plants that would grow healthily without much added extraneous chemicals, such as fertilizers, insecticides, and herbicides, or in a certain case would withstand a specific herbicide, by manipulating the genes.

For example, clover and some bean plants produce their own nitrogen fertilizers (ammonia, NH_3). You may have seen small round things attached to the roots of clover. These small nodules contain an enzyme called nitrogenase that converts nitrogen (N_2) in the air into ammonia. If a gene of this enzyme was incorporated into a plant gene system and if the plant was made to produce this enzyme, then the requirement for nitrogen fertilizer would be greatly reduced for such a plant. This is the kind of thinking behind the efforts to create "GM" plants. This particular (gene) modification has not yet been successful.

Alternatively, a gene may be introduced into a useful plant so that it can tolerate an extraneous chemical such as herbicide. Then the plant can grow in the presence of the herbicide while other plants, e.g., weeds can be eliminated by the applied herbicide. Of course, one may simply incorporate an insecticide-producing gene itself into the gene system of a plant, so that the necessity of applying insecticide may be lessened for the plant.

As in the use of chemicals in conventional farming, the problems arise from the paucity of experimental data on the overall effects of such a gene manipulation. More detailed studies need to be conducted before a widespread use of such a technique, though, in reality, some GM plants are now widely planted in many countries.

Before some examples are discussed, let us focus on the widely held conviction or rather justification for the use of GM technology. That is: "the gene manipulation has been practiced in the form of cross-species-breeding and selection over the millennia, and this new technique is no different, and it only speeds up and enhances such an effort. Besides, the biological evolution itself is such a gene manipulation by nature." Natural crossbreeding has its limitation because it can happen only among similar organisms. On the contrary, GM technique can incorporate e.g., a bacterial gene into a plant; this is not usually feasible in nature. (Though, it could have happened in nature in the event of formation of eukaryotic organisms by symbiotically incorporating prokaryotes. For example, mitochondrion is believed to be a former aerobic prokaryote incorporated in a eukaryotic cell).

Natural selection would have eliminated a crossbred species if the product species would not fit. Nature in general would have selected a best-suited species in the niche. "Being best-suited" implies a species that would survive best under the circumstance, and, ideally, also implies a species that would provide the best quality for the farmer who was attempting to crossbreed. On the other hand, a species that is obtained by artificial gene manipulation is not directly subject to the elimination process by nature (the ecosphere), and it may eventually become dominant or extinct if left to the natural selection pressure. This process may take decades or much longer or even shorter. However, it is likely that artificial interventions may preclude the action of natural selection, thus forcing the GM species to be artificially dominated by, e.g., using the same GM seed every growing season. Meanwhile how a GM plant affects the ecology is uncertain. The plant-growing process is difficult to be confined spatially because of cross-pollination, etc, and hence the gene artificially embedded in a GM plant may be transmitted to other plants in the ecosphere.

Another issue is that the newly incorporated gene would not only produce a substance that must be usually beneficial to the plant growth, but that it may also produce other substances or interfere with the production of other substances and may not necessarily be beneficial for human consumption. These need to be carefully investigated and evaluated before a GM plant is allowed to be widely grown. The reality is always less than ideal; for example, the process that is supposed to incorporate a particular gene may inadvertently introduce factors (genes and others) other than intended. And these other factors unintentionally incorporated may be harmful, when the product of the GM plant is consumed. This is a technical issue; techniques are not always perfect.

(Note: Another basic problem is economical; in other words, a few large corporations seem to be intent on controlling the world agriculture through attempts to become the sole distributors of the GM plants, i.e., seeds. However, this is not a scientific issue).

A few examples will now be discussed. One of the most widely used GM plants is soybean resistant to herbicide "Roundup." Roundup is a trade name of glyphosate, which is effective and widely applied. If a useful plant like soybean is made resistant to Roundup, the herbicide can be used very easily and worrilessly to eliminate all other plants including weed that grow in the same lot. Glyphosate (its isopropylamine salt) is an inhibitor of an enzyme 5-enolpyruvylshikimate-3-phosphate synthase, which is necessary for the formation of aromatic amino acids such as tyrosine, tryptophan, and phenylalanine. Hence glyphosate will kill the plants that come into contact with it. Some microorganisms including *Agrobacterium* CP4 have a version of the enzyme that is not inhibited by glyphosate. Hence a plant with the gene for this particular version of the enzyme will tolerate glyphosate, whereas others will be killed by the herbicide. So the gene of a plant has been genetically modified to incorporate the gene of this enzyme. This is the principle. Roundup-resistant crop has been developed for soybean, corn, sugar beet, and others.

Glyphosate is supposed to be nontoxic to mammals and humans, as the shikimate pathway is not present in them. Glyphosate inhibits the binding of phosphoenol pyruvate to the active site of the enzyme, and phosphoenol pyruvate is one of the pivotal metabolites present in all organisms, plants, and animals. Hence the residual glyphosate in the food consumed by humans has the potential to effect the metabolic pathways. It has been reported that indeed some toxicities are associated with glyphosate (see for example: http://www.percyschmeiser.com/Toxic.htm). Glyphosate has been shown to damage some beneficial microorganisms in soil, to interfere with uptake of nutrients by plants and also affect negatively the symbiotic nitrogen fixation. These factors were not taken into account while promoting Roundup. The commercial formula of Roundup contains chemicals other than the main ingredient glyphosate; they are a surfactant and others. A surfactant is used to facilitate the incorporation of glyphosate into plant cells. Many surfactants are known to cause disruption of reproductive processes in many animals including human beings. Indeed a substantial increase in miscarriage, infant malformation, and others has been reported among the people in some areas in Argentina where a huge amount of Roundup was sprayed.

Another issue is that repeated use of a same herbicide Roundup may enhance the development of a resistance against it among some plants including the weeds that are intended to be the target of Roundup. Indeed the so-called superweeds that are resistant to Roundup have already developed in some areas. This will necessitate an increased use of the herbicide and eventually may also require use of other herbicides. This will negate the original intention of this GM plant.

Another popular GM plant is "Bt"-cotton and others. "Bt" stands for "*Bacillus thuringiensis*," which is known to produce several toxins to kill some insects. A toxin is modified in its DNA sequence by introducing introns, polyA signals, promoters, and enhancers. The modified DNA is then incorporated into the gene of cotton (or others). The thus-modified cotton will produce the toxin that is a little altered to make it more active and more soluble in the plant cells. The toxicities of Bt-toxins in the natural state have been evaluated in mammals and environments, but that of the artificially modified in the GM crops has not been tested. It has been assumed that the toxins produced in the GM plants should be equivalent to the natural ones, because the toxin-encoding domain of the DNA is the same both in the natural and the artificial one. Nature is capable of producing Bt-resistant strains of insects through evolutionary processes, and indeed reports have already been published that indicate emergence of Bt-resistant insects.

In general, GM plants may work profitably for a while, but it is very likely that Nature would eventually defeat their advantages.

22.3   Synthetic Chemicals in General

Synthetic or artificial chemicals are those that have been synthesized by human beings. Naturally occurring compounds can also be synthesized artificially. If such a compound is pure, it should be chemically the same as the corresponding natural one. The basic problems with the synthetic non-natural compounds stem from the fact that they have not been present in Nature until humankind synthesized them in recent centuries. Nature has been dealing with all the naturally occurring chemicals, and as a result, Nature has been able to maintain its integrity in the presence of these compounds.

The ways Nature deals with chemicals are physical (force and heat) and/or chemical. Particularly important is the biological processes that are chemical process at molecular level. Chemicals interact with the biological systems and are modified to the forms appropriate for them and, if toxic, may be rendered harmless in some cases. The biological and ecological systems, as a whole, know how to deal with those chemicals inherent in Nature, though individual organisms may or may not be able to cope with them all.

In order to sustain oneself, an organism produces chemicals harmful to other organisms that compete with it for its survival. These chemicals often kill the target organisms; these are naturally occurring chemical weapons. A few examples are discussed in Chap. 17. Many antibiotics used in human civilization are chemical weapons produced by microorganisms to kill other microorganisms (Chap. 7). The target organism may counteract such a chemical, by evolving a mechanism to render it harmless.

Humankind, having learned how to synthesize compounds, has produced an enormous number of non-natural chemicals in the last one-and-a-half century. Nature, the ecological system and the organisms in it including the human body, has had no experience to deal with them when they first appeared on the earth. Some of them may be harmless to most organisms and simply be tolerated by them. Others can become harmful to some organisms beyond a certain threshold level. Yet others can be toxic even at a low level; they cannot be rendered harmless as the organisms do not have mechanisms to detoxify them.

Some of these examples have been discussed in the main body of this book, but another important issue has not been dealt with. That is the issue of "synergy." When two or more compounds act simultaneously, the combined effect may be much greater than the sum of the effects of the individual compounds. In reality, it is quite rare that a single agent acts at a single location and at a single instant. There are always a multitude of compounds present even in a single cell, into which two extraneous compounds are supposed to enter. The two compounds may interact with one another, maybe forming another compound, and each of these compounds interact with the compounds present in the cell in a variety of ways. Or one compound elicits an effect and the other compound enhances or reduces the effect. There are a number of ways that two or more compounds exert "synergistic" effects. Chemistry has been focused on the effect of a single compound at a single location and at a single time (instant and scale). Multiple and simultaneous effects of several compounds have not been well researched and understood yet because of their complexities.

Huge numbers and quantities of artificial compounds have been created and dispersed on this earth, and their combined effects are only now becoming manifest, but humankind as a whole does not have a clear picture of their extents and effects. Understanding of this whole issue requires more than chemical knowledge.

Acknowledgement

The author gratefully acknowledges the permission for the use of copy-righted material from the sources shown below.

Fig. 4.4. Fig. 28.5 (p852) in D. Voet and J. G. Voet, *"Biochemistrt,* 2nd ed*"* (J. Wiley and Sons, 1995)

Fig. 4.6. Modified from Fig. 7.53 (p240) in R. H. Garrett and C. M. Grisham, *"Biochemistry"* (Saunders, 1995)

Fig. 4.8. Figure on p878 in T. Cech, *Science*, 289 (2000), 878

Fig. 5.6. Fig. 19.12 (p717) in P. Atkins and L. Jones, *"Chemistry Principles*, 3rd ed" (W. H. Freeman and Co, 2005)

Fig. 7.11. Figs. 31.22 and 31.23 (p880, 881) in J. M. Berg, J. L. Tymoczko and L. Stryer, *"Biochemistry* 5th ed" (W. H. Freeman and Co, 2002)

Fig. 8.2. Figure on p23 in D. A. Mcquarrie and P. A. Rock, *"Descriptive Chemistry"* (W. H. Freeman and Co., 1985)

Fig. 11.1. (the left hand side): Fig. 6.1 (p55) in D. A. Mcquarrie and P. A. Rock, *"Descriptive Chemistry"* (W. H. Freeman and Co., 1985)

Fig. 11.5. Fig. 2 in E. Stolyarova, *et al.*, *Proc. Nat. Acad. Sci. (USA),* **104** (2007), 9209–9212

Fig. 11.8. Fig. 3 in L. T. Scott, *et al., J. Am. Chem. Soc.*, **118** (1996), 8743–8744

Fig. 11.9. From L. C. Venema, *et al.*, *Appl. Phys.* **A66** (1989), S513–S515

Fig. 13.1. Fig. 2.1 (p23) in B. Mason, *"Principles of Geochemistry*, 2nd ed" (J. Wiley and Sons, 1966)

Fig. 13.2. Fig. 2 (p120) in W. M. Irvine and A. Hjalmarson, Comets, Intersteller Molecules and the Origin of Life in *"Cosmochemistry and the Origin of Life"* (C. Ponnamperuma, ed., D. Reidel Publ., 1983)

Fig. 14.1. After Fig. 3.8 (p47) in B. Mason, *"Principles of Geochemistry*, 2nd ed" (J. Wiley and Sons, 1966)

Fig. 14.3. After Figs. 9.6, 9.7 (p350, 351) in N.N. Greenwood and A. Earnshaw, *"Chemistry of the Elements*, 2nd ed" (1997, Elsevier)

Fig. 14.4. Fig. 9.5 (p349) in N.N. Greenwood and A. Earnshaw, *"Chemistry of the Elements*, 2nd ed" (1997, Elsevier)

E. Ochiai, *Chemicals for Life and Living*,   281
DOI 10.1007/978-3-642-20273-5, © Springer-Verlag Berlin Heidelberg 2011

Fig. 17.3. Fig. 1a in D. A. Dougherty and H. A. Lester, *Nature*, **411** (2001), 252–255

Fig. 17.5. From W. Zhang *et al., Proc. Nat. Acad. Sci. (USA)*, **95** (1989), 12088–12093

Fig. 19.11. Fig. 3.27 (p106) in P. Atkins and L. Jones, *"Chemistry Principles*, 3rd ed"* (W. H. Freeman and Co, 2005)

Fig. 21.1. The cover of R. Petrucci and R. K. Wismer, *"General Chemistry with Qualitative Analysis"* (Macmillan, 1983)

Fig. 21.3. Fig. 1(a) in R. R. Myer, *et al., Science*, **289** (2000), 1324–1327

Fig. 21.4. Fig. 1 in Z. Zhang, *et al., Science*, **302** (2003), 846–849

Fig. 21.6. Fig. 6 in D. P. Kern, *et al., Science*, **241** (1988), 936–944

Fig. 21.7. Fig. 1(c) in B. C. Stipe, *et al., Science*, **279** (1998), 1907–1909

Fig. 21.8. Fig. 1 in P. Saulet and C. Joachim, *Chem. Phys. Letters*, **185** (1991), 23

Fig. 21.9. Fig. 3c in J. Mou, *et al., FEBS Letters*, **371** (1995), 279–282

Fig. 21.10. Fig. 3.44 and 3.45 (p61, 62) in J. M. Berg, J. L. Tymoczko and L. Stryer, *"Biochemistry* 5th ed"* (W. H. Freeman and Co, 2002)

Fig. 21.11. Fig. 9.19 (p238) in J. M. Berg, J. L. Tymoczko and L. Stryer, *"Biochemistry* 5th ed"* (W. H. Freeman and Co, 2002)

Fig. 21.12. Fig. 24.4 (p668) in J. M. Berg, J. L. Tymoczko and L. Stryer, *"Biochemistry* 5th ed"* (W. H. Freeman and Co, 2002)

Index